797,885 Books
are available to read at

Forgotten Books

www.ForgottenBooks.com

Forgotten Books' App
Available for mobile, tablet & eReader

ISBN 978-1-330-67934-0
PIBN 10091290

This book is a reproduction of an important historical work. Forgotten Books uses state-of-the-art technology to digitally reconstruct the work, preserving the original format whilst repairing imperfections present in the aged copy. In rare cases, an imperfection in the original, such as a blemish or missing page, may be replicated in our edition. We do, however, repair the vast majority of imperfections successfully; any imperfections that remain are intentionally left to preserve the state of such historical works.

Forgotten Books is a registered trademark of FB &c Ltd.
Copyright © 2017 FB &c Ltd.
FB &c Ltd, Dalton House, 60 Windsor Avenue, London, SW19 2RR.
Company number 08720141. Registered in England and Wales.

For support please visit www.forgottenbooks.com

1 MONTH OF FREE READING

at

www.ForgottenBooks.com

By purchasing this book you are eligible for one month membership to ForgottenBooks.com, giving you unlimited access to our entire collection of over 700,000 titles via our web site and mobile apps.

To claim your free month visit: www.forgottenbooks.com/free91290

* Offer is valid for 45 days from date of purchase. Terms and conditions apply.

English
Français
Deutsche
Italiano
Español
Português

www.forgottenbooks.com

Mythology Photography **Fiction**
Fishing Christianity **Art** Cooking
Essays Buddhism Freemasonry
Medicine **Biology** Music **Ancient Egypt** Evolution Carpentry Physics
Dance Geology **Mathematics** Fitness
Shakespeare **Folklore** Yoga Marketing
Confidence Immortality Biographies
Poetry **Psychology** Witchcraft
Electronics Chemistry History **Law**
Accounting **Philosophy** Anthropology
Alchemy Drama Quantum Mechanics
Atheism Sexual Health **Ancient History**
Entrepreneurship Languages Sport
Paleontology Needlework Islam
Metaphysics Investment Archaeology
Parenting Statistics Criminology
Motivational

Date Due

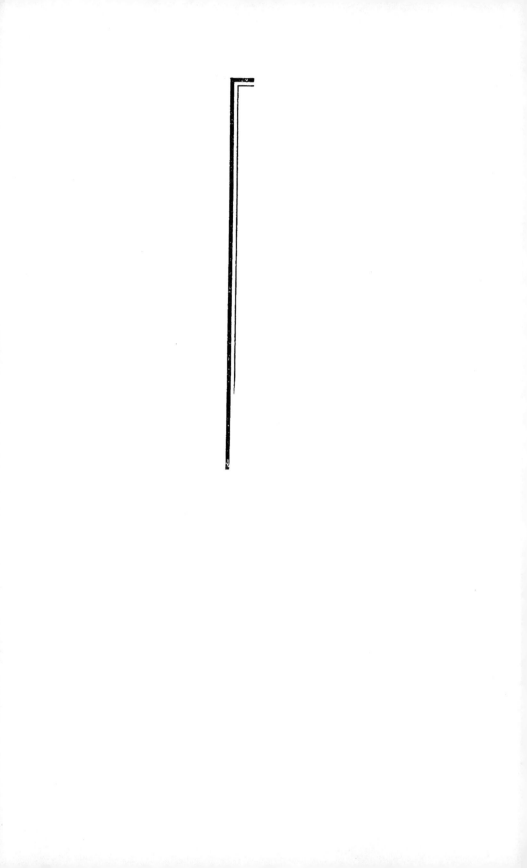

ANTIRRHINUM MAJUS LUTEUM RUBRO-STRIATUM,
With Bud Variation.

A-C.—CHRYSANTHEMUM SEGETUM PLENUM,
D.—CHRYSANTHEMUM SEGETUM.

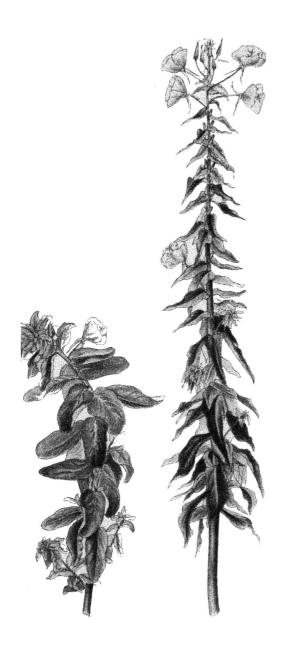

O. LATA.   O. LAMARCKIANA.   O. NANELLA.

HYOSCYAMUS PALLIDUS × NIGER.
A.—H. niger. B.—H. pallidus.

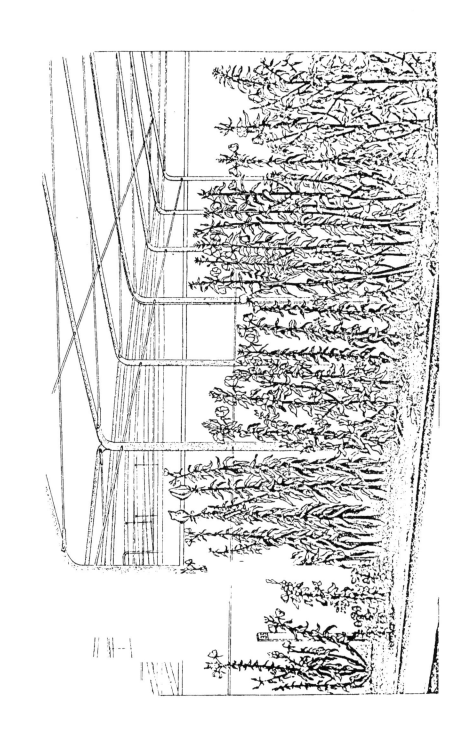

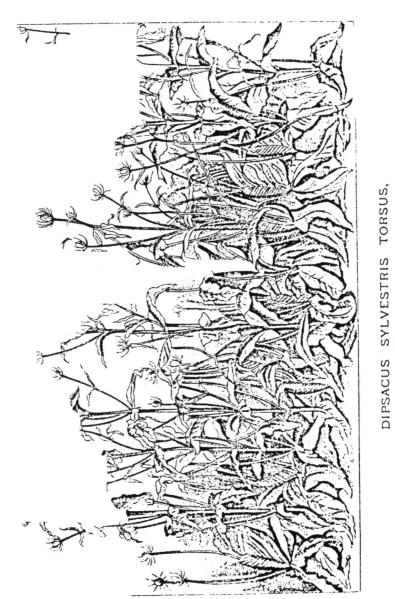

DIPSACUS SYLVESTRIS TORSUS.
With Atavists

THE
MUTATION THEORY

EXPERIMENTS AND OBSERVATIONS
ON THE
ORIGIN OF SPECIES IN THE VEGETABLE
KINGDOM

BY

HUGO DE VRIES
PROFESSOR OF BOTANY AT AMSTERDAM

TRANSLATED BY
PROF. J. B. FARMER AND A. D. DARBISHIRE

VOLUME II
THE ORIGIN OF VARIETIES BY MUTATION

WITH NUMEROUS ILLUSTRATIONS AND SIX COLORED PLATES

CHICAGO
THE OPEN COURT PUBLISHING COMPANY
LONDON AGENTS
KEGAN PAUL, TRENCH, TRÜBNER & CO., LTD.
1910

COPYRIGHT BY
THE OPEN COURT PUBLISHING CO.
1910

CONTENTS.

PART I.
THE ORIGIN OF HORTICULTURAL VARIETIES.

	PAGE
I. THE SIGNIFICANCE OF HORTICULTURAL VARIETIES IN THE THEORY OF SELECTION	3
1. Variability in Garden Plants	3
2. The Doctrine of the Increase in Variability in One Direction Brought About by Selection	9
II. LATENT AND SEMI-LATENT CHARACTERS	18
3. Eversporting Varieties	18
4. Half Races and Half Curves	26
5. Trifolium Pratense Quinquefolium, An Eversporting Race	36
III. THE DIFFERENT MODES OF ORIGIN OF NEW SPECIES	56
6. Horticultural and Systematic Varieties and Elementary Species.	56
7. Progressive, Retrogressive and Degressive Formation of Species	65
IV. THE SUDDEN APPEARANCE AND THE CONSTANCY OF NEW VARIETIES	76
8. Examples of Constant Races	76
9. Sterile Varieties	88
10. Instances of Races which Have Arisen Suddenly in Nature	95
11. Horticultural Varieties which Have Arisen Suddenly.	99
V. ATAVISM	104
12. Atavism by Seeds and Buds	104
13. Vilmorin's Suggestion as to the Origin of Striped Flowers	113
14. Antirrhinum Majus Striatum	120

Contents.

		PAGE
15.	Hesperis Matronalis	136
16.	Clarkia Pulchella	144
17.	Plantago Lanceolata Ramosa	148

VI. EXPERIMENTAL OBSERVATION OF THE ORIGIN OF VARIETIES.. 161
- 18. The Origin of Chrysanthemum Segetum Plenum 161
- 19. Double Flowers and Flowerheads 194
- 20. The Origin of Linaria Vulgaris Peloria 201
- 21. Heritable Pelorias 220

VII. NON-ISOLABLE RACES 227
- 22. Trifolium Incarnatum Quadrifolium 227
- 23. Ranunculus Bulbosus Semi-Plenus 243
- 24. Variegated Leaves 265
- 25. Alternating Annual and Biennial Habit 291

VIII. NUTRITION AND SELECTION OF SEMI-LATENT CHARACTERS. 307
- 26. Increased Nutrition Favors the Development of the Anomaly 307
- 27. The Influence of External Conditions and of Manuring 315
- 28. The Periodicity of Semi-Latent Characters 323
- 29. The Choice of Seeds in Selection 332

PART II.

THE ORIGIN OF EVERSPORTING VARIETIES.

I. TRICOTYLOUS RACES 343
- 1. The Occurrence of Tricotyls as Half Races and Intermediate Races 343
- 2. Tricotyls, Hemi-Tricotyls and Tetracotyls 356
- 3. The Influence of Tricotyly on the Arrangement of Leaves 365
- 4. Tricotylous Half Races 379
- 5. Tricotylous Intermediate Races Do Not Arise by Selection 393
- 6. The Isolation of Tricotylous Intermediate Races 417
- 7. Partial Variability of Tricotyly 444
- 8. The Influence of External Conditions on Tricotyly .. 450

II. SYNCOTYLOUS RACES 457
- 9. Hemi-Syncotyly, Syncotyly, Amphi-Syncotyly 457
- 10. Helianthus Annuus Syncotyleus 466
- 11. Improvement of a Hemi-Syncotylous Race 476
- 12. Atavistic Races 481

Contents.

	PAGE
13. The Influence of External Conditions on Hereditary Values	485
III. THE INCONSTANCY OF FASCIATED RACES	488
14. The Inheritance of Fasciations	488
15. Half Races with Heritable Fasciation	502
16. Eversporting Varieties with Heritable Fasciation	508
17. The Significance of the Atavists	514
IV. HERITABLE SPIRAL TORSIONS	527
18. Spiral Disposition of the Leaves	527
19. Rare Spiral Torsions	537
20. Spirally Twisted Races	543
21. The Significance of the Atavists	554

PART III.

THE RELATIONS OF THE MUTATION THEORY TO OTHER BRANCHES OF INQUIRY.

I. THE CONCEPTION OF SPECIES ACCORDING TO THE THEORY OF MUTATION	567
1. Systematic Biology and the Theory of Mutation	567
2. Progressive, Retrogressive and Degressive Mutations.	569
3. The Theoretical Distinction Between Species and Varieties	578
4. The Practical Conception of Species	589
5. The Parallel Between Systematic and Sexual Relationship	592
II. THE RANGE OF VALIDITY OF THE DOCTRINE OF MUTATION	599
6. The Significance of the Available Evidence	599
7. The Explanation of Adaptations	606
8. Vegetative Mutations	614
III. THE MATERIAL VEHICLES OF THE HEREDITARY CHARACTERS	631
9. Darwin's Pangenesis	631
10. Intracellular Pangenesis	639
11. The Pangenes as Bearers of the Hereditary Characters	643
IV. GEOLOGICAL PERIODS OF MUTATION	651
12. The Periodicity of Progressive Mutations	651
13. Iterative Formation of Species	661
14. The Biochronic Equation	663
INDEX	675

LITERATURE.

LIST OF THE AUTHOR'S PAPERS BEARING ON THE THEORY OF MUTATION.

(See the List at the Beginning of the First Volume.)

a. Intracellular Pangenesis. Translated from the German by Prof. C. Stuart Gager. Chicago: The Open Court Publishing Co., 1910.

b. Fluctuating Variability and Mutability.

Eine zweigipfelige Variationscurve. Roux' Archiv für Entwicklungsmecianik der Organismen, 1895, II, Heft I.—Archiv. Néerl., 1895.

Sur les courbes Galtoniennes des monstruosités. Bull. scient. France et Belgique, 1898, T. XXVII, p. 395.

Over het omkeeren van halve Galton-curven. Botanisch Jaarboek, Gent, 1898, X, p. 27.

Ueber Curvenselection bei Chrysanthemum segetum. Ber. d. d. bot. Ges., 1899, Bd. XVII, Heft 3.

De zaadkweekeryen te Erfurt. Het Nederlandsch Tuinbouwblad, 1891, p. 327. Gladiolus nanceianus, ibid., VIII, Jan. 1892.—Tulipa Greigi, ibid., May 1892. —Caladium, ibid., July 1892.—Caladium's van ALFRED BLEU, ibid., July 1892. —Dubbele Seringen, ibid., Sept. 1892.—Grootbloemige Canna's I and II, ibid., Dec. 1892.—Amaryllis, ibid., IX, Sept. 1893.

c. Spiral Torsions.

Ueber die Erblichkeit der Zwangsdrehung. Ber. d. d. bot. Ges., 1889, VIII, p. 7.

Eenige gevallen van Klemdraai by de Meekrap. Bot. Jaarboek, Gent, 1891, III, p. 74.

Monographie der Zwangsdrehungen. Jahrb. f. wiss. Bot., 1891, XXIII, pp. 13-206, Plates II-XI.

Bydragen tot de leer van den Klemdraai. Bot. Jaarboek, 1892, IV, p. 145.

Eine Methode, Zwangsdrehungen aufzusuchen. Ber. d. d. bot. Ges., 1894, Bd. XII, Heft 2.

On Biastrepsis in Its Relation to Cultivation. Annals of Botany, 1899, XIII, p. 395.

d. Fasciations and Other Anomalies.

Sur un spadice tubuleux du Peperomia maculosa. Archiv. Néerl., 1891, T. XXIV, p. 258.

Over de erfelykheid der fasciatiën. Bot. Jaarboek, Gent, 1894, VI, p. 72.
Over de erfelykheid van synfisen. Bot. Jaarboek, 1895, VII, p. 129.
Erfelyke monstrositeiten in den ruilhandel der Bot. Tuinen. Bot. Jaarboek, 1897, IX, p. 66.
Een epidemie van vergroeningen. Bot. Jaarboek, 1896, VIII, p. 66.
Sur la culture des monstruosités. Cps. rs. de l'Acad. des Sc., Paris, 1899.
Sur la culture des fasciations des espèces annuelles et bisannuelles. Revue générale de botanique, 1899, T. XI, p. 136.
Ueber die Abhängigkeit der Fasciation vom Alter bei zweijährigen Pflanzen. Bot. Centralblatt, 1899, LXXVII.
Ueber die Periodicität partieller Variationen. Ber. d. d. bot. Ges., 1899, XVII, Heft 2, p. 45.
Over het periodisch optreden van anomalien op monstreuze planten. Bot. Jaarboek, 1899, XI, p. 46.
Sur la périodicité des anomalies dans les plantes monstrueuses. Archiv Néerl., Série II, T. III.
Over verdubbeling van Phyllopodiën. Bot. Jaarboek, 1893, V, p. 108.
Ueber tricotyle Rassen. Ber. d. d. bot. Ges., 1902, Bd. XX, Heft 2.

e. Unit-Characters.

ADAM's Gouden regen (Cytisus Adami). Album der Natuur, 1894.
Hybridizing of Monstrosities. Journ. Roy. Hortic. Soc., 1899.
Sur la fécondation hybride de l'albumen. Cps. rs. de l'Acad. de Paris, 1899 and Ref. Biol. Centralbl., 1900.
Sur la fécondation hybride de l'endosperme chez le Mais. Revue générale de botanique, 1900, T. XII, p. 129.
Sur la loi de disjonction des hybrides. Cps. rs. de l'Acad. de Paris, 1900.
Das Spaltungsgesetz der Bastarde. Ber. d. d. bot. Ges., 1900, Bd. XVIII, Heft 3.
Ueber erbungleiche Kreuzungen. Ber. d. d. bot. Ges., 1900, Bd. XVIII. Heft 9.
Sur les unités des caractères spécifiques. Revue générale de botanique, 1900, T. XII, p. 257.
The Law of Separation of Characters in Crosses. Journ. Roy. Hortic. Soc., 1901, XXV, Part 3.
On Artificial Atavism. Proceed. Americ. Hortic. Soc., 1902.
La loi de MENDEL et les caractères constants des hybrides. Cps. rs. de l'Acad. de Paris, 1903.
Anwendung der Mutationslehre auf die Bastardirungsgesetze. Ber. d. d. bot. Ges., 1903, Bd. XXI, p. 45.
Befruchtung und Bastardirung; ein Vortrag, 1903. Leipsic, Veit & Co.

PART I.

THE ORIGIN OF HORTICULTURAL VARIETIES.

I. THE SIGNIFICANCE OF HORTICULTURAL VARIETIES IN THE THEORY OF SELECTION.

§ 1. VARIABILITY IN GARDEN PLANTS.

DARWIN based his theory of selection, in great part, on the well-known horticultural principle that new varieties are obtained by seeking for small deviations with subsequent isolation and selection. Variations which at their first appearance almost escape observation can be worked up by the skill of the gardener; in doing so variability is seen to increase, and in favorable cases, very rapidly. In this way a new form arises, which answers the purposes and rewards the labors of the breeder.

We have all heard how beautiful double varieties have resulted from the appearance of single flowers in which only one stamen and this often only partially was transformed into a petal.

In the first volume we dealt with this practice more than once, and pointed out how liable it is to give rise to misunderstanding when applied to the elucidation of the problem of specific differentiation (Vol. I, § 23, pp. 176-185). The object of the present Part is to collate the relevant data and to show what light they throw on this all-important problem. Of course we can only go so far as the incomplete and scanty character of the material will allow.

The development of the statistical treatment of variation which took place after DARWIN's time, allows of an altogether different conception of the phenomena than was possible some fifty years ago. It was shown that the fluctuation of characters is due to their development to a greater or less degree. But the character in question does not vary in any other than these two directions. The variation is linear (Vol. I, p. 118). It increases or diminishes but creates nothing new. New characters can arise, so to speak, alongside of it, but they arise independently of the fluctuation of the old ones.

This applies to the case before us. *The variations which the horticulturist looks for and then works up are not variations of the old characters;* such may indeed give rise, by selection, to improved races, but not to new types (Vol. I, p. 82). The required deviations are anomalies, as in the example of the origin of double flowers, just cited. When such an anomaly arises we may be sure that the new character already existed in the internal organization of the plant. Where it springs from and how it arose is a matter of indifference to the breeder: he has got it and can work it up. In other words: "The first condition necessary for raising a novelty is to possess it" (Vol. I. p. 185).

In this connection two cases are distinguished in practice according as one is dealing with apparently invariable forms, or with forms exhibiting a high degree of fluctuating variability. In the former case all that has to be done is to isolate the novelty and to free it of possible impurities introduced by crossing. If this can be done without much difficulty the variety is perfect and constant from the beginning and needs only a few years of multiplication before it can be put on the market (Vol.

I, pp. 77-80). Many white flowered varieties afford good examples of this kind of novelty.

But it is very different with the second case. A novelty which exhibits fluctuating variability in a high degree seldom makes its first appearance in a full state of development. As a rule it is very slightly developed at first. The novelty is betrayed, as the expression is, by a quite small trace or indication. From the scientific standpoint we have to regard this as a *minus*-variant, i. e., as an extreme variant in the *minus* direction of the new character (Vol. I, p. 51). And it is plain that the seeds of such a variant of the new variety will, when sown in the garden, soon give the mean value of the character in question.

This process is, as we can easily see, fundamentally a phenomenon of regression (*Vol.* I, Figs. 18 and 19, pp. 73 and 84); but to the breeder it is a progressive change, and by no means an inconsiderable one, since on it the success of his operations largely depends. This apparent paradox, however, has been a great obstacle to the understanding of these phenomena. But, to us, it explains in a very clear way the initial and rapid increase in variability; for it is obvious that an approximation to the mean value will take place much more easily and rapidly than a departure from it.

The breeder can now either rest content with this "regressive advance"; or he can endeavor to raise the new form above its mean value by choosing plus-variants as seed-parents. But in the latter case the value of the new form remains dependent on the continuance of selection (*Vol.* I, p. 80).

Notes dealing with this process of breeding are not rare in horticultural literature, but they are generally short and lack precision being much inferior in this respect to the accurate accounts that are given of artificial crossings. I shall bring together the most important facts that I have been able to find, in the following section (§ 2).

In order to penetrate more deeply into these phenomena I have endeavored to apply this method to a series of cases. With the help of control experiments, and by keeping detailed records, I succeeded in finding out how such novelties usually develop themselves. Just as happens in practice, I was successful with some cases but not with others. And the correspondence between my results and the experience of breeders seems to me to be so complete that my experiments may simply be taken as instances of the method under discussion.

I propose to distinguish, therefore, between highly variable and only slightly variable novelties. The latter are generally assumed to be instances of *single variations* which arise suddenly. In the case of these I shall, therefore, only have to discuss their origin and the question of their constancy. (Chapter IV of this Part.) Much more important from the critical standpoint are the varieties with a high degree of fluctuating variability, i. e., those very cases which passed for instances of the origin of new characters by artificial selection (Chapters II and VIII). As examples of this I refer to variegated leaves and to double and striped flowers.

If we now compare, from a theoretical standpoint, this high variability with the normal examples which we dealt with before (Vol. I, pp. 47-52 etc.) we shall see that the two are not exactly the same. In variegated leaves the yellow alternates with the green, in semi-double flowers the petaloid stamens alternate with normal ones and so forth. Therefore we are not here dealing with the variable development of a single quality, but with the simultaneous operation, or rather with the conflict, of two qualities. For in proportion as the one or the other of them prevails the plant will be more or less variegated, double and so forth. One of the characters is the old, normal one, pertaining to the original species. The other is the new, abnormal one pertaining to the variety in process of formation, in fact the anomaly. And the conflict of these two antagonistic types affords at least a partial explanation of this extraordinary variability.

The green color itself is only very slightly variable, and the pure yellow or golden varieties, in which the green is entirely absent, are equally uniform (*varietates aureae*, for example *Pyrethrum Parthenium aureum*). Of "double" forms there are two types; the ordinary highly variable more or less double sorts, and on the other hand the sterile varieties which exhibit this peculiarity to its full extent in all their flowers (see *Ranunculus acris petalomana*, Vol. I, Fig. 40, p. 194). In this case the types with a high degree of fluctuating variability might be considered as a connecting link between two almost invariable forms, the normal single and the petalomanous types.

If we regard this principle as an explanation of the case in point we arrive at the conception of *intermediate forms with two antagonistic characters striving for the mastery, and possessing a remarkably high degree of variability as a result of this struggle*. The extent of this variability differs from case to case: in the most extreme examples whole organs or even whole individuals can come exactly to resemble one of the types between which they oscillate. Pure green or, on the other hand, pure yellow or even white leaves or seedlings are by no means rare in variegated varieties. But the resemblance is only superficial. The green *minus* variant of the variegated type does not belong to the original species, nor the yellow *plus* variant to the golden variety; as may often be seen by sowing the seeds of such extreme types.

I propose to call such varieties *intermediate races,* and if neither of the two antagonistic characters preponderates too much over the other, balanced races or *ever-sporting varieties*[1] (see § 3).

If we attempt to make a statistical study and graphical description of the variability in such intermediate forms we must obviously not expect such simple and straightforward curves as those which describe the variability of normal characters (*V*ol. I, p. 48). In principle we may expect to obtain figures which simultaneously exhibit the two magnitudes—that is to say *compound curves* such as have been studied by LUDWIG, BATESON, PEARSON, DAVENPORT and others. It is evident that they will present very different forms according to the mutual proportion of the two characters (see below §§ 3-5). At the same time it is clear that in such cases selection may lead to special results which will often be due to the impossibility of transgressing the characters of the two limiting types (see § 5 and Fig. 3).

The two following generalizations may be derived from the facts we have been discussing.

1. *Some horticultural varieties owe their existence to a single new character.* These are usually not more variable than the original species and as a rule just as constant from seed. Very frequently the novelty consists in the loss or latency of a character of the parent spe-

[1] See *Species and Varieties, Their Origin by Mutation,* p. 309.

cies. In cases where the origin of such a novelty is satisfactorily known it always happened suddenly. For the combination of several characters in the same variety see *Vol.* I, p. 197.

2. *On the other hand some horticultural varieties are compound types which owe their existence to the association of two (or more) antagonistic characters.* The two characters tend to exclude one another more or less completely and struggle for the upper hand; from this there results a very high degree of variability in their manifestation (as in variegation, stripes, doubleness and so forth). These forms usually first appear as *minus* variants, i. e., with a slight degree of development of the abnormality in which condition they are sought for and isolated and subsequently improved by selection. The artificial production, therefore, of such a form is not a sudden one but a process of gradual improvement. Their first origin however remains unknown.

§ 2. THE DOCTRINE OF THE INCREASE IN VARIABILITY IN ONE DIRECTION BROUGHT ABOUT BY SELECTION.

One of the most attractive parts of the doctrine of selection is that according to which variability may be increased by selection. Many observations, especially in horticulture, seem to support this view; which, if it were true would afford an almost irrefutable argument in favor of the prevailing belief in the omnipotence of natural selection (*Vol.* I, p. 119).

Varieties are said to be incipient species. By selecting the individuals which deviate most from the type of the species it is believed to be possible to attain first to variations and then to varieties. To these is ascribed the tendency to become fixed and so to become races: in the same way these races would later be transformed into species. This is the generally accepted view.

This view is based, as I attempted to show in the first part of the first volume, on an unwarranted extension of DARWIN's theory of selection. DARWIN argued from the results obtained in horticulture; but these, at least as described in the works of the best authorities, do not seem to me to justify such an extension.

According to the prevailing view, man has the power to produce any desired amelioration in any species ready to hand. All characters vary and all that need be done would be to isolate the extreme variants and to breed further from them. The process takes some time of course but in many species the experiment is already lasting about half a century. But the advances which have been made, and which are of the very greatest practical importance, do not tally with this assumption. On the contrary we learn from them that for much that has been attained much has proved unattainable.

The comparative studies of systematists show us that almost everywhere there exist unperceptible transitional stages between the smallest differences and perfectly distinct species. This forms a weighty argument for, but no proof of, the prevalent view. For we have to reckon here with transgressive variability ($Vol.$ I, Part II, § 25, p. 430), which tends to blur the boundaries of related groups.

I have indicated in the foregoing section (§ 1) the principles on which in my opinion an elucidation of the process in question must be based. If a small anomaly is found in a wild or cultivated species, and a new and constant form is raised from this by selection, the whole sequence of events may have the appearance of having been gradually brought about by the free will of the experimenter; whereas as a matter of fact the result was attained mainly by good luck.

If we look through the literature of horticulture we shall soon see that this illusion has not taken in the really efficient breeder. He knows perfectly well that neither the beginning nor the end of such an experiment is under his control. It is only between these two limits that everything depends on his skill.

The first indication of an anomaly in a pure species appears by chance; and it is a well-known rule in horticulture that the breeder should always be on the lookout for such chance occurrences. It does not matter how small the deviation is so long as it is an anomaly (p. 4). When such a deviation has once been found it lies with the breeder to perfect it and bring it to its full development. But the ever present, more or less considerable, fluctuating variations of normal characters are of no use for this purpose; by their means many varieties may be made better and prettier, but they can give rise to nothing really new.

The best horticultural authorities are in agreement on this point. CARRIÈRE for example says: *"L'horticulteur ne peut faire naître les variétés,"* and in greater detail in reference to double flowers: *"Le point de départ des fleurs doubles est en dehors de notre puissance comme de nos calculs; nous ne pouvons rien, ou à peu près rien, sur le fait initiatif; nous ne pouvons que le saisir lorsqu'il se présente; nous ne pouvons pas le provoquer; c'est un effet, dont la cause nous est inconnue."*[1] A well-known

---

[1] E. A. CARRIÈRE, *Production et fixation des variétés dans les végétaux*, 1865, p. 64 and p. 15.

English breeder, WILLIAM PAUL, says:[1] "He who is seeking to improve any class of plants, should watch narrowly and seize with alacrity any deviation from the fixed character..... However unpromising in appearance at the outset, he knows not what issues may lie concealed in a variation." SALTER also said that the greatest difficulty lies in finding a small initial deviation; but when this has once been found all the rest lies within our power, however small the variation may be. And DARWIN, who cites this,[2] always emphasized its great importance whenever he had occasion to refer to it.

In other words, which we have already often quoted: The main condition necessary to produce a novelty is to be in possession of its first step.

And yet as is well known the attempt is not by any means always successful. Sometimes the variation disappears without leaving a trace behind; in which case of course all further efforts to deal with it are in vain.

Unfixable deviations of this kind are, according to my experience, the occasional manifestation of latent characters. What the breeder wants to find are those cases in which the chance anomaly has already become a heritable although hidden race. If this has happened the anomaly will, in the first place, easily manifest itself, if the conditions of life are not quite unfavorable and in the second can rapidly be developed to the level of a good horticultural variety.

So far as the available data enable us to judge, breeding experiments of this kind always follow the same course. Hosts of examples can be found. Extensive

---

[1] *Contributions to Horticultural Literature*, 1892. *Nature*, Vol. 46, p. 583.

[2] *Variations of Animals and Plants*, II, p. 249. See also Part I, p. 267 et seq.

sowings repeated year after year avail nothing if chance does not play its part. *Anemone coronaria plena* arose in the nurseries of WILLIAMSON in England as a single plant, which exhibited a slight petaloid broadening of one of the stamens.[1] From the seed of this specimen a race has been started, the flowers of which became fully double in the course of a few generations. The double varieties of roses, *Campanulas,* and many other garden plants have arisen in the same way. I saw a bed of mignonette (*Reseda odorata*) some of which had double spikes, in a nursery at Erfurt. The spikes were fasciated, the flowers were broader and the whole plant fuller, more compact and handsomer than the species. The plants of this bed had been produced from the seed of two fasciated specimens which had accidentally appeared the year before. The normal were weeded out and the abnormal saved and allowed to set seed with a view to putting a new variety on the market.

In cases such as this, selection has a twofold object. In the first place the variety must be isolated,—that is purged of the impurities resulting from free crossing. It must also be actually improved by selection. The first indications of doubling are, as we have seen, single supernumerary petals, or in composites single supernumerary ray florets on the disc; the first indication of a new color is often very pale; slit leaves and petals are indicated by quite small invaginations, combs (Vol. I, p. 191) appear as small outgrowths. All these qualities had to be improved by selection up to the level of the mean of the character and then even perhaps beyond.

An improvement of this kind, when once started, is effected not only rapidly but with increasing swiftness.

DARWIN, *loc. cit.*, II, p. 269.

Hence the illusion of an increase in variability. The explanation is simply this that, as shown in the preceding section (§ 1), we first find a *minus* variant of the new character, which, in accordance with the law of regression, approaches not the character of the old species but the mean value of the new variety, as soon as it is isolated. And this takes place easily and swiftly since the new variety in this case behaves like an improved race on the cessation of selection or under reversed selection (Vol. 1, § 14, p. 122).

The progress made by this improvement and through the purification from the results of crossing is often so rapid that it can be expressed in terms of a geometrical series. This generalization does not attain to the rank of a law, but my meaning will become clearer by citing an example. HOFMEISTER sowed the seeds of plants of *Papaver somniferum polycephalum*,[1] which he had found growing between normal examples of the species. By selecting the fruits which were richest in supernumerary carpels, but without isolation, he effected the following increase in the number of abnormal examples in the succeeding generations:

Year:	1863	1864	1865	1866	1867
Percentage of abnormal plants:	6%	17%	27%	69%	97%
Geometric series:	8	16	32	64	(100)

These figures, as we see, do not differ considerably from a geometric series. I do not lay much stress on the fact, but I have myself more than once obtained similar series of figures in breeding experiments.

The limits that can be reached are as little under the control of the breeder as the starting-points that had to

[1] *Allgemeine Morphologie*, p. 565. See our Fig. 27 on p. 138 of the first volume; also HOFFMANN, *Bot. Zeitg.*, 1881, p. 397, and VERLOT, *Production et fixation des variétés*, p. 88.

be waited for. This is most forcibly brought out by the fact that numerous horticultural varieties are still at exactly the same level as they were at the time of their introduction. The most vigorous selection continued over long periods of time has only rarely succeeded in effecting a further improvement in the same direction. We are familiar with hosts of variegated plants, but *Aurea* varieties are very rare. Flowers with petalomany are sterile, and the plants can only be multiplied by vegetative methods. But it is quite clear that this difficulty is by no means the cause of their rarity. Amongst composites we occasionally find isolated heads without tongue florets, but how small is the number of discoid varieties. I once found an example of *Coreopsis tinctoria* in my cultures, which exhibited only some spare ray florets, but although I isolated the plant, the abnormality did not reappear from its seed. *Catacorolla* (an outward doubling of the corolla so as to form lappets) occurs almost only as a commercial race in *Gloxinia superba*. Fistulous composites are rare; there is room on the market for monophyllous and laciniate varieties of many species, if only we could make them. But so long as chance does not put them into our hands, all our labor is in vain.

Nevertheless, all plants no doubt possess numerous latent characters. Any culture carried out on a sufficiently large scale, or continued for several years, will give convincing proof. In fact it is often very difficult to keep races free from anomalies. *Agrostemma Githago, Raphanus Rhaphanistrum* and many other species contain an almost inconceivable number. Amongst garden plants desirable novelties must obviously be rare now because they must have been already found and put on the market;

useless and indifferent anomalies are common enough, especially in extensive cultures.

When a new horticultural variety has been isolated and "fixed," that is to say improved and rid of impurities by a few years' cultivation no considerable further improvement in the same direction is to be expected. Only two ways of progress are still at hand. These are to wait for the chance appearance of a new abnormality in the same strain, or to combine the new character with others by crossing. The former method is dependent on chance and therefore often unsatisfactory. The second is almost sure to succeed, and thus it is always chosen. Each new character is immediately transferred to numerous other varieties of the species and a corresponding number of novelties obtained in this way. Thus LEMOINE transferred the double flowers of a single double lilac to several dozen varieties, and the *Cactus Dahlia* was, very soon after its introduction, obtainable in almost every shade of color and doubleness. Ordinarily this process is described in the opposite way—that is to say, it is claimed that the properties of the old varieties are transferred to the new type. In this way there appears a vast series of varieties forming a new group co-extensive with the older forms of the original species. Thus a single new character can double the number of varieties. *Petunias, Zinnias* and *Fuchsias* are familiar examples of the application of this method in former times, *Gladiolus, Begonia* and many others of its recent application. The ostrich-feather *C*hrysanthemum (with ciliated petals) arose about thirty years ago in a single variety (*Alph. Hardy*), but can now be obtained in large numbers of forms.

The doctrine of the onesided increase of variability

by selection is based, therefore, as far as the available data enable us to decide, *on the existence of strains with heritable but hitherto latent characters.* Such races are highly variable, and their existence is betrayed when they first are met with, by trifling anomalies which however can easily be worked up by selection. As a result *they rapidly depart from the type of the species but only because they approach their new type*: and as soon as this has been reached by isolation or exceeded by selection it is just as difficult to effect any further improvement as in ordinary improved races. *These varieties cannot be evoked at will;* we have to wait till they chance to appear. Nor when once fully developed can we improve them further. *Nothing but chance*—that is to say some unknown factor—*can as yet overstep these two limits; selection can effect no more than the most transparent illusion of any thing approaching complete control.*

II. LATENT AND SEMI-LATENT CHARACTERS.

§ 3. EVERSPORTING VARIETIES.

Before I proceed to deal with the results which have been obtained, in horticulture, with these highly variable varieties it is desirable, in order to clear up the conceptions involved, to fix our attention on the various stages which may be interpolated between a species and a simple and constant variety derived from it.

We will start from the fact that the chance appearance of an anomaly by no means always opens up the way to the acquisition of a novelty. One example out of many will suffice. Pitchers (Figs. 16, 106, and 109, *Vol.* I, pp. 61, 470, 484) are usually found as quite rare and isolated variations,[1] but in some species of plants, such as *Magnolia* and *Tilia,* tolerably frequently. But a variety as rich in these structures as, for example, *Trifolium pratense quinquefolium* is in 4- and 5-merous leaves does not exist, although it would obviously attract attention and pay the trouble of breeding experiments.[2]

This shows that an anomaly discovered by chance may be the expression of a latent character which cannot be brought to its full state of development. Besides this

---

[1] Over de erfelykheid van synfisen, *Kruidkundig Jaarboek,* Gent, 1895, p. 129.

[2] A variety of *Ficus religiosa,* with all its leaves changed into pitchers, has since been introduced into Europe by Mr. PRAIN, the Director of Kew-gardens. (Note of 1910.)

extreme but very common mode of appearance two other cases are possible, according to my experience:

1. When the seeds of an abnormal individual are sown the anomaly is repeated from time to time in a few or more individuals, remaining rare or only appearing in a feeble state of development. Selection may improve it, but only to a very inconsiderable extent.

2. Under favorable circumstances the anomaly may increase rapidly both in the degree of its development and in the number of individuals which present it. A so-called constant race is formed in the course of a few generations. It is subject to a high degree of fluctuating variability in respect to the character in question and is largely dependent on cultivation.

I propose to term the first type of characters *semi-latent* and to distinguish amongst latent characters between the genuine completely latent ones and those which occasionally come to light or the *semi-latent* ones. This term refers to the behavior of the character in the race as a whole; a semi-latent character may remain latent in many individuals and organs and be active in others. A true latent character on the other hand only very rarely becomes active.

If we study these three cases statistically, trying to plot the variation of the anomaly in the form of a curve (p. 8) we generally obtain the following results:

First case. The genuine latent characters appear so rarely that they do not afford sufficient material for a curve.

Second case. Semi-latent characters must be studied in combination with their antagonistic active characters, and are expressed by half curves (Fig. 1, p. 28), from which a two-sided curve may be derived by selection

(Fig. 2, p. 34), the apex of which however does not become very distant from that of the half-curve.

Third case. The characters are first expressible by half curves because they are *minus* variants; but after isolation the curve very soon becomes a two-sided one with a new apex. The new variety reaches its full development and is maintained without further selection.

A schematic presentation of the conflict between two antagonistic characters is given below:

	The normal character is:	The anomaly is:
I.	active	latent.
II.	active	semi-latent.
III.	An equilibrium is maintained.	
IV.	semi-latent	active.
V.	latent	active.

I do not of course suppose that no further cases are possible, that there may not for example be various stages of semilatency. The facts at our disposal do not admit of any such definite statement. On the other hand it must be stated that the scheme I have given covers the cases which have been so far collected; we shall soon see large numbers of examples of the main cases, whereas of others I have not yet found any.

In the above table I obviously represents the normal, original species, and V the slightly variable and constant variety derived from it. The three other numbers represent the three intermediate forms of which the two first (II and III) correspond to actuality whilst the fourth merely follows from the scheme. I am rather doubtful as to its occurrence in nature.

It is necessary to introduce special names for the first two intermediate forms. I shall therefore call them both intermediate races, one of which—No. II— I shall call a *half race,* and No. III a *middle race.* The word race is obviously not used here in the sense of an improved rare (Vol. I, p. 80) but simply means a heritable form. Instead of middle race I shall usually employ the more convenient term of eversporting variety.[1]

Two examples to which reference has already been made will serve to illuminate the foregoing discussion.

EXAMPLES.

		VARIEGATED LEAVES.	DOUBLE FLOWERS.
I.	Original species.	Green.	Simple.
II.	Half-race.	Rarely variegated.	Occasional petaloid stamens.
III.	Eversporting variety.	*Var. variegata.*	*Var. plena.*
V.	Constant variety.	*Var. aurea.*	*Var. petalomana.*

The parallelism of these two groups rests on the assumption that the same character appearing in a state of full development would give rise to the constant golden and to the fully double varieties;[2] and that it is by their mixture with the antagonistic character that variegated and half-double varieties arise. The object of this assumption is solely to present the matter more clearly; for in cases of segregation the characters behave slightly differently (see p. 124).

There are many examples of half races and eversporting varieties; the former constitute a very considerable part of the material of teratology and afford suitable material for the experimental study of monstrosities. The same holds good for many eversporting varieties, and I shall have to recur to this point in the second part of this volume with especial reference to twisted stems and fasciations. Half races as a rule exhibit their

[1] See *Species and Varieties, Their Origin by Mutation,* Chapters XI-XV, pp. 309-459.

[2] See § 19 and especially § 24 (on variegation).

abnormality too seldom to be of any use, or at any rate to be of more than of secondary value, in horticulture. On the other hand the eversporting varieties highly contribute to the diversity among horticultural plants. Numerous varieties with variegated leaves, with striped or double flowers, with double heads amongst the composites, belong to this group. The *Formae cristatae* of many ferns, the combs in the flowers of *Primula sinensis, Cyclamen persicum, Begonia* etc., the polycephaly of *Papaver,* the catacorolla of *Gloxinia superba,* and a series of other more or less rare instances may also be adduced.

It is, obviously, not necessary that all the forms named should exist for every pair of antagonistic characters. In many cases the intermediate races are absent and in others one or two of them. It is, likewise, not necessary that the pure type corresponding to a certain intermediate race should exist. We can, in such cases, very often reconstruct it by the help of analogy. The following are instances which will be described more fully later on in this part, in which the corresponding constant variety is still failing.

SPECIES.	HALF-RACE.	EVERSPORTING VARIETY.
Trifolium pratense	wild four-leaved clover	*T. p. quinquefolium.*
Trifolium incarnatum	*T. i. quadrifolium*	unknown.
Ranunculus bulbosus	*R. b. semiplenus*	unknown.
Chrysanthemum inodorum	unknown	*C. i. plenissimum.*
Chrysanthemum segetum	*C. s. grandiflorum*	*C. s. plenum.*

Caltha palustris furnishes another instance; it exhibits in nature a half race with supernumerary petals and is represented on the market by a uniformly double sterile variety exhibiting petalomany. *Camellia japonica* presents the two types of doubling in different varieties.

The remarkable fistulous and monophyllous varieties, so well known as examples of partial atavism, are further instances of eversporting types (*Vol.* I, Fig. 38, p. 193, and Fig. 15 of this volume), together with the viviparous grasses (*Poa alpina vivipara, Poa bulbosa vivipara,* etc.) and a number of other viviparous forms such as *Agave vivipara* and so forth.[1] If the constant variety corresponding to a certain intermediate race does not exist, this latter is usually classed as a variety in the case of middle races, but as a heritable anomaly in the case of half races.

It is, further, very probable that many natural species which attract attention by the high degree of variability of some particular character are really in a way intermediate races, i. e., that they owe their multiformity to the co-existence of two antagonistic characters. Instead of entering further into this very attractive subject I shall content myself with citing the case of *Acacia diversifolia* which owes its name and its nature to the conflict between the two characters of bipinnate leaves and flattened petioles without leaflets (phyllody of the stalks).

The question of the constancy of these intermediate races is a very important one. I propose to deal with it when referring to individual cases in detail; and the only general statement I shall make now is that both constant and inconstant intermediate races exist. On the one hand there are those cases in which an overstepping of the limits between these two races is apparently as rare as the mutations by which new species arise, and

[1] See GOEBEL, *Organographie,* I, pp. 153-159; E. H. HUNGER, *Ueber einige vivipare Pflanzen.* Diss. Rostock, 1887. Bot. Jahresber., 1888, T. XVI, I, p. 421, and also, especially, CLOS, in *Actes du congrès international de botanique,* Paris, Sept., 1900, p. 7.

in which at least, in spite of every precaution and care, I have not yet succeeded in obtaining the one race from the other. (*Trifolium incarnatum quadrifolium, T. pratense quinquefolium, Ranunculus bulbosus semiplenus.*) On the other hand are those races which when cultivated on a sufficiently large scale give rise every year to individuals which seems to overstep the otherwise fixed limits of the race. These are therefore inconstant intermediate races. I regard this phenomenon as one of atavism, at any rate in those cases where, as in my own observations, they revert from an eversporting variety to the type of the parent species without however acquiring the constancy of the latter. Atavistic phenomena of this kind are well known in striped flowers and variegated leaves; and I have also found very striking examples of it in *Linaria vulgaris peloria* and *Plantago lanceolata ramosa* (§ 20 and § 17).

Besides the cases which fall into the two categories just discussed, I succeeded in finding a third in which one intermediate race arose from the other very rarely and only in isolated cases. I have seen two cases of this so far. One was the origin of *Linaria vulgaris peloria* from *L. v. hemipeloria* (§ 20); the other was the formation of the double *Chrysanthemum segetum plenum* (Plate II), from *C. s. grandiflorum* with 21 instead of 13 tongue-florets (§ 18). *Linaria vulg. peloria* is probably an intermediate race, on account of its inconstancy; whereas *L. vulg. hemipeloria* (with stray peloric flowers) is obviously a half race. The origin of the former from the latter presumably occurs in nature from time to time. My *Chrysanthemum segetum plenum* is a novelty in the horticultural sense of the term, being just as double as the double varieties of other composites; so far as I

Eversporting Varieties. 25

know it has not as yet arisen anywhere else. It constitutes an eversporting variety like a number of other double composites which are analogous to it; and arose in my experimental garden, not from the original species, but from a variety known in the trade as *C. s. grandiflorum,* which forms a first step towards it in respect of the number of its tongue florets, and is therefore to be regarded as a half race.[1]

Let us now briefly summarize the foregoing discussion:

1. There exist both in the cultivated state and in nature a series of forms which are either not constant or highly variable, a state of affairs which is probably due to the interaction of two antagonistic characters.

2. Of these two characters one is to be regarded as normal, that is to say, as belonging to the parent species; the other as the abnormal.

3. Where the former preponderates, teratological half races with their half curves are the result.

4. If the two maintain an equilibrium, there are formed what I have called middle races, intermediate races, or eversporting varieties, of which many examples are to be found amongst garden varieties and "heritable" teratological races.

5. The high degree of fluctuating variability of the eversporting variety, its occasional discovery in nature and in cultivation, and the possibility which it affords of the working up of striking novelties by means of isolation and selection, afford an explanation of the major-

[1] The numerous apices of the curves describing variation in the number of rays in composites, which have received no explanation so far, tend however to make the application of this conception difficult.

See also the origin of *Dahlia variabilis fistulosa* in my cultures (§ 11, p. 100.

ity of the phenomena which led DARWIN to his theory of the slow transformation of species. For at that time it was believed that the inception of this process was to be sought in the variation of a character already existing, whereas as a matter of fact the variation in question is independent of the fluctuation of the existing characters.

6. The origin of a constant variety or a new species could be easily imagined to occur in this way: First a half race would arise from a pure race, then from this half race a middle race and lastly, from this latter, a new constant form. But this would be pure fancy, since it is without any basis of fact. Besides in many cases the intermediate stages are entirely wanting.

§ 4. HALF RACES AND HALF CURVES.

The study of anomalies must be based on the theory that external factors can only be efficient in altering the form of the plant if the power to react to them (or the potentiality for the change) is already present.[1] "The induction of malformations by external causes is no more than the manifestation of latent potentialities," says GOEBEL.[2]

Every plant possesses a whole host of such latent potentialities. A single plant of *Plantago lanceolata* may be *ramosa, stipitata,* and *bracteata*; it may have split leaves and pitchers composed of one or two leaves; and it may exhibit abnormal twisting and forked ears, or present a whole series of other anomalies. The seeds of a single self-fertilized plant will very often give rise to

[1] See *Intracellulare Pangenesis,* p. 194.
[2] GOEBEL, *Organographie,* p. 158.

a whole series of malformations. Many cultivated plants, such as *Cyclamen, Pelargonium* and *Fuchsia,* are particularly productive of such abnormalities.

The internal factors may either be latent or semi-latent. In the former case the characters are either not manifested, or only exceptionally, as in the pinnate leaves of the red clover (Fig. 46) and as in the numerous cases of pitchers which have been found once, or only at long intervals, in the same species. In the second case they appear more or less regularly, often yearly, and in many specimens. For example I have observed the formation of pitchers on *Magnolia obovata* in the various botanical gardens which I have visited; and this species as well as its near allies bears pitchers with us every year.[1]

In both cases these potentialities are heritable. This is proved by their frequence in the case of the semi-latent characters and rendered extremely probable in that of the latent ones by their occasional reappearance.

Latent and semi-latent characters constitute what we may call the outer range of the forms of a species. The inner range of forms consists of the normal characters of a species which are exhibited during its normal life or are only induced by such common stimuli as wounds, mutilations, darkness, or the uncovering of subterranean organs and so forth. They are part of the innermost essence of the species. But the countless latent characters belong just as much to the essence of the species, especially when they have formed part of the inner range in some remote ancestor and are therefore atavistic. And it is just this outer range which presents the best indica-

[1] *Over de erfelykheid van Synfisen,* Bot. Jaarb. d. Gesellsch. Dodonaea, Gent, 1895, p. 129. In the course of ten years I have observed about 100 pitchers on *Magnolia.*

tions of descent and therefore of systematic relationship. It fully deserves and repays the attention given to it, as CELAKOWSKY'S admirable papers show. It is to be hoped that others will, following the lines laid down by HEINRICHER, undertake the task of rendering these characters more amenable to study by cultivation, and so bring an increased number of them to light.

The manifestations of latent characters are so rare that they scarcely ever lend themselves to statistical study (p. 19). When they recur from time to time they are seen to be extremely variable, since as a rule even the rarest anomalies are not quite the same each time they appear. It is easily seen in such cases that the variability is a unilateral one; but the construction of curves usually fails owing to the sparsity of the material.

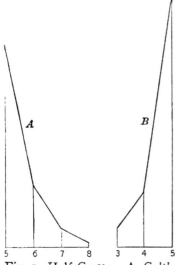

Fig. 1. Half Curves. A, *Caltha palustris*. Curve of the number of petals in 416 flowers. B, *Weigelia amabilis*, Curve of the slips of the corolla in 1145 flowers.[1]

The half races are much more favorable in this respect. Here the deviations are by no means so very rare. As a rule the normal character still preponderates, but material sufficient for statistical study can often be found without difficulty. It shows clearly that the variation is a unilateral one. The apex of the curve is the mean of the normal character, and the deviations all lie on the same side. And in ordinary cases they are less numerous

[1] *Ber. d. d. bot. Ges.,* Vol. XII, 1894, p. 197, Plate X.

Half Races and Half Curves. 29

the further they deviate from the type of the species. Fig. I gives a couple of examples at *A* and *B*. *A* gives the number of petals of *Caltha palustris* in a locality not far from Hilversum; the flowers, where the species is pure, are pentamerous. But in this place there occurred flowers with 5-8 petals in the following proportions:

Flowers with	5	6	7	8 Petals.
Relative number	72%	21%	6%	1%

Weigelia amabilis, also, has normally pentamerous flowers; but it often varies in a minus direction. I found in 1145 flowers on three bushes in our garden (Fig. 1 B):

Number of slips in the corolla	3	4	5
Number of flowers	61	196	888

Half curves differ from the half of a normal curve because the height of the mean, i. e., the number of normal cases, is too great. Such curves do not display the variability of the character given by the highest ordinate, but that of another character which is concealed in the normal flowers.[1]

Half- or unilateral curves are widely distributed in nature. Where they occur they point to the existence of half races. Nevertheless middle races can, under certain circumstances, as we have already pointed out (p. 20) exhibit half curves; just as, on the other hand, the half

---

[1] Half curves are therefore compound curves. Their apex corresponds to the mean value of the normal character; their flank is the expression of the semi-latent character. If the normal character, in the material at our disposal, does not vary it has no curve of its own, which accounts for the absence of a flank on the other side. This for example is the case for curves based on numbers, when the normal number is constant or practically constant as in the case of the three-leaved clover or pentamerous flowers. If the normal character is distinctly, though slightly, variable, as in the case of data based on measurements, the half curve has a flank on the other side, but it is very steep. I do not propose to pursue this point any further here, since it is merely my object to show that half curves are only a special case of asymmetrical curves.

curve of a half race can be tranformed into a bilateral curve by selection and high nutrition. I shall recur to this point shortly.

The well-known researches of FRITZ MÜLLER with *Abutilon* give instances of half curves.[1] MÜLLER obtained the following figures from a culture with seeds from flowers with six petals: 145 with 5 petals, 103 with 6, and 13 with 7. Of more recent investigations we may mention those of BATESON and PERTZ with *Veronica Buxbaumii* according to which the normal cases always composed 70-90% of the culture in spite of the selection of the extreme variants in petal-number as seed-parents, the remaining 30-10% being composed of abnormal cases in a rapidly diminishing series.[2] The fruits of *Aquilegia* are pentamerous, but 6-, and still more rarely 7-merous, ones occur. The fruit of the cotton is also pentamerous, but I have found several tetramerous and occasional trimerous ones. *Papaver Argemone* has tetramerous flowers, but specimens with 5, and less often with 6 petals, also occur; by sowing seeds from the latter I was not able to obtain any increase in the number.

Duplications of leaves, concrescence of umbel-rays in *Umbelliferae,* of the fruit stalks of *Cruciferae,* of the fruits themselves in the *Compositae* and so forth, the adnation of an axial bud with its axillary branch and a number of other anomalies behave as half races. The abnormal cases, which are of course infinitely rarer than the normal ones, become rarer in proportion as they depart from the normal. It is unnecessary to give a longer list here, I may just mention the catacorolla on the outer

[1] HERMANN MÜLLER, *Die Befruchtung der Blumen*, p. 450.

[2] W. BATESON and MISS D. F. M. PERTZ, *Notes on the Inheritance of Variation in the Corolla of Veronica Buxbaumii.* Proceed. Cambridge Phil. Soc., Vol. X, Pt. II, p. 78.

side of the corolla in a half race of *Linaria vulgaris* which I have studied for a few generations, and for which the half curves have recently been plotted and investigated by GARJEANNE.[1]

It is well known that every species has a tendency, as the expression is, to vary in certain definite directions; in these the deviations occur fairly frequently, in others either not at all or very seldom. The number of anomalies is by no means an unlimited one for a given species, but strictly limited. One expression of this phenomenon is the fact that one species tends to produce and repeat one particular abnormality, and another species, another. This general fact, with which we are familiar in vague expressions of this kind, can be made the starting point of valuable experimental investigations. For what are we to understand by "tendency" in these cases? In my opinion simply the existence of a half race or sometimes even of an eversporting variety. These two types of races are, so far as my experience reaches, perfectly distinct, and in numerous cases amenable to experimental study; they are things with nothing intrinsically vague about them although they are sometimes blurred in their manifestation, under a superficial examination, on account of the high degree of fluctuating variability which they exhibit.

If we take a plant which presents this tendency to a particular anomaly and cultivate its progeny, isolating it with an eye to this tendency, we shall usually find that we are dealing with an intermediate race of the kind of which we have spoken. I shall refer to an instance in the following section (§ 5); but this will be only one out of

[1] A. J. M. GARJEANNE, *Beobachtungen und Culturversuche über eine Blüthenanomalie von Linaria vulgaris.* Flora, 1901, Vol. 88, p. 78; with Plates IX and X.

many. It is frequently uncertain, at first at any rate, whether besides the half race, the "species" itself exists in pure condition, that is to say, a race in which the character in question is not semi-latent but latent. But when, as is so often the case, the species is widely distributed but the half race is only observed locally, we are evidently fairly safe in assuming the separate existence of both.

Anomalies which are very common in nature point to the existence of eversporting varieties; those which are rare, to half races. In the former case they are often reckoned among the characters proper to the species, as for instance the remarkable lateral fruitlets on the fruits of *Tetragonia expansa,* which were included by DE CANDOLLE in his diagnosis of the species, in his *Prodromus*.[1] Other well-known instances are the incomplete apetaly of *Ranunculus auricomus*,[2] as well as the branched ears of *Lolium perenne ramosum* which seem to be relatively common everywhere in my own country. LENECEK[3] records lime-trees with 20-30% of their leaves transformed into pitchers; and with us trees with single pitchers, and others which produce large numbers of them every year are met with from time to time (Vol. I, Fig. 106, p. 470).

In many cases we know both the half race and the middle race of the same, or of closely related, species. For example, there grows very commonly here a form of *Plantago major* (*f. bracteata*) which bears more or

[1] A. DE CANDOLLE, *Prodromus Regni Vegetabilis.* See also EICHLER, *Blüthendiagramme,* II, p. 120.

[2] WINTER, *Journ. of Bot.,* Vol. 35, 1897, p. 406. This form also grows in our garden and in our country in the wild condition.

[3] O. LENECEK, *Mitth. d. naturw. Vereins,* Vienna, 1893. Found not far from Leitmeritz.

less numerous green bracts on the lower parts of the spikes. The well-known *Plantago major rosea* of our gardens, all of the bracts of which are green and fairly large, constitutes the complementary, and constant, eversporting variety. Besides *Papaver somniferum polycephalum* (Vol. I, Figs. 27-28, pp. 138-139) which is to be regarded as an eversporting variety, there are polycephalous half races of *P. commutatum* and several other species which in my cultures behave in quite a different manner from the former, in response to selection. Besides the favorite *Varietates cristatae* of our cultivated ferns we occasionally find, in nature, wild species with a split leaf. *Celosia cristata,* the cockscomb, is an exceedingly interesting eversporting variety,[1] besides which fasciated half races in numerous other genera are known.[2] But I must refrain from the citation of further instances.

Just as a species can as a rule be distinguished from its nearest allies by two or several characters, so a half race can manifest as semi-latent anomalies two or more characters which are latent in the species in question. Nor is this by any means rare. In the case of characters which deviate in the opposite direction from the type of the species, "double half-curves" may be formed which have two unequal flanks. The number of petals of *Hypericum perforatum* varies in this way, in this neighborhood, round a mean of 5; on the one side going frequently to 4 and rarely to 3, and on the other side rarely to 6. The corolla of *Campanula rotundifolia* often varies from 5 to 6 and 7, and rarely from 5 to 4 and 3.[3]

[1] See the second part of this volume.
[2] *Botanisch Jaarboek,* Gent, 1894, p. 72.
[3] See also *Ber. d. d. bot. Ges.,* Vol. XII, 1894, p. 202, where further examples will be found.

Selection and nutrition have as usual a great effect on half races. I shall not deal exhaustively with this point until the end of this part, but will give here a brief discussion of the general principles underlying it in order to prepare a proper understanding of the question.

Our discussion of the phenomena of fluctuating variability in the third part of the first volume led us to the conclusion that selection and nutrition usually operate in the same manner on the individual characters of plants.

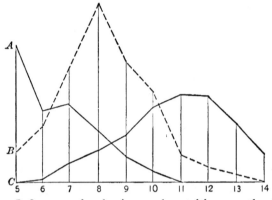

Fig. 2. Influence of selection and nutrition on the half race *Ranunculus bulbosus semiplenus*. A, Half Curve after several years of culture. B, Curve of the 12 best individuals (i. e., those richest in petals). C, Curve of the best plant.[1]

Positive selection and plenty of food enhance the development of a character, whilst selection in a minus direction or defective nutriment operate in the opposite direction.

Polycephaly in *Papaver somniferum* behaves in this way,[2] and, as we shall see later on, all the other anomalies which have been tested do so, as well. Half curves can thereby be transformed into unilateral ones (Fig. 2), either

[1] *Ber. d. d. bot. Ges.*, Vol. XII, 1894, Plate X, Fig. 4.
[2] Vol. I, Part I, pp. 135-143.

by making a special curve from plants which exhibit the largest number of abnormalities, or by making a curve from a race bred from such plants. But an improved race of this kind remains dependent on selection and high nutrition, and is soon lost if these are withheld.[1] One instance will suffice. *Achillea Millefolium* has white flowers, but occasional specimens have red ones. From this I have bred a race with red flowers, which sometimes even attain the deep red of dark wine. After four years of stringent selection, all the plants in successful cultures were more or less red. But if the plants were grown too close or were on poor land, more than half were white, and when I made further sowings without selection the proportion of reds rapidly reverted to its original small amount. On the other hand there is on the market the well-known *Begonia semperflorens atropurpurea Vernon* as a constant dark brownish red variety.

Eversporting varieties revert rapidly under *minus*-selection, but it is seldom possible to eradicate their character altogether as I experienced when working with the adnation of the lateral branches to their main stems in *Aster Tripolium* and *Bidens grandiflora*, and as I shall describe later in greater detail in the case of *Celosia cristata*. (See Part II of this volume.)

In conclusion, we see that in nature as well as in cultivation (especially in the case of horticultural varieties and other anomalies) intermediate forms between the original species and its constant variety are often met with. The two commonest are the half race and the middle race or eversporting variety. The former has a half curve, the latter a bilateral one. Both occur in

---

[1] Vol. I, Part I, § 14, p. 122.

numerous species and genera, either together, or separately. Both are easily influenced by high nutrition and selection, but are usually quite distinct and only apparently connected by transitional forms.

§ 5. TRIFOLIUM PRATENSE QUINQUEFOLIUM, AN EVER-SPORTING RACE.

Four-leaved clovers are notoriously rare in nature, but it is perfectly easy to have many hundreds of them, provided a hereditary race can be obtained. Isolated ex-

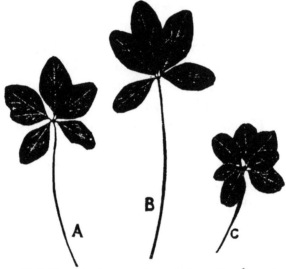

Fig. 3. *Trifolium pratense quinquefolium*, five-leaved and seven-leaved leaves of clover. The left leaf, A, shows a transition towards the 6-merous leaf in the splitting of one of its leaflets.

amples of this race seem to occur sporadically in nature; all that has to be done is to find, to isolate, and to multiply them. (Fig. 3.)

In the following section I shall describe the history of a particular race. I shall do so largely with a view to emphasizing the contrast between a middle race and a half race. In a half race the latent or semi-latent character is very seldom visible, perhaps in one leaf or on one plant amongst many thousands, and after several years of selection it is only on isolated individuals that two or three specimens of the anomaly may be found.

In the middle race, or eversporting variety as I call it in contradistinction to the half race, the anomaly is by no means so rare. Most of the leaves consist of from 4-7 leaflets, and plants without such scarcely occur at all even in the absence of any selection. Trifoliate leaves are not wanting; indeed no plant is without them, particularly in its early stages and on weak branches.

On the other hand pure five-leaved or pure seven-leaved races do not as yet exist: I mean forms which do not revert.[1] There is no ground for supposing that we may not succeed, some time, in obtaining at least a constant seven-leaved variety. But for this to happen the right combination of unknown causes must chance to offer itself (see § 2); and this has not yet occurred in the case before us.

When a variable race has been found in nature the next step is to isolate it. And if, as is the case with red clover, isolated individuals of the species are sterile, two or three of them must be cultivated together, or if this is not possible one or several generations must be grown as a rule in order to purify the race of the effects of crossing. But this is easily effected. Further, the character can be improved by selection within the limits of variability in the new race, just as in the case of pure specific characters. When once the furthest point in this direction has been reached, and this usually occurs after a

---

[1] Or at least revert as rarely as four-leaved individuals occur in the ordinary clover, which are in reality also partially atavistic.

few generations, further improvements are only to be expected from a corresponding amelioration of the conditions of cultivation. In this way I succeeded in the beginning in improving my four-leaved clover, but after 1895, in spite of continuous and stringent selection, no further improvement has been observed. I shall therefore confine myself to a description of the first seven generations.

These were:

1st Generation.	1886–89.	Two plants from Loosdrecht.
2nd "	1890.	Four plants with some four- and five-leaved leaves.
3rd	1891.	36% abnormal leaves per plant.
4th[1]	1892.	S With isolated abnormal seedlings. C With 60% seedlings of which the first, second or third leaf was tetramerous.
5th	1893.	C With 55% seedlings with compound primary leaf.
6th	1894.	C With 96–98% seedlings with compound primary leaf.
7th	1895.	C With 95–97% seedlings with compound primary leaf.

To proceed to a more detailed account I begin with the examples collected in the field.[2] I found them near Loosdrecht on the edge of a road which was covered with grass. They bore several tetramerous and one pentamerous leaf and seemed therefore to afford better opportunities than the usual find which often is only a single four-leaved clover leaf in a meadow. I transplanted them to my garden, where they lived for another three years. Here the anomaly not only reappeared but in-

---

[1] The result for this year is a double one. *S* (spring) refers to the crop of 1892 itself. *C* (crop) to the record of the seedparents in terms of the seedlings raised from their seeds (see p. 40): Similarly with the subsequent years.

[2] *Over het omkeeren van halve Galton-curven*, Kruidkundig Jaarboek, Gent, Vol. X, 1898, pp. 27-54 with Plate I.

creased, on account, doubtless, of the improved conditions of life. In July and September, 1889, I counted 46 tetramerous and 19 pentamerous leaves amongst a large number of normal ones. But there was no sign of a 6- or 7-foliate leaf on these two parent plants of my race.

I saved seed from them in the autumn of 1889 and sowed it next spring on a bed in my experimental garden. I obtained something over one hundred plants of which about one-half showed at least one four-leaved leaf. The rest were removed either in July before they flowered, or whilst they were in flower. On September 1, I selected the four plants which bore the largest number of abnormal leaves, and destroyed the rest. These four bore 64 tetramerous and 44 pentamerous leaves. Of the destroyed plants the best had only an average of about 5 abnormal leaves per plant. This year again there were no instances of 6- or 7-foliate leaves.

In 1891 I obtained the third generation from the seeds of these four selected plants, sowing in the garden as before. It consisted of 300 plants on which I examined 8366 leaves when they were beginning to flower. Of these 1117 or 14% were tetra- or pentamerous. Leaves with 6 or 7 leaflets were not observed; they were first seen in August and September of that year. The number of plants with at least one quadrifoliate leaf also exhibited an advance. There were about 50% of them in 1890, but now there were nearly 80%. These plants had on an average about four tetramerous and as many pentamerous leaves. At the beginning of August I chose the twenty best individuals and destroyed all the rest. I only harvested seed from the nine best plants among them and in the following spring only sowed the seeds of a single seed-parent which seemed to me to be the very best of all, 36% of its leaves being composed of more than 3 leaflets.

In the spring of 1892 I sowed the seed in pans in the greenhouse attached to my laboratory instead of in the beds as before. The advantages of this were (1) that more seeds germinated and (2) that the examination of the seedlings was greatly facilitated. They stayed in the pans until the unfolding of the third leaf, were then looked through, and the best ones transplanted into pots with manured garden soil. Amongst the several hundred seedlings there were 18 in which the quadruplicity was already manifest among the first leaves. Only these specimens were planted out; during the summer they bore a large number of tetra- and pentamerous leaves; and some 6- and 7-foliate ones, which appeared now for the first time in considerable numbers.

With this, the isolation of the five-leaved race of clover was brought to an end. The elaboration of the ordinarily latent or semi-latent character had been fully accomplished. The race could, like any other, be improved by selection but it could not be expected to change its character any further in the process.

Of course I did not omit to effect this further improvement. But there was no point in paying further attention to the characters of the adult plants, since differences could now only be found in them by a statistical examination of all their leaves. And it was found to be practically impossible to carry out this scrutiny with the necessary detail, for the plants soon become too big to be grown in pots. Therefore in order to make curves it is necessary to defoliate the plants, and this can not be done until after the choice of the seed-parents, whose leaves must obviously neither be removed nor even damaged.

For these reasons it is desirable to effect the selection in the seedling stage, or at any rate before transplanting. This process had already been begun in the spring of 1892 and needed therefore only to be perfected by continned selection. And the result justified my expectations.

In the spring of 1893 I sowed the seed of the 18 plants of the year before, already referred to, separately for each seedparent. I recorded the seedlings when the third leaf had unfolded. If all the leaves were normal, I straightway weeded out the plant; but if one or more of its leaves had a supernumerary leaflet I preserved it. Of the 3409 seedlings which I examined 2471 were normal and 938 were not, i. e., about 30%.[1] Of course the remaining 70% must also be abnormal, but the anomaly was not yet recognizable in the seedlings. Some of them which I transplanted produced, as adult plants, leaves with from 4 to 7 leaflets in large numbers.

I determined the percentage production of abnormal seedlings in this manner for 16 of the 18 seed-parents; the values were distributed over them as follows:

	10–20%	21–30%	31–40%	41–50%	51–60%	61–70%
Seed-parents:	1	7	3	2	2	1

I further chose from this series a seed-parent producing 60% abnormal seedlings. It had itself had in its early stages a compound primordial leaf, which fact also marked it out for the continuation of the race. It will be found in the table on p. 38 under 1892 C.

Amongst the seedlings from the seeds of this parent

[1] *Botan. Jaarboek,* Gent, Vol. X, p. 37, where the two figures have been transposed by an oversight.

several occurred with trifoliate (instead of single) primordial leaves (Fig. 4). I only selected these as seed-bearers, for transplanting, and I effected a considerable simplification in my cultures by adopting this mark as a criterion for all further selection of stock plants. For the definitive selection could now be made 2-3 weeks after sowing, and it was not necessary to pay any further attention to the development of the character; this was fully insured. Nevertheless I took care by means of further experiments to satisfy myself that there exists a fairly close relation between a large number of 4-7-merous leaves on a plant and a high percentage of abnormal seedings produced by it.

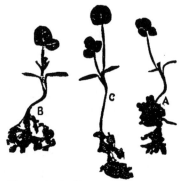

Fig. 4. *Trifolium pratense quinquefolium.* A, Seedling with a trifoliate primordial leaf. B, C, Seedlings with single and bimerous primordial leaves; these two latter types were regarded in my race as atavistic.

In July 1893 I only saved the 12 best plants raised from the seed of the plant of 1892 with 60% abnormal offspring. With the exception of two plants they all bore not only 4-6-foliate leaves, but even some 7-merous ones. The four best had 27, 30, 33 and 34 of this latter type. There were no leaves with more than seven leaflets.

The plant with 34 7-merous leaves also produced the highest percentage of abnormal seedlings, as shown by the result of the sowing in the following spring. Of 209 seedlings produced, 51 had a bimerous, and 61 a trimerous, primordial leaf, i. e., 55% of abnormalities. It was therefore chosen as seed-parent (see p. 38). It should be remarked that in previous years seedlings with a compound primordial leaf had either been entirely absent or at any rate very rare.[1]

In the summer of 1894 I only bred offspring from the plant with 55% of abnormalities in its seedlings, and of these only the twenty best, with compound primordial leaf and the next leaf tetra-pentamerous. These only did I allow to flower and to bear seed. The result was recorded by means of the same characters in the following spring. For eleven plants it was 70-90%, for five others 91-96%, and for the two best 98-99% seedlings with compound primary leaf. And the higher the number the greater was the percentage of trifoliate, as opposed to bimerous, primordial leaves.

The same high percentage was obtained in the culture of the next year, 1895, in the seventh generation of my experiment. Since then the race has remained constant under the same conditions of selection.

I have employed this constant and highly abnormal race for a series of observations and experiments, to the more important of which I shall now refer,[2] for they are well qualified to afford us some insight into the nature of such a race. *This race exhibits a high degree of variability, which is due to the possession of a semi-latent character besides that which it has obviously inherited from the parent species. The extent to which this parental heritage, the normal trifoliate leaf, is developed depends on the conditions of life of the plant.* And, speak-

---

[1] See the remarks in § 22 relating to the size of the seed in *Trifolium incarnatum*. In the five-leaved clover, especially in later years, practically all the seedlings had compound primordial leaves, so that this character had nothing to do with the size of the seed.

[2] For a detailed account see the oft-cited paper in *Kruidkundig Jaarboek*, Vol. X.

ing generally, favorable conditions favor the characters of the race, and unfavorable ones those of the species (see below, § 26).

This is only a special case of the well-known principle: *Every injury increases the tendency to atavism.*[1]

In the first place let us consider the *periodicity*. The number of multipartite leaves increases with the individual strength both on the whole plant and on the separate branches. And if, at the end of growth, weakness supervenes this number again decreases.

Let us examine Fig. 5. It is a photograph of a strong young branch which was removed on August 1, 1900. The lowest leaf was nearly withered; it was small and had the inversely egg-shaped form of the leaflets which is characteristic of the leaves of the young red clover, It consisted of only 3 leaflets. The two following leaves were markedly larger and stronger, of a more elliptical form and tetramerous. Then follows a 6- and then a 7-merous leaf, after which the leaves again return to the simpler types.

The branch photographed was chosen for its regularity; and yet a pentamerous leaf is absent from the ascending series. Most of the branches, even on the best plants, were less regular: indeed it often happened that tetramerous leaves were succeeded by some trimerous ones, and so forth.[2]

What has been stated concerning the lateral branches is also true of the rosette of radical leaves whose axis

---

[1] That is, reversion of the race to its parent species, for the character of the race is itself, morphologically speaking, a reversion to a more remote ancestor.

[2] For exact figures the reader is referred to: *Ueber die Periodicität der partiellen Variationen*, in Ber. d. d. bot. Ges., 1899, Vol. XVII, p. 48.

is, of course, the main stem of the whole plant. Here also the number of leaflets per leaf first increases, on the average, and then decreases, with many fluctuations however. The branches themselves exhibit a certain periodicity since the lower ones contain a smaller quantity of

Fig. 5. *Trifolium pratense quinquefolium*, 1900, showing the periodicity of the anomaly on a branch. Beginning from below the leaves have 3—4—4—6—7—5—leaflets.

abnormal leaves than those next above them, whilst the highest of all are poorer again.

If therefore the conditions are favorable to a branch in its earliest stages it will develop more 4-7-merous leaves. And it is obvious that such leaves will extend both above and below the maximum of the period in direct proportion to their number. Whence it again follows that the better nourished the plant is, the earlier will the abnormality appear. And this is true both of the individual branches and of the rosette of radical leaves, and therefore of the whole plant.

From these conclusions the converse rule may be deduced that the earlier a seedling produces its first tetramerous leaf, the greater will probably be the number of abnormal leaves on the adult plant. The most abnormal plants will probably be those which in the seedling stage had a compound primary leaf. Experience has proved the truth of this rule throughout my experiments.

If we now take another glance at the table on page 38 we see that the character recorded has gradually shifted in the course of generations and as a result of selection. The more the improvement advanced the earlier could selection be effected. In the third generation I kept 300 plants in the beds to be selected from; since the fourth generation I have carried out the selection in the seed-pans and only planted out the few best (e. g., 10-20) to act as seed-parents.

It is possible, therefore, within the limits of such a race, on the one hand to effect an increase in the number of multipartite leaves, and on the other to reduce it by reversed selection. In both cases we go as far as possible from the mean of the race, without, however, succeeding in overstepping its definite boundaries. Let us see what selection is able to effect in the two cases, and let us begin with the former. It is the question of intensifying the anomaly to its extreme limit.

A striking peculiarity of my race is the fact that leaves

with more than seven leaflets have never, or only extremely rarely, been produced. As a matter of fact a duplication of the leaves by splitting, which is so common among other plants,[1] occurs in my race also, and if it affects a pentamerous leaf, makes a 10-merous one of it. But that is the expression of another latent character which we are not concerned with here. Apart from these I have not yet found in my cultures, in spite of the most careful search, a single instance of a leaf with more than 7 leaflets.

The character of my race is the quinquefoliate leaf which is usually in the majority; the remaining types are grouped round it in accordance with QUETELET's law, so far as the tendency to symmetry permits this. For it is clear that this tendency does not favor the regularity of the curve of variation. The increase in the number of leaflets from 3 to 4 takes place by the lateral splitting of one of the lateral leaflets (see Fig. 3 A), one of the lateral veins becoming the primary vein of the new leaflet. Transitions such as that figured are certainly fairly rare, but all degrees of them, down to a splitting of the small partial stalk of the leaflet, occur from time to time. If only one leaflet is split, the leaf becomes asymmetrical; but if the two lateral leaflets split, the whole may remain symmetrical. The duplication can extend to the terminal leaflet and turn a vein of this, either on one side or on both sides, into the primary vein of a new leaflet. In this way the 6- and 7-merous leaves arise; the former are asymmetrical, the latter symmetrical.

The statistical examination of large numbers proves that the symmetrical leaves predominate over the asymmetrical ones. The plant seems to prefer to retain its

[1] DELPINO, *Teoria generale della Fillotassi*, 1883, p. 197.

symmetry even in the anomalies. This is brought out in the curves by the relative shortness of the ordinates corresponding to 4 and 6 (Fig. 6).

Let us return to the processes of selection. The mean of the race is a pentamerous leaf, which varies within fairly narrow limits, never (or hardly ever) less than

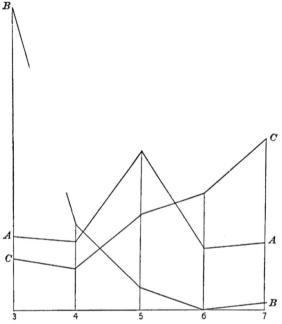

Fig. 6. *Trifolium pratense quinquefolium.* A, Normal curve of the number of leaflets in the leaf. B, Curve of an atavistic individual. C, Curve of the maximum degree of abnormality, 1894.

three or more than seven leaflets being produced. Selection can therefore be either in the direction of the 7 or the 3. In both cases the original symmetrical curve becomes unilateral. But in the former case the improvement of the race is pushed on as strongly as possible, in the latter the reverse happens until it can hardly be distinguished from the ordinary instances of the rare four-leaved clover.

A glance at the table on page 38 will show that my race was only very slightly developed at first, and had to be brought to its normal type by isolation and selection. But in spite of this selection it is not so constant that it does not occasionally give rise to atavistic individuals. On the other hand individuals with a maximum development of the character of the race are from time to time produced. And these extremes are sometimes both found within the limits of a single culture.

I observed this in 1894 with plants which had been raised from the seeds of a single individual in the third generation (1891, p. 38). The seed-parent in question had survived the winter and did not ripen its seed until the second year. In July, 1894, there was a large number of strong plants of the same age, of which I chose the seven best for a detailed examination of their leaves. Some of the oldest leaves were already withered, the youngest not yet unfolded; these were not recorded. Each of these seven plants was plotted in the form of a curve, one (Fig. 6 A) gave the normal curve of the race, another (B) was atavistic, whilst all the rest had their highest ordinate at 7. I have only given the mean value for these five (C).

These three groups gave the following percentage of leaves with the number of leaflets written above them:

Number of leaflets:	3	4	5	6	7	Number of leaves counted
A. Normal example:	17	16	37	14	16	172
B. Atavistic example:	75	19	5	0	1	216
C. Extreme variants:	12	9	22	17	40	97[1]

[1] Mean number per plant.

These figures are presented graphically in Fig. 6. It will be seen that the normal curve is a symmetrical one slightly depressed, however, over the ordinates of the even numbers as a result of that symmetry which we discussed above. The two other lines form half curves; in both of them the apex coincides with one extreme. The curve B, of the atavistic individual, is almost the same as the curve which was the dominant one in the first years of my experiment when there were, as yet, no 6-7-foliate leaves (p. 38). It is an ordinary half curve of variation, which is characteristic of the half races of semi-latent anomalies. The curve C is, however, reversed; it displays the predominance of the racial character over the antagonistic one which is that of the original species. It also shows the preference for symmetrical leaves.

If atavistic individuals are used as seed-parents the character of the race can be observed to vanish more or less completely in a short time. I carried out an experiment of this kind in the years 1896-1898, after the race had reached its maximum development in 1894-1895 as described on page 38. Within the space of three generations this race has retrogressed so far that the plants could no longer be recognized as belonging to it. For the purposes of this reversed selection I chose, from the plants which had borne a large number of 5-7-merous leaves in 1895, those seedlings of which the primary leaves were single and the first leaves trifoliate. With a few exceptions they had all developed occasional tetrapentamerous leaves by the middle of June. Three of the exceptional ones were isolated before flowering, they subsequently developed a few multipartite leaves. But when their seeds germinated it was seen that they were not only not poorer in seedlings with compound leaves but even slightly richer; they were therefore not chosen for the continuation of the experiment. I chose the seeds of three plants of 1896 which had given rise to no more than 2-3% seedlings with compound primary leaves. Atavistic seedlings only were transplanted, but in the following summer (1897) even those bore some tetra-pentamerous leaves, almost without exception. On the other hand 6-7-merous leaves were almost entirely absent, and the race had thus returned to the condition described by the unilateral curve of the first year of the experiment (1891-1892). Some plants produced nothing but trifoliate leaves during the whole of the summer and the following spring.

In 1898 I made another culture of atavists from the seeds harvested in 1897. This was therefore the third atavistic generation. But two thirds of the generation raised still consisted of plants with some tetra-pentamerous leaves, and therefore possessed this character in a far higher degree of development than ordinary red clover. This stringent, thrice occurring reversed selection had therefore considerably reduced the development of the anomaly but had not succeeded in destroying or even in concealing the fact that the culture belonged to the pentamerous race.

I also made an experiment on the influence of external conditions on the development of multipartite leaves. There are two ways of dealing with experiments of this kind; we may either subject the different parts of the same plant to diverse conditions of life or similar samples of seed to diverse treatments from germination onwards. In the former case we determine the effect on the grown plant. This is however seldom great, inasmuch as the plant is most sensitive in its early stages. In this form of the experiment we can, so to speak, only investigate the last vestiges of its former susceptibility. Far more striking results are to be expected from experiments with seedlings; but here a great uniformity in the samples of the seeds is necessary for the results to be reliable. It is not sufficient to mix the seed, but it is advisable to harvest seed from two or three or still better from a single seed-parent of known and pure ancestry. It is even better to allow the influences that are to be investigated to operate during the development of the seed on the parent plant.

In accordance with these considerations, therefore, I cut one of my plants into two parts, one of which I transplanted into poor sandy soil but the other into good garden soil, and allowed them to set seed. I was thus able to study both the direct effect on the plant and also the indirect effects on the succeeding generation. (See *Vol.* I, Part III, pp. 521-522.)

The experiment, which was carried out during the years 1892-1894, was made with a single individual which arose from the stock plant for 1891, mentioned on page 38 and marked S. This plant had, when it germinated in 1892, a bimerous primordial leaf, and in the same year bore seeds which, when sown in the spring of 1893, gave rise to about 40% seedlings with a tetramerous leaf. As soon as this was visible in the seedpan the choice was made and the parent plant, which I had kept through the winter in a bed, was cut in two and transplanted into the above mentioned kinds of soil. Both halves grew well, although not with equal luxuriance; they flowered in July, were pollinated from the various plants around them composing the main culture of that

Trifolium Pratense Quinquefolium.

year, and set seed in August. At this time I examined an equal number of leaves on the two halves and obtained the following result:

Number of leaflets:	3	4	5	6	7
On garden soil:	12	25	34	20	18
On sandy soil:	18	19	35	19	17

The behavior of the two halves was identical; the difference in the soil exerted no visible effect. Moreover the seeds on the two halves were of about the same size and produced in roughly equal numbers. The two sets were harvested separately and sown in the following spring (1894) in pans. When the young plants had about 3 leaves they were examined. Calling a plant with a tetra- or a pentamerous leaf "abnormal" the result was:

Seeds from garden soil 30% abnormal
"   "   sandy soil   24%   "

The experiment involved 150 and 200 seedlings. The abnormal ones were further sorted according to the composition of their primary leaves.

Leaflets	1	2	3	Totals
Seeds from sandy soil	24	10	13	47
" " garden soil	16	12	13	41

Both counts therefore gave a difference in favor of the better nourished seeds. For further investigation I selected those which appeared most abnormal from both series, i. e., the seedlings with a trimerous primordial leaf, and planted them out under similar conditions. In July when each plant had twenty or more stems, I pulled them up, selecting for examination the ten best plants from each group; i. e., those ten, the leaves of which numbered about 100 per plant. The leaves were recorded separately for each individual, and as there happened to be practically no difference between the several individuals in each group, I calculated the mean for the two sets in percentages.

Number of leaflets per leaf	3	4	5	6	7
From seeds from garden soil	14	13	25	16	32
" " " sandy soil	39	13	23	10	15
Difference	−25	0	+2	+6	+17

The effect of the treatment in the previous year is now perfectly plain. The curves for both groups have become unilateral but in the case of the better nourished ones the apex is at seven, and for the others at three leaflets per leaf.

Conversely we may conclude that, in the experiment described on page 47 and graphically exhibited in Fig. 6, the atavists were produced by poorly nourished and the maximal variants by highly nourished seeds. And the following generalization about anomalies seems to be justified: that the nutrition of the seed on the parent plant is the most important factor influencing the development of the anomaly (Vol. I, pp. 521-522).

Let us now briefly summarize the results of this experiment. I began by finding in the field two plants belonging to a five-leaved race, which however as the result of indifferent nutrition for several generations only developed tetra-pentamerous and no 6-7-foliate leaves. By better cultivation and by the continued selection of the most abnormal individuals, no doubt those which happened to have been best fed, a race was evolved in the course of a few generations with a number of leaflets per leaf varying between 4 and 7 round a mean of five. After this selection had been repeated four or five times maximal variants were produced the majority of the leaves of which were 7-foliate. At the same time there were still "atavists" in the seventh generation the apex of whose curve was over 3 leaflets. The atavists however really belong to their race as is shown by the fact that even after repeated selection in an atavistic direction they produce far more quadrifoliate leaves than the normal red clover (or more exactly, the corresponding wild half race of the red clover).

The better the seeds are fed on the parent plant the greater is the development of the anomaly on the individuals produced by them. Poor seeds give rise to atavists, good ones to extreme variants.

My experiment extends over ten generations. It gives no support to the view that the five-leaved race was, so to speak, caught in the act of developing its character, or that it could give rise to a higher type without further mutation. It is a highly variable, but constant variety.

III. THE *DIFFERENT* MODES OF ORIGIN OF NEW SPECIES,

§ 6. HORTICULTURAL AND SYSTEMATIC VARIETIES AND ELEMENTARY SPECIES.

The opinion has of late been often expressed, by VON WETTSTEIN in particular, that there is no ground for the assumption that all species have arisen in the same way.[1] There is no difficulty in applying this view to the theory of mutation, although one of the chief objects of this book is to show that ordinary or fluctuating variability does not provide material for the origin of new species. But this does not exclude the possibility of different modes of origin of new species. The simultaneous origin of species in groups, in definite periods, such as I have described in the case of *Oenothera Lamarckiana,* must constitute for me the main type of this process, until the origin of species has been experimentally studied in other cases. Such experiments would have to study the phenomenon before and during the first appearance of the new type. Inferences drawn from data obtained after its appearance can hardly be considered as decisive.

This essential type explains in my opinion in the first

[1] R. v. WETTSTEIN, *Der Saison-Dimorphismus als Ausgangspunkt für die Bildung neuer Arten im Pflanzenreich,* Ber. d. d. bot. Ges., Vol. XIII, 1895, p. 303; and particularly the same author's *Descendenztheoretische Untersuchungen;* I. *Untersuchungen über den Saison-Dimorphismus im Pflanzenreich;* Denkschr. d. Mat. Naturw. Classe d. k. Akad. d. Wiss., Vienna, 1900.

place the progressive origin of species, that formation of new characters to which in the main the evolution of the plant kingdom is due. On the other hand there is a whole series of other types which are now, so far as it is possible to judge, mainly confined to the lateral branches of the phyletic tree. With regard to these however we must content ourselves at present with indirect methods of investigation.

DARWIN'S statement that varieties are incipient species is well known. So also are the words of one of the most famous authorities[1] on horticulture, VERLOT: *Toute variété a d'abord existé à l'état de variation.* These two generalizations are evidently based on phenomena entirely different from those with which we have become familiar in *Oenothera.* They constitute, so to speak, the other extreme of the series.

I propose therefore now to investigate the manner in which "variations" in the sense of so-called structural abnormalities or anomalies (and not the individuals which exhibit variation in accordance with QUETELET's law) arise, and how they result in the origin of "species." But here we come across an obstacle on the very threshold of the inquiry in the manifold meanings of the word variety.[2] It will soon become clear that horticultural and systematic varieties are to be considered as categories of entirely different values. But both DARWIN's and VERLOT's sentences just quoted are based on data obtained from horticultural varieties; and we must now

[1] B. VERLOT, *Production et fixation des variétés*, 1865, p. 100.

[2] The general conception of this term is that formulated by CARRIÈRE in the following words: "*On nomme variété tout individu qui, par quelque caractère que ce soit, se distingue d'un ou de plusieurs autres avec lesquels on le compare et qu'on considère comme appartenant à un même type spécifique* (*Production et fixation des variétés*, 1865, p. 6).

inquire how far their transference to systematic varieties is justified.[1]

The origin of horticultural varieties will therefore be submitted to a critical and experimental examination. But before we do this I think it advisable to consider first the meaning which is attached to the term variety in systematic works, and secondly the various ways in which species can arise. And we shall find that whilst there is no question that the mode of origin of horticultural varieties is often analogous to that of so-called "good" species, this parallel is by no means so common as the present form of the doctrine of descent would lead one to believe.

To begin with systematic varieties: Here we find we can draw a pretty natural line between what we called elementary species on the one hand, and real systematic varieties on the other.

In connection with this antithesis I think it desirable, after what has already been said on this topic in the first volume,[2] to lay especial stress on the fundamental difference between these two conceptions. LINNAEUS and his pupils describe the elementary species as varieties; JORDAN, DE BARY, and others who argue from experimental data, refer to all forms as species.

The terms "species" and "variety" have become so familiar that it is no longer possible to effect any radical change in their definition. For their exact meaning we have to refer to the works of LINNAEUS himself. His

[1] For some interesting observations relating to the origin of new forms, see the papers by F. KRASAN in ENGLER's *Botanische Jahrbücher*, Vol. XIII, Pts. 3-4; Vol. XXVIII, Pts. 1, 2 and 5, and also his *Mittheilungen über Culturversuche mit Potentilla arenaria*, Graz, 1900.

[2] See Vol. I, § 7, "Species, Subspecies and Varieties," especially pp. 169-172.

conception of them is now common property, and in my opinion our best course is to interfere with that conception as little as possible.

There can be little question that the difference between variants and variations is becoming more and more widely recognized. Variants are what we call individual deviations; they are instances of fluctuating variability. The characters which distinguish them disappear under suitable cultivation and are therefore to be regarded as inconstant. In systematic works they are not as a rule given a place, or merely briefly mentioned, or, lastly, treated as a *Forma*, which is the lowest subdivision of the system; e. g., *Forma alpestris, Forma aquatica*. But this can only be done when the relationship of the form is sufficiently known; lack of material in the case of exotic plants, or incomplete investigation of indigenous species of course would make this impossible, and such forms have therefore often first been described as varieties or even as species.[1] In many cases of course the true relationship is still unknown and the systematic grouping, therefore, to be considered as provisional; as for instance in the case of *Anthyllis Vulneraria alpestris, Limosella aquatica caulescens, Carlina acaulis caulescens,* and so forth.

BONNIER's researches on Alpine plants, discussed in detail above (Vol. I, p. 146), have demonstrated that some of these differences are not even instances of individual but of partial variability. From the two halves of a single individual can be grown the form characteristics of the plains and the *Forma montana*.

[1] For example *Ranunculus aconitifolius* L. in alpibus minor, caule 3-5 floro; *R. aconitifolius altior* KOCH, caule multifloro, fol. laciniis longius acuminatis, in montibus humilioribus = *R. platanifolius* L. mant. 79 (KOCH, *Synopsis*, p. 12).

But the large number of cases of forms preliminarily described as varieties but which possibly may be only variants, is one of the most considerable obstacles in this inquiry.

LINNAEUS himself followed two distinct rules in subdividing his species. According to the one the species was regarded as the type from which the varieties were derived; according to the other, however, the species was regarded as a collective group which embraced a certain number of units of equal value. The separation is sharp and definite and LINNAEUS was obviously perfectly conscious of its reality. In the derived varieties the series begins with β followed by γ, δ, ϵ etc.; it is taken for granted that the type or *Forma genuina* represents the a. In a homonomous series there is no such *Forma genuina*, and the series of varieties therefore begins with a.

Let us consider the two cases separately and let us begin with the second.

LINNAEUS's homonomous varieties, a, β, γ etc., are sometimes arranged in groups, and sometimes not (as in *Teucrium Polium, Lavandula Spica,* etc.). In the former case the species falls into two or several subspecies, each of which again may include one or several varieties. For instance *Euphorbia exigua* has two subspecies *acuta* and *retusa*, the first of which consists of one and the second of two varieties. *Beta vulgaris* has the well-known subspecies *rubra* and *Cicla*; the first of these embraces five, the second two varieties. In these species there is no *Forma typica* or *Forma genuina*. The variety which is named first has no other priority over the others.

In such cases the species is a group of similar composition to that of a genus and of a family; since in these no particular species or genus is regarded as the prototype from which the rest would be merely derived forms. Species of this kind are therefore obviously and avowedly collective species.

LINDLEY, A. P. DE CANDOLLE, ALPHONSE DE CANDOLLE and other eminent systematists consider the collective species without *Forma typica* to be the only really existing type. Species must be subdivided in exactly the same way as genera, says the last named of these authors in his *Phytographie*.[1] LINDLEY splits up his species of roses on the same principle; *Rosa rubiginosa* into 8, *R. spinosissima* into 9 varieties, etc. DE CANDOLLE deals with the difficult and numerous subgenera and elementary forms of *Brassica* in the same way in the second volume of his *Systema Vegetabilium*.

DE CANDOLLE calls the units, which in such cases are treated as varieties, *"les éléments de l'espèce"*;[2] they are related to the species as these are to the genera and as the genera to the families.

But the majority of botanists regard varieties as forms which have been derived from the species. For them the species is the type, the real entity, from which the varieties have arisen by small changes. They follow the course taken by LINNAEUS who based his diagnoses, in the vast majority of cases, on one of the forms of a species and arranged the rest in a lower grade under this. The origin of the varieties from the species was simply inferred from *a priori* premises as I have already shown in the first volume, this origin having only been directly observed in isolated cases of horticultural products; for the majority and certainly the most important

---

[1] ALPH. DE CANDOLLE, *La Phytographie ou l'art de décrire les végétaux*, 1880, pp. 74-82. Much of the argument set forth in the text is due to this excellent work.

[2] *Loc. cit.*, p. 80.

cultivated varieties are as old or even older than cultivation itself.

If we examine a number of such derived forms in any systematic work or flora, it immediately becomes evident that the same kind of differences recur in the most widely separated families, genera and species. Everywhere varieties present series of parallel forms. The recurrence of white flowered varieties in numerous species with blue or red flowers is so familiar a phenomenon, that often all reference to them is omitted. LINNAEUS himself knew that nearly all such species had a white variety. If the color of a flower is compound, and if one of the components is lacking, a white flower with a dark center often results and is known as a *Var. bicolor* (for example *Cynoglossum officinale bicolor, Agrostemma coronaria bicolor*); or the dark patches are absent as in *Gentiana punctata concolor,* which case is exactly analogous to that of *Arum maculatum immaculatum.*

Often too, the clothing of hair is lacking either on the whole plant or, when only certain parts are densely hirsute in the "species," on these. The nomenclature of the series of parallel forms, under this heading, is particularly rich in terms which all indicate the same property, as for example: *Papaver dubium glabrum, Biscutella laevigata glabra, Arabis ciliata glabrata, Arabis hirsuta glaberrima, Veronica spicata nitens, Amygdalus Persica laevis, Eritrichium nanum leiospermum, Paeonia corallina (peregrina) leiocarpa,* etc.

Thornless forms are usually termed *inermis*; they occur in *Ranunculus arvensis, Genista germanica, Robinia Pseud-Acacia* and many others. The *Varietas ciliata* occurs in *Cytisus prostratus* and in *C. spinescens*, also in *Lotus corniculatus,* etc. A dense clothing of hair is the

Horticultural and Systematic Varieties. 63

distinguishing feature of *Solanum Dulcamara tomentosum, Veronica scutellata pubescens, Melissa officinalis villosa, Galeopsis Ladanum canescens, Vicia lutea hirta, Lotus corniculatus hirsutus,* etc.

The patches of color at the base of the petals are often absent in *Papaver orientale,* in *Erodium cicutarium* and many other plants. Such names as *ochroleuca, purpurascens, integrifolia, serratifolia, angustifolia, latifolia* denote varieties each one of which may recur in several unrelated species. Finally I may mention the red berries which occur as a varietal character in *Empetum nigrum* and characterize the red variety of the gooseberry; and the yellow berries of *Atropa Belladonna lutea* and *Daphne Mezereum album* which are only selected examples from a long series of such varieties.

All these forms differ from their species in the fact that a particular one of their characters is either developed to a greater extent (*hirsuta, ciliata, purpurascens,*) or on the other hand very slightly developed or entirely absent.

The absence of a character may also be a case of extreme rarity in the vegetable kingdom such as the strawberries without runners, and the peculiar *Pinus Abies aclada,* with its tall absolutely unbranched stem, which has been figured by SCHRÖTER.[1] *Fragaria vesca monophylla* (Vol. I, Fig. 38, p. 193), *Robinia Pseud-Acacia monophylla, Fraxinus Ornus monophylla,*[2] and a monophyllous form of *Melilotus coerulea* (Fig. 12 on page 87) belong to the same category.

The varietal names enumerated above almost always occur, in systematic works, in series which begin not with

---
[1] C. SCHRÖTER, *Die Vielgestaltigkeit der Fichte,* 1898, pp. 52-53.
[2] A. BRAUN, *Verjüngung,* § 332. Here also the earlier literature will be found; and some facts concerning *Rubus Idaeus monophyllus.*

α but with β and are therefore considered as having been derived from a *Forma typica* or *genuina* and not as being of equal value with this. Evidently the principle on which they are classified is borrowed from a consideration of horticultural varieties. This proceeding, however, is only justified in the relatively rare cases in which horticultural varieties can be demonstrated to be younger than the species. Besides this the geographical distribution of the forms in question is often employed to decide which is the species, and which are the varieties. If one particular form is wide-spread and another only local or sporadic in its appearance it is obvious that the former will be regarded as the older and therefore as the species. Often this fits in conveniently with the fact that the species was discovered earlier than the variety, so that instead of disturbing the classification in the system all that had to be done was to range the variety under the species.

The exigencies of space prevent me from going into further detail here. What I have already said may suffice to show that the systematic term "variety" means two fundamentally different things to LINNAEUS and the later systematists:

1. *Homonomous Forms,* amongst which even LINNAEUS could not select one as a type for the others; "Eléments de l'espèce" (DE CANDOLLE) or elementary species.
2. *Derived Forms,* which are distinguished from the type of the species only by the decreased or increased development of a particular quality; or by its complete absence: True varieties.[1]

[1] Amongst these, again, the simple invariable types are to be distinguished from the intermediate or eversporting races. (See §§ 3-4.)

I think it undesirable that these two types of subdivisions of the species should continue to be denoted by the same term. The simplest plan would be to refer to the former as elementary species and only to the latter as varieties, and I hope that this limitation of the terms will come into general use.

The question, however, is a purely systematic one and belongs to the department of descriptive science. For as soon as it is treated from the experimental standpoint the whole difference disappears. Many of the best varieties prove, when tested by sowing, to be as constant as elementary species, so that a separation on the basis of constancy is out of the question.

* * *

On the basis of the foregoing discussion I treat the homonomous subdivisions of the LINNEAN species as elementary species and eventually denote them with binary names. In the case of derivative varieties, however, I prefer to make no definite choice; I regard for example *Chelidonium laciniatum Miller* and *Chelidonium majus laciniatum* as equally justifiable. And when for instance several species in the same genus have white flowered or glabrous varieties, a binary nomenclature would obviously be much too cumbrous.[1]

§ 7. PROGRESSIVE, RETROGRESSIVE AND DEGRESSIVE FORMATION OF SPECIES.

A glance at the phylogeny of the vegetable kingdom reveals the fact that all species cannot have arisen in the same way. Progressive development is due to the con-

---

[1] For instance if specific names like that of *Agrostemma nicaeensis* for *Agrostemma Githago pallida* were generally used for white flowered varieties.

tinual formation of new characters, to increasing differentiation. Nevertheless the great multiformity of species within the orders and families is only in part due to this progressive process, but to a large extent to an infinite variety of combinations of characters already existing. This is combined in innumerable cases with instances of regression; that is, with the absence of characters which are otherwise proper to the group to which the species belongs. *Sium* and *Berula* have, for example, simple pinnate leaves within the group of the *Umbelliferae* with doubly pinnate leaves; and the assumption is that they have arisen from the latter by a simple loss. Similarly *Primula acaulis* stands in the middle of a group containing the *Primulas, Androsace* etc. with umbellate inflorescences, and the same inference is drawn as to its origin. The same is true of a host of other cases, and even for whole groups. For instance DELPINO holds, as is well known, that the Monocotyledons have arisen from the lower *Dicotyledons* by the loss of a whole series of characters.

Cases such as these are spoken of as instances of retrogressive metamorphosis. And it is probably not too much to say that there are possibly more species on the face of the earth at present that have arisen on retrogressive than on progressive lines.

The question is often debated whether, in retrogression, the characters absolutely disappear or only become invisible, or latent. There is much evidence for the latter view, derived largely from the great variety of atavistic structures (youth forms, subvariations on the lower internodes of lateral branches, the form of the leaf in suckers, the effects of parasites, anomalies, reversions to the ancestral form by bud-variations, etc.). Latency is without doubt the general rule. That an actual internal loss may also occur is probable on general grounds, but very difficult to demonstrate in a given case. For every positive result points to latency, and nothing but a negative result after exhaustive investigation could warrant the conclusion that a character had absolutely disappeared.

The multiformity of species within the larger groups is also due to a phenomenon which DARWIN calls parallel variation. I refer to the repeated appearance of the same new character in related or remote groups.[1] Climbing and tendril-bearing plants, parasites, saprophytes and insectivorous plants, decussate phyllotaxy, are a few names from a vast number of instances. One of the greatest difficulties for the systematist, the question as to the mono- or polyphyletic origin of many characters is a problem of a similar nature. For example, are the siliqua and silicula in the Cruciferae, or is the position of their embryo to be regarded as an indication of mono- or polyphyletic origin? Do the Sympetalae with an inferior ovary originate from other Sympetalae or from epigynous Choripetalae? Have the Gymnosperms arisen once or oftener from the vascular Cryptogams? We do not know, because, on such points, the highest authorities are not in agreement. And so long as these differences of opinion exist it will be difficult to approach the question as to the cause of the parallel formation of specific characters—whether they arise from a common latent source, or afresh each time—with any hope of success.

The origin of systematic and horticultural varieties

---

[1] On this point see also my *Intracellulare Pangenesis*, English translation by Prof. C. Stuart Gager (Chicago, The Open Court Publishing Co., 1910).

is evidently due, in the vast majority of cases, to retrogressive development or latency, as I have already attempted to show.

There is a close analogy between the formation of these varieties and that of certain species. The origin of varieties (such as *Var. hirsutissima, spinosissima* and *ciliata*) as the result of the intensification of characters is a much rarer phenomenon. This form of variety, which seems to be of very little importance in the evolution of the vegetable kingdom, may be called subprogressive, and the phenomenon of its origin subprogressive formation of species.

The parallel, retrogressive, and subprogressive modes of origin have this in common that they only provide new combinations and do not contribute new units or any essentially new elements to the progressive evolution of the vegetable kingdom. In this respect they stand in sharp contrast to progressive formation of species.

There is another series of phenomena to be mentioned here, of still less significance in the phylogeny of plants. The first of these is the manifestation of old, latent characters. A whole series of anomalies are so widely distributed in the vegetable kingdom, or at least among flowering plants, that it is almost impossible not to assume a common cause for them. This cause must be a latent character that has arisen in some common aucestor and therefore must be of great antiquity. The commonest and best known example of a widely distributed anomaly of this kind is that of fasciation, instances of which in almost any desirable number of species can be collected in the course of a few years. It appears that almost every species amongst Coniferae and Monocotyledons, but especially among Dicotyledons, can exhibit fasciations.[1] As a constant horticultural variety it occurs in *Celosia cristata*; but as a specific character, so far as I know, it does not occur. This is, however, true of the almost equally widely distributed split leaves (as in *Boehmeria biloba*), of adherences (*Solanum*), of flowers on leaves (*Helwingia rusciflora* and others) and of numerous other anomalies of which CASIMIR DE CANDOLLE has given a valuable general account.[2] He calls them "*Variations taxinomiques*"; whereas anomalies which do not occur as specific characters such as fasciation, twisting, virescence and sterile varieties, are designated by him as ataxinomous. I take the following further instances from his list: Connation of opposites leaves, which occurs normally in *Dipsacus, Lonicera* and others, or of the cotyledons (amphisyncotyly, normal in *Sicyos*); Pitchers, normal in *Sarracenia* etc., and in the peltate leaves, as for instance *Eucalyptus citriodora*; foliacious excrescences on the leaves, normal in *Senecio sagittifolius* from Uruguay, and on petals, normal in *Petaquia saniculaefolia* and as a sport in *Clarkia elegans*; Synanthy in *Lonicera,* and so on.

For our purposes, however, the question is not which anomalies can also occur as specific characters, but conversely which specific characters can also occur as anomalies in other species. For our task is to deal with the problem of species and especially to provide an answer to the question how far their characters can be derived from more or less widely distributed latent qualities which have existed for a long time in the vegetable kingdom or in particular groups of it. With this end in view

---

[1] See Vol. I, Figs. 34 and 35 on pp. 182-183.

[2] C. DE CANDOLLE, *Remarques sur la tératologie végétale,* 1896, pp. 5-6.

70 *The Different Modes of Origin of new Species.*

I shall supplement the examples named with a few more; they serve to show how general this parallelism between anomalies and specific characters is. Thus, for example, *Polygonum viviparum* and *Agave vivipara* bear adventitious buds or bulbils normally in the inflorescences; but I found them also as an anomaly in *Aloë verrucosa* and *Saxifraga umbrosa*. A spiral involution is normally exhibited by the flowerstalks of *Vallisneria* and *Cyclamen*, and it occurs as a variety in the stalks of *Juncus spiralis*, and as an anomaly in *Scirpus lacustris* of which latter a beautiful instance came under my notice. Hypocotylous buds are, for example, normally present in *Linaria* and *Linum*; they occur as an anomaly in *Siegesbeckia*[1] according to BRAUN, and I have also observed them in *Phaseolus multiflorus*. The numerous flowerbuds on the leaf stalk of *Cucumis sativus* as described by CASPARY[2] are analogous to the buds scattered on the internodes of *Begonia phyllomaniaca*. The bulbs of *Gladiolus* carry their lateral corms on stalks; I observed the same mode of connection as an anomaly in *Hyacinthus orientalis*. MASTERS has collected a series of teratological cases[3] of buds on leaves, which may be regarded as parallel to the normal instance of the same phenomenon furnished by *Bryophyllum*.

We see therefore that a large number of specific characters are analogous to taxinomous anomalies. The latter recur in related forms, but much more frequently in more or less remote groups. In so far as they are due to a common cause, they point to the widespread

[1] A. BRAUN, *Verh. d. bot. Vereins Brandenb.*, XII, 1870, p. 151.
[2] CASPARY, *Ueber Blüthensprosse auf Blättern*, Schriften d. phys Gesellsch., Königsberg, 1874, p. 99 and Table II.
[3] MASTERS, *Vegetable Teratology*, p. 170.

existence of latent characters. I shall refer to this mode of formation of species as degressive. In it, that which arises is always something new, and often something strikingly new, but usually without any clear relation to the progressive development on the main lines of evolution. They form, rather, lateral improvements of types already existing.

Degressive formation of species is therefore due to the activation of long established latent characters. Of these, as GOEBEL has shown in his *Organographie,* there are two types to be distinguished.[1] Either the character in question was active in the earlier ancestors, or it was not. In the former case we have an instance of reversion or atavism, and moreover a genuine systematic reversion, at least inasmuch as the ancestral relation can be demonstrated. In the other case we have only the development of a specific character from a taxinomous anomaly.

It is perhaps hardly necessary to state that the application of this criterion of grouping can only be effected at the present moment in a relatively small number of cases. The information at our disposal does not as yet meet the demands of such a system. On the physiological side, however, the question of prime importance is only the distinction between the chief groups; so that we will only lay stress on that point here .

Bearing this limitation in mind we can summarize what we have already said, as follows:

THE ORIGIN OF NEW SPECIES.

A. By the formation of new characters: Progressive specific differentiation.

B. Without the formation of new characters.

[1] K. GOEBEL, *Organographic,* Vol. I, p. 170.

72 The Different Modes of Origin of new Species.

B_1. By characters already existing becoming latent: Retrogressive specific differentiation, cases of atavism.

B_2. By the activation of latent characters: Degressive specific differentiation.

    a. From taxinomous (latent) anomalies.

    b. As genuine atavism.

B_3. From hybrids.

* * *

This list does not of course claim to be complete. There is no doubt a whole series of further types which can be more or less easily ranged under or parallel to these subdivisions. On the other hand it is at once clear that the distinction between *A* and *B* is, in the present state of our knowledge, the important thing, and moreover that it will suffice as a basis for experimental inquiry. But before I proceed to illustrate this antithesis I will offer some remarks on the last section (B_3).

New species can arise from hybrids but specific characters cannot arise by means of hybridization; or, we may say that with regard to the production of mutations, hybrids behave just like ordinary species, except that according to the prevalent view they are slightly more prone to it. The existence of a vast number of species, however, is due to the various combinations of characters which also exist in closely allied or in remote species. And it is evident that, by crossing, characters can be combined which have not appeared in the same genealogical line but in distinct though allied species. Thus for example by crossing *Oenothera rubrinervis* with *O. nanella* I obtained an *O. rubrinervis-nanella* which has remained constant for many generations without segregation and without reversion. And a host of new species have doubtless arisen on similar lines.

Coming now to the discussion of the difference between our two groups *A* and *B,* we draw a distinction between *progressive* specific differentiations or the origin of new specific characters on the one hand, and *retro- and degressive* specific differentiation, which consists in the activation or latency of potentialities already in existence.

It is obvious that a premutation is necessary for progressive but not for retro- and degressive differentiation. For in the case of the former the new potentialities must first arise before they can become visible externally, whilst in the case of the two latter we are only dealing with potentialities already existing. I propose therefore to apply the results obtained with *Oenothera Lamarckiana* and the conclusions regarding the premutation period to which we arrived, to the further elucidation of this question.[1] It is of course a purely speculative discussion that we are embarking on, but one which will, in my opinion, materially help in clearing the ground. And I may therefore say, in anticipation, that this theory is supported by the experiments to be recorded in this section and most strikingly by the history of my *Linaria vulgaris peloria* (see § 20).

I have already stated, in Vol. I, Part II, that I regard the mutational period in *Oenothera Lamarckiana* as a type of the mode of origin of species in general; that is to say, of the essential form of that process, the progressive type.[2] We often find in the vegetable kingdom analogous groups of closely related species which are

[1] See Vol. I, Part II and especially § 31, p. 490.
[2] Vol. I, p. 259.

usually ranged as elementary species of larger species, but sometimes regarded by the best authorities as "good" species. The group most closely related to our experimental series is that of *Oenothera biennis* or the subgenus *Onagra*;[1] more remotely connected are the groups of *Hieracium, Rosa,* etc., or of *Draba verna, Viola tricolor* etc. Such groups appear to us as the relics of past periods of mutation. The new forms which arise from such periodical mutations are to be regarded as homonomous subdivisions of the older species or as elementary species.

It is natural in such periods not only that new specific characters should arise, but that old established latent ones should reappear more easily than at other times; and among the mutations of *Oenothera Lamarckiana* our *O. nanella* is undoubtedly analogous to typical horticultural dwarf varieties, and *O. laevifolia* to those systematic varieties which arise by the loss of a character.

These latter, however, and similar retro- and degressive changes are an entirely different matter. The essential condition for their production is always present, and all that is needed is the external stimulus to induce the mutation. This, it appears, need not occur periodically, nor affect several characters at the same time. New horticultural varieties appear at irregular intervals of time, and here and there in the area of cultivation of the species. But it is equally certain that we have to do in such such cases exclusively, or almost exclusively, with retrogressive and degressive changes.[2] Analogy and parallelism are universal, and their effects sometimes go so far that the characters of the species fall into the background. Double flowers look so much alike that one

[1] See Vol. I, p. 439; and § 31, p. 490.
[2] I am excluding from this consideraton the effects of crossing.

often cannot tell, even from the best illustrations, to which genus or family a given instance belongs.

I shall therefore throughout this Part attempt to describe the origin of horticultural varieties as exhaustively as possible. In the present state of our knowledge they form in my opinion the pattern of retrogressive and degressive formation of species; just as the mutations of *Oenothera* were the pattern of progressive changes. Together they give us some idea of the main lines along which specific differentiation takes place in nature, at the present time as well as in the past.

In conclusion: Progress on the main lines of descent results from the production of new characters; but the extraordinary variety of forms results from the occasioual disappearance of characters already existing, or from the activation of latent ones (retrogression, degression, atavism).

IV. THE SUDDEN APPEARANCE AND THE CONSTANCY OF NEW VARIETIES.

§ 8. EXAMPLES OF CONSTANT RACES.

Horticultural varieties are generally constant; exceptions to this rule are usually noted expressly in the textbooks. Most varieties are not only constant from seed but also pure. By constant is meant that in ordinary cultivation they produce no more impurities than are unavoidable (that is to say, at most 3%). Absolute purity means that when isolated under experimental conditions the seeds reproduce their own variety without exception. Constancy in this case is complete, but it is seldom of practical interest to bring either the old established sorts or the novelties to this pitch of purity, or even to find out how closely they approach it.

This has, however, been repeatedly done by scientific investigators and especially by DARWIN and HOFFMANN.[1] Insufficient familiarity with the danger of chance crossings robbed the results of the older investigators of much of their value as evidence, except of course in those cases where the race proved constant. The large number of observations of instances of complete constancy were ob-

---

[1] See the *Rückblick auf meine Culturversuche* of the latter author in the *Botanische Zeitung*, 1881, and the literature cited there. IHNE and SCHRÖTER have given a complete list of HOFFMANN's papers in the obituary of him in *Berichte d. d. bot. Gesellsch.*, Vol. X, 1892, p. 18 of the last part.

viously made under conditions which excluded the effects of crossing.

In spite of the existence of these experimental data, it is still the general view that varieties are inconstant forms. That which distinguishes them from true species is supposed to be their faculty of giving rise to occasional and not even rare reversions to the type of the species. This, moreover, is supposed to be a proof of their assumed relation to their species.

Every one of course is free to choose his own definition of a variety. But he who makes inconstancy an essential part of the definition will have to exclude a very large number—and perhaps the most important—of our horticultural varieties, and regard them as elementary species.

I have spent much time in the endeavor to test the constancy of horticultural and also of wild varieties with a view partly of directly satisfying myself as to their purity and partly of finding inconstant forms for subsequent experiments. I have usually started from seed but sometimes, in the case of perennial varieties, from bought plants. Whenever possible the visits of insects were excluded and the plants artificially pollinated. But in the great majority of cases pollination has to be left to bumblebees and moths, and we must be content in providing complete isolation.

The most important point is the extent of the experiment. Absolute constancy can obviously never be directly demonstrated. The space and time needed for other experiments seldom permit the bringing to flower of more than a few thousands of plants of one sort. And even if this is continued for several years the possibility of the occurrence of rare cases of atavism (e. g., once in a million) can not be excluded. The object of experimentation cannot therefore be to demonstrate absolute constancy. The best plan usually is to be content with a few hundred individuals; it is even often impossible to get sufficient seed for more. Experiments on a smaller scale should only serve to confirm the results obtained in other cases; but even if they only do this they are, in my opinion, by no means without value.

Fig. 7. *Bidens tripartita.* Type without ray florets.

The nearest that we can get to demonstration of absolute constancy is to make observations on races which grow in vast quantities in certain districts and are nevertheless true to their type. In these cases the constancy is so striking as to induce many systematists to regard the form as a species. Amongst the better known examples of this are the *Discoidea* forms of many composites.

MOQUIN TANDON regarded the *Discoidea*, i. e., the form without tongue-florets, as the *Peloria* of the Composites.[1] They are generally regarded as having arisen from the *Radiata* forms. Sometimes the discoid form is

Fig. 8. *Senecio Jacobaea* L. (f. *radiata*).

commoner than that with rays, and then the *Discoidea* form may be found described in systematic works as a species and the *Radiata* as the variety; as for instance in *Bidens tripartita* (Fig. 7), and *B. cernua*,[2] although *B.*

[1] *Tératologie végétale*, p. 179.
[2] KOCH, *Synopsis Florae Germanicae*, p. 309.

grandiflora, B. bipinnata, and *B. atropurpurea* are well-known species with ray florets. *B. tripartita* and *B. cernua* are very common in Holland and I have often tried to find or to obtain examples with ligulate florets,

Fig. 9. *Senecio Jacobaea discoideus*, KOCH.

but as yet in vain. By this fact both forms are proved to be constant as thoroughly as a proof can be. In other countries, however, the *varietates radiatae* are known to occur. Similarly *Senecio Jacobaea* has a *Forma radiata* and a *F. discoidea*[1] (Figs. 8 and 9) both of which grow in this country and are absolutely constant. The *Discoideus* grows in thousands of specimens in the dunes in the province of North Holland; but the *Forma radiata* grows equally abundantly in South Holland; both are amongst the commonest and most widely distributed species of our flora. For twenty and more years I have had them under observation, and never saw any trace of admixture or reversion; the two varieties were always absolutely pure in the respective localities. Of late, however, there have been some cases of intermingling near the limits of their areas, probably as the result of seed transportation. The two sorts can therefore be regarded as absolutely constant.[2] *Matricaria Chamomilla discoidea*[3] has proved equally constant in my experimental garden, but MURR mentions the occasional occurrence of heads with rays.[4] In 1897 I raised from the seeds of a single plant of *M. discoidea* 575 plants, all of which were without ligulate florets. On these I only harvested the seeds of the weakest branches of the higher orders and raised 460 plants in 1898, all of which again were with-

[1] See Vol. I, p. 196.

[2] A valuable summary dealing with this point is given by. J. MURR, *Strahllose Blüthen bei heimischen Kompositen*, Deutsche Bot. Monatsschr., Vol. 14, 1896, pp. 161-164. See also *Botan. Jahresber.*, T. 24, 2, p. 11, where rare instances of forms with rays belonging to normally discoid species and rayless flowers on normally radiata forms, are given. I cite *Senecio Jacobaea* as an instance of the latter, in opposition to the observations given in the text. An attempt to discriminate half races amongst these forms (See § 3, p. 18) would probably lead to valuable results.

[3] For an account of the rapid spread of this form in Norway see JENS HOLMBOE, *Nogle Ugraesplanters Invandring i Norge*, 1900. *Nyt Magaz. f. Naturv.*, Vol. XXXVIII, p. 187 (with map). The variety is there also fully constant.

[4] J. MURR, *loc. cit.*, pp. 161-164.

out ligulate florets. From these plants I only harvested the poorest possible seed on the latest branches after cutting away the main stem and the stronger branches; but from this seed, as before, I obtained nothing but *Discoidea* (750 plants in 1899).

Flowerheads without, or almost without, rays also occasionally occur in races usually normal in this respect. Examples of this have occurred in my experimental garden in *Chrysanthemum coronarium, Coreopsis tinctoria, Dahlia striata nana* and others.[1]

In the first volume I cited numerous examples of constant varieties[2] and showed[3] that many of them were certainly one or two centuries old; in fact as old, or probably even older than, the cultivation of their species itself. The varieties are generally as constant as the wild elementary species, of which *Draba verna* and *Viola tricolor*[4] were cited as examples. Belonging to the same group are the two remarkable types, which HERMANN MÜLLER has distinguished in *Iris Pseudacorus*, of which the one with narrow openings to the flower is adapted for pollination by *Rhingia*, whilst the other is adapted for pollination by bumble bees;[5] IRWIN LYNCH has recently compiled a very complete and valuable list of

---

[1] Further examples are given by MURR, *loc. cit.*

[2] See p. 196. Examples are afforded by GAILLON-strawberries (Vol. I, Fig. 7, p. 34) and by *Chelidonium laciniatum* (Vol. I, Fig. 36, p. 190).

[3] On page 183 of the first volume will be found a list of the varieties known to MUNTING (1671) and still cultivated.

[4] See Vol. I, Figs. 3 and 4, pp. 22 and 23. For the constancy of the elementary species of *Viola tricolor* see also V. B. WITTROCK, *Viola Studier* in Acta Horti Bergiani, Vol. II, No. 1, 1897 (Cultures extending over three years).

[5] H. MÜLLER, *Die Befruchtung der Blumen*, p. 68.

constant varieties, based on data given by gardeners and botanists as well as on his own observations.[1]

It is a common opinion amongst gardeners that white flowered varieties are among the most constant. They are very plentiful and easy to control. From the cases as yet investigated it may be assumed that hybrids between them and the colored species will be colored also, and therefore soon and easily discovered; so that in the purification or fixation of these varieties the hybrids are usually removed soon and altogether, which is a very important thing for approaching constancy. Several investigators have tested the purity of white varieties. For instance HILDEBRAND[2] worked with white Hyacinths, *Delphinium Consolida, Matthiola incana* and *Lathyrus odoratus*; HOFFMANN with *Linum usitatissimum album*;[3] HOFMEISTER for thirty years with *Digitalis parviflora alba*;[4] PREHN with *Scabiosa alba*,[5] etc.

I myself have made similar observations. I started by buying a few plants of each of some varieties of perennial species, allowing them to flower on isolated spots and then saving and sowing their seed. Wherever the isolation was complete all the offspring, with a single exception (*Aquilegia chrysantha*), were white flowered. The following were the species tested in this way (I give in parentheses after each one the number of plants which were raised from their seed and observed in flower): *Campanula pyramidalis alba* (26), *C. persicifolia alba* (1044),

---

[1] IRWIN LYNCH, *The Evolution of Plants*, Journ. Roy. Hort. Soc., Vol. XXV, Pt. 1, pp. 34-37, Nov. 1900.

[2] HILDEBRAND, *Die Farben der Blüthen*, p. 70.

[3] HOFFMANN, *Botan. Zeitung*, 1876, p. 566. See also the very complete list of constant white varieties given by CARRIÈRE, pp. 12-13, and the literature cited there.

[4] HOFMEISTER, *Allgemeine Morphologie*, p. 556.

[5] J. PREHN, *Schr. Naturw. Vereins Holstein*, Vol. X, 1895, p. 259.

Catananche coerulea alba (5), *Hyssopus officinalis albus* (198), *Lobelia syphilitica alba* (537), *Lychnis chalcedonica alba* (401), *Polemonium dissectum album* (126), *Salvia sylvestris alba* (296). The following white varieties of annual species I also found to be perfectly constant: *Chrysanthemum coronarium album* (400), *Godetia amocna*, white Pearl (15), *Linum usitatissimum album* (779), *Phlox Drummondi alba* (50), *Silene Armeria alba* (617). Among wild species I subjected especially *Erodium cicutarium album*, which is common in Holland, to a severe test. In this form the pigment characteristic of the species is lacking both in the leaves and in the flowers. I found the variety constant through five generations in my experimental garden, not a single red plant appearing although the sowings were conducted on a very large scale. Later I collected seeds of the *var. alba* from another locality and found it also to be constant (43 specimens).

Other color varieties usually prove equally pure if the seeds of plants that have been isolated are sown. In some cases this fact is so generally known that they have, on this ground, been raised to the rank of species, as, for instance, *Anagallis* (*arvensis*) *coerulea*. In 1897 I had 25 examples of this variety flowering on an isolated spot, and from these in 1898 I had 866 plants which were without exception blue. *Tetragonia expansa*, whose leaves and flowers are normally reddish brown, has given rise to a pure green form which has been raised to the rank of a species under the name of *T. crystallina*. This I also found perfectly constant. In 1898 I sowed about 600 fruits obtained from a culture of 1897. Each fruit contains from 6 to 10 and often more seeds, which germinate sooner or later, some of them not until after a few years have elapsed. In the course of the first summer 3975 seeds germinated, during the second 1082, during the third 88, and during the fourth 90. All the 5235 seedlings thus raised were green without a trace of the red pigment, and belonged therefore to the *T. crystallina*. In this case, therefore, the seeds which germinate late are just as constant as those which germinate early.[1]

In other cases where the constancy is just as complete but happens to be less well known, the sorts in question are "only" regarded as varieties. Some of these forms even seem to be wholly unknown in botanical circles,[2] as for instance, *Silene Armeria rosea* whose color is intermediate between that of the species and that of the white variety and which is not a hybrid but an old established perfectly constant sort and just as "good" as the other two. In 1898 I had about 4000 plants raised from the seed of isolated plants of 1897 of the *Var. flore roseo* in flower; they were all of the same color as the parent plants. The same result was obtained on a smaller scale in subsequent years. *Clarkia pulchella carnea* behaved in the same way (50 specimens). I also found the pale flowered *Agrostemma Githago nicaeensis* constant (for 10 years), and *Hyoscyamus* (*niger*) *pallidus* (40 spec.) and *Agrostemma coronaria bicolor* (349 spec.). Further examples of the same phenomenon are afforded by the yellow *Chrysanthemum coronarium*, the varieties of the flax with white and with yellow seeds, many varieties without the dark patches at the base of the petals, which are characteristic of the species, as in *Papaver somniferum Danebrog*, *Papaver commutatum*, *Madia elegans* (Fig. 10) and others.

[1] This is not the case with *Trifolium incarnatum quadrifolium* (See § 22).
[2] See *Bot. Zeitung*, 1900, p. 234.

Another interesting constant variety is *Chelidonium majus latipetalum*. (Fig. 11), for the possession of which I am indebted to Prof. J. W. MOLL in Groningen. It differs from *C. majus* in its petals which are so broad that their edges overlap so that they form an unbroken

Fig. 10. *Madia elegans.*

Fig. 11. A, B, *Chelidonium majus latipetalum.* C, D, *Chelidonium majus.*

crown instead of an open cross. I found it to be constant through several generations.

The constancy of the fasciated variety of *Myosotis alpestris; Victoria* with its broad, many-petalled central flowers, is likewise well known, as is also that of *Linaria vulgaris tricalcarea*,[1] of many glabrous forms like *Lychnis vespertina glabra*, of thornless types like *Datura Stramonium inermis*,[2] etc., all of which I have tested personally. Space does not permit of the compilation here of anything like a complete list of constant varieties.

Fig. 12. *Melilotus coerulea monophylla*. Each leaf consists of a single blade but is more or less deeply incised. One of the lateral leaflets in the middle leaf on the right of the figure appears to be quite free. A, a bract from the inflorescence; here the leaves are least incised.

[1] J. H. WAKKER, *Linaria vulgaris*, Nederl. Kruidk. Archief, 1889, with plate X.
[2] See Fig. 5 on page 31 of Vol. I.

I shall conclude by referring to a race of *Melilotus coerulea,* the possession of which I owe to the kindness of Prof. M. W. BEYERINCK (p. 63). Its leaflets[1] are fused to a single blade in which the three main nerves still diverge from the base. The blade moreover has three distinct tips, the depth of the indentations between these being subject to considerable fluctuating variability. Not rarely the three parts are only united from the base to halfway up or less, and sometimes they are separated almost to the base and in rare cases even entirely so. All these forms may occur on the same plant. But there was no reversion in my experimental sowings; every plant exhibited this monophylly to a greater or less extent.

§ 9. STERILE VARIETIES.

One of the greatest difficulties presented by the current doctrine of selection lies, as I have pointed out more than once in the first volume of this work, in the fact that the gradual origin of species, which is presupposed by it, has never been observed. In every case in which observations have been made sufficiently close to the origin of a new form, they indicate a sudden change. We do not find those gradual transitions which the doctrine of selection would lead us to expect. The new form may be highly variable, and in that way the limits between it and the parent species may sometimes overlap; but, as I have already shown (*Vol.* I, § 25, p. 430) transgressive variability of this kind only provides a morphological transitional series and not a genetic one.

My object in the present chapter is to bring together

[1] This form has been described by WYDLER, *Flora,* 1860, p. 56, and occasionally since.

a list of further instances partly from the literature and partly from my own observations in order to place my conclusions on a broader basis of facts.

The difficulty of this task is increased by the fact that it often seems impossible to show how those cases, in which other investigators believed that they had detected transitional series, are to be explained on the theory of mutation. This is especially so where the authors have simply relied on comparative investigations. The results of these can usually be explained, no doubt, by the supposition of transgressive variability, but a proof can only be given if the phenomena in question are investigated by statistical methods.

In strong contrast to these doubtful cases, however, there is a long series of observations in which the absence of transitions is practically certain. Perhaps the most striking of these are the sterile varieties which constitute one of the most serious obstacles to the current doctrine of selection, at any rate as regards its exclusive application. DARWIN himself repeatedly cited them as objections and examined them minutely.

Fig. 13. A flower of *Lilium candidum plenum*. The thalamus is changed into a long stalk on which the narrow perfectly white petals are spirally arranged.

In the case of the vast majority of sterile varieties we know neither how, when nor where they arose. They are propagated by vegetative methods and have been from time immemorial. But they differ from their supposed parent species so markedly that they take rank with the best varieties. Nobody supposes that they have arisen gradually.

The first instance that I give is *Lilium candidum plenum*, a form which is on the market. It is a well-known variety, the bulbs of which are offered every year by dealers in bulbs, in their catalogues. Instead of flowers it has long stalks clothed with petals (Fig. 13). The stalk is the elongated thalamus; the petals are narrow and dead white, and of the color and structure of the petals of other white lilies. In each flower they continue to appear for several weeks; the lowest may be brown and withered before the uppermost have unfolded. Figure 13 shows a fairly short flower; they are often twice as long. Stamens and carpels are never formed; the apex consists of a compact bud of the youngest petals.

We do not know how the variety first arose. It was first described by G. VROLIK in 1827, after he had already seen it flowering for 20 years in the Botanical Garden in Amsterdam.[1] It is therefore nearly a century old. In horticultural literature it is not referred to until much later, about 1840.[2]

Another well-known sterile garden plant is the green Dahlia (*Dahlia variabilis viridiflora*). The flowerheads

---

[1] G. VROLIK, *Over een rankvormige ontwikkeling van witte leliebloemen.* Verhandelingen der eerste klasse v. h. k. Nederl. Instituut van Wet. te Amsterdam, Part I, 1827, pp. 295-301, with one table. The spike with five flowers figured there, is still preserved in our collection.

[2] See MÉRAT, *Ann. Soc. d'hortic. de Paris*, 1841-1845, and VERLOT, *loc. cit.*, 1865, p. 91.

are destitute of flowers; and the thin, transparent bracts are transformed into small green leaves. The variety is much cultivated in gardens, partly as a curiosity and partly because their green "flowers" do not wither but remain fresh on the plant; which renders it of a decorative effect until far into the autumn.[1] The variety arose in a crop of seedlings about the middle of the last century in Boskoop in Holland, and since then has been grown from tubers. It occasionally bears isolated red ray florets but, so far as I know, never sets seed.

Some years ago I obtained what seems to be a new and hitherto undescribed form of green *D*ahlia through the kindness of Messrs. ZOCHER & Co. in Haarlem. It is not known whence this form came because it was at first taken for the type of green *D*ahlia we have just been considering. It differs from this however in the fact that the green heads are not of the normal form and size but transformed into long green leaf-bearing spikes like that figured in Fig. 14 with the exception of the clump at the top.

This form produced elongated flowers of this kind in great numbers in the nursery garden; but it could never, so to speak, bring its growth to a conclusion. They grow until the autumn and often longer, and frequently attain a length of 30 centimeters and more. They behaved in exactly the same way in my garden until last year when I manured them heavily. Then there appeared from a few of the green "flowers" in late autumn a little head at the uppermost end (Fig. 14). This unfolded, but consisted of green bracts only; it contained neither flowers nor seeds. The plant is therefore perfectly sterile.

Another variety closely analogous with this is the

[1] See the literature in PENZIG's *Teratologie,* II, p. 71.

92 *Sudden Appearance and Constancy.*

Fig. 14. Elongated Green Dahlia, a new variety.

wheat ear carnation (*Dianthus Caryophyllus spicatus*). In this case we find instead of the flowers small green ears which are formed of green bracts arranged crosswise. This sterile form does not appear to be in general cultivation, although it is perennial; nevertheless it turns up here and there in crops of seedlings especially of mixed varieties. I cultivated a plant arisen in this way; most of the ears were sterile as usual, but some of them produced a flower at their top. From these I obtained several germinative seeds.[1]

The green rose has been known from time immemorial, but the green *Pelargonium zonale* is a modern product. In both cases the petals and stamens are transformed into green leaves. They are said to be perfectly sterile, and are only propagated by cuttings.

Many double flowers never produce seed, and this is especially true of those forms which do not develop structures intermediate between stamens and petals, but are described as instances of Petalomania.[2] Ra-

[1] After having been sown these seeds have repeated the wheat-ear variety (Note of 1909).

[2] K. GOEBEL, *Beiträge zur Kenntniss gefüllter Blüthen, Jahrb. f. wiss. Bot.*, Vol. XVII, pp. 217-219, and elsewhere.

nunculus acris (Vol. I, Fig. 40, p. 194), *Caltha palustris, Anemone nemorosa, Hepatica triloba, Tropaeolum majus flore pleno, Clematis recta, Barbarea vulgaris floribus plenis* and many others are alleged in horticultural literature to be perfectly sterile. Double varieties of composites also afford instances such as *Achillea Ptarmica, Ageratum mexicanum* (some varieties), *Pyrethrum roseum*, etc. Others, such as *Anthemis nobilis,* are known to bear seed from time to time and so do not belong here.

Viburnum Opulus, Hydrangea hortensea, Muscari comosum plumosum and others have become sterile by the transformation of their reproductive flowers into sterile ornamental ones. Bananas and other seedless fruits have already been dealt with in the first volume (p. 195).

Many varieties of the sugar-cane never set seed, such as the Cheribon cane which is the richest in sugar. This variety, which extends over vast regions, consists of a single individual; that is to say, it is derived from one single unknown stock plant and has always been propagated by cuttings or so-called *Bibits* only.

Robinia Pseud-Acacia inermis[1] is also said never to flower; and is only propagated by grafting.

If this sterility affects annual species or such as cannot be permanently reproduced by vegetative means, the sterile form must vanish sooner or later. Such forms hardly deserve the name of variety, and are usually spoken of as monstrosities. But, in regard to their origin, they are perfectly analogous with the sterile forms of which we have already treated. In the first volume (p. 195, Fig. 41), I gave the sterile maize as such an instance. More remarkable still is the unbranched Fir (*Pinus excelsa aclada* or *monocaulis*) which SCHRÖTER

[1] DE CANDOLLE, *Physiologie,* II, p. 735.

has described in his excellent monograph. The whole plant consists of a single branchless stem, which is merely slightly swollen at the limits of every year's growth; the needles remain adherent for a long time.[1] This form has appeared in diverse localities. SCHRÖTER records 4 examples from Italy, one from Baden, several from Westphalia, Mittelfranken and Bohemia, and some from Mariabrunn near Vienna. The majority of these plants reached a height of 1-2 meters, a few of them as much as 5-6 meters; some of them are still alive

RIMPAU has described an instance of sterile rye.[2] Ears of this rye appeared almost every year during a period of more than ten years; they were often much and sometimes excessively branched, especially in years and localities where the rye was very thin. But as ears of this kind occurred on plants which also bore normal ones, the repeated appearance of the anomaly may perhaps be due to inheritance.

And lastly, instead of giving a further record of the numerous existing sterile varieties, let me refer to *Nitella syncarpa,* which has recently been described by A. ERNST, and which bears, instead of oogonia, incompletely developed antheridia which never produce spermatozoids.[3] The examples in question were observed near Zürich, and were completely sterile.

[1] See P. 63 and C. SCHRÖTER, U*eber die Fichte* (*Picea excelsa Link*) Vierteljahrsschr. d. nat. Ges. in Zürich, Jahrg. XLIII, 1898, Parts 2 and 3, pp. 50-53, Fig. 18. This valuable work contains a very full review of the varieties, forms, and monstrosities of this highly "variable" tree.

[2] *Deutsche landwirthschaftliche Presse,* Berlin, October 4, 1899, where photographs of monstrous ears of rye are given.

[3] ALFRED ERNST, *Ueber Pseudo-Hermaphroditismus bei Nitella syncarpa.* Flora, 1901, Vol. 88, Part I, with Plates I-III.

§ 10. INSTANCES OF RACES WHICH HAVE ARISEN SUDDENLY IN NATURE.

In nature, elementary species are, as is well known, not connected with their closest allies by transitional forms. Nevertheless fluctuating and transgressive variability frequently bring about the appearance of continuous series, which however on closer examination especially by statistical methods dissolve into perfectly distinct component units.[1] In very many cases these transitional forms are absolutely lacking, and the separation of a particular form as variety, subspecies, elementary species, or even species, is mainly founded on their absence.

The absence of transitional stages in the case of forms which have been long familiar and are widely distributed obviously tells little concerning their mode of origin. Results are more likely to follow from the investigation of those cases in which the types in question are local in occurrence and in which, therefore, if transitional stages should occur, one would expect to find them in the locality inhabited by the plant. In some instances an exhaustive and minute study of the geographical distribution of certain varieties has led to the discovery of the center from which distribution took place. ASCHERSON and MAGNUS obtained a result of this kind with the pale fruited varieties of the European species of *Vaccinium* and some related Ericaceae.[2] In inquiries of this kind, the absence of transitions at the present time points to the conclusion that they may never have existed.

---

[1] See Vol. I, Part II, § 25, p. 430.

[2] P. ASCHERSON and P. MAGNUS, *Verhandl. d. k. k. zool.-botan. Gesellschaft in Wien*, 1891, p. 677.

In contradistinction to these more or less distributed varieties, there is a whole series of records scattered through the literature of cases in which a new form has been found on a particular spot under circumstances which warrant the conclusion that it has arisen exactly there and fairly recently. In such cases transitional forms are always lacking, a fact which proves pretty conclusively that such have not been produced in the origination of the form. In Part II of the first volume two cases afforded by *Oenothera Lamarckiana* were described in detail: I refer to the appearance of *O. brevistylis* and *O. laevifolia* on the original locality at Hilversum. Both species proved, when tested, to be perfectly constant from seed, without any atavism; and transitional forms were not seen in the field. If these species had arisen where I found them, their origin must have taken place between the year of the introduction of the species and the first year in which I discovered them; that is, between 1870 and 1886 (See Vol. I, p. 266).

The most important and accurate observation of such an occurrence is that which has recently been recorded by SOLMS-LAUBACH, and deals with a species newly arisen from *Capsella Bursa Pastoris*.[1] This was found by Professor HEEGER in the market place near *Landau* amongst the ordinary Shepherd's purse and called *C. Heegeri* after him, by SOLMS. It occurred in 1897 and 1898 in very small numbers and only on this one spot. In its vegetative parts it is exactly like *C. Bursa pastoris,* from which it only differs in the form of its fruits. But the differentiating characters are of the rank of some of

[1] H. GRAF ZU SOLMS-LAUBACH, *Cruciferen-Studien*, Botanische Zeitung, 1900, Heft X, Oct., 1, 1900, pp. 167-190, Plate VIII.

those which serve to separate genera amongst the *Cruciferae*.

The fruits of *Capsella Heegeri* are oval, and about as thick as they are broad. The seeds are notorrhizous. The valves lack the firm anatomical structure, characteristic of the normal valve, but are soft and full of sap, a condition which may be considered as due to arrested development. On the weaker branches in the autumn, deviations from this type occur which revert more or less to that of *C. Bursa*; moreover the flowers and young fruits may develop into malformations, as the result of the attacks of *Cystopus candidus,* which closely resemble those of *C. Bursa pastoris.*

The seeds of isolated plants of *C. Heegeri* gave rise solely to the parent type (382 examples) without reversion to *C. Bursa.*

There can therefore be scarcely any doubt that *C. Heegeri* is a good elementary species which arose from *C. Bursa* in 1897, or a few years previously, somewhere near Landau. It is moreover a species which is distinguished from its nearest allies by characters of far greater systematic importance than those which separate many species of known origin.

I myself found a *Stellaria Holostea apetala* not far from Wageningen in Holland under similar circumstances (1889), and also in the same year the well-known *Capsella Bursa Pastoris apetala*[1] near Horn in Lippe. But I did not succeed in obtaining seed from either of them. In 1888 I collected some seed of *Lychnis vespertina* not far from Hilversum and obtained some perfectly glabrous plants by sowing it. The new variety *L. v. glabra* proved fully constant as soon as I was able

[1] See PENZIG, *Teratologie*, I, p. 267.

to isolate it, and has maintained itself up to the present day without ever reverting.

So far as published data go, forms which have suddenly appeared in nature, or have not previously been noticed, prove constant, provided that cross-pollination is guarded against. In the opposite case they will prove themselves pure as soon as they can be isolated. One of the oldest cases in point is the constancy of *Ranunculus arvensis inermis* which was established by HOFFMANN.[1] The majority of records refer to trees of which the larger number of varieties, if not all, according to DARWIN himself, have arisen suddenly,[2] such as the weeping oak, the weeping white hawthorn, etc.[3] A single specimen[4] of *Fagus sylvatica aspleniifolia* was found in a wood in *L*ippe-*D*etmold and could be multiplied from seed. According to LOUDON, *Taxus baccata fastigiata* was found in 1780 growing wild in Ireland;[5] but no pure seedlings of it have been obtained since only one specimen was observed (a female one).

The above list of cases is not a rich one; but it makes no claim to completeness. The observations in point are, with few exceptions, relatively incomplete inasmuch as there is always the possibility that the first discovery of the new species or variety may have been preceded by a long period of evolution. If we assume this to be true, the absence of transitional forms and the constancy of the

[1] HOFFMANN, *Bot. Zeitung*, 1878, p. 273, where several other examples will be found.

[2] DARWIN, *Variations*, I, pp. 461-463.

[3] Further examples are given by BRAUN, *Verjüngung*, p. 333 (the sudden origin of red-leaved varieties of *Quercus, Corylus*, etc.).

[4] RATZEBURG, cited by BRAUN in *Abh. d. k. Akad. Berlin*, 1859, p. 217.

[5] L. BEISSNER, *Handbuch der Nadelholzkunde*, 1891, p. 169. A great number of further examples is given in this work.

new form are the only arguments for its sudden appearance.

§ 11. HORTICULTURAL VARIETIES WHICH HAVE ARISEN SUDDENLY.

It is a matter of common knowledge that horticultural varieties have very often arisen by sports. But opinions differ on two points. One is an empirical one and relates to the question of constancy; the other relates to the meaning of the word variety. The two points are narrowly bound up with one another. If the new form is not constant and pure from seed but frequently reverts to the parent species it is usually supposed to be derived from that species and is treated as a subdivision of it. But if the new form is as constant as the parent species, the empirical means of demonstrating its relationship are lacking, and the conclusions are drawn from historical data and based on analogy; a proceeding which, as we all know, often leads to differences of opinion.

Besides the historical records the main point in such cases is always the proof of the constancy from seed. But inasmuch as the interest of the practical man only extends to the question whether the variety can be conveniently multiplied by seed and is not concerned with the possibility of occasional reversions, such information, especially in older cases, can only be accepted with caution.

With this reservation, I propose to give a brief review of some of the better known instances. But before I do this I will call attention to a very beautiful variety which I have not yet found described nor seen in trade-catalogues, but which has appeared in my own cultures.

Fig. 15 represents a single Dahlia, whose ray florets are all transformed into long and broad tubes which are open above. The same thing occurs in many other composites, for instance in *Chrysanthemum segetum fistulosum, Corcopsis tinctoria fistulosa,* etc. On the analogy of these cases I propose to call this new Dahlia, *Dahlia variabilis fistulosa.* This variety arose from a crop raised

Fig. 15. *Dahlia variabilis fistulosa,* a new variety which has appeared in my cultures.

from the seeds of *D. var. Jul. Chrétien,* a dwarf single Dahlia with red flowers of the color of red lead, the tubers of which I had bought in 1892 in *Lyons.* From the seeds which I saved in that year from this variety, I raised in 1893 several plants of which one had a white flower. I only sowed seeds of this in 1894.[1] It was

[1] I have unfortunately not yet succeeded in fertilizing Dahlias artificially by their own pollen.

in the crop thus raised that the plant which bore the flower head shown in Fig. 15 appeared. The color was dark carmine red, not that of red lead. The flower heads were all fistulous from the beginning of June until well into October; but the later flowers manifested the abnormality in varying degrees. Either the base only of the tube was closed; or only some of the ray florets had the form of a tube. The plant had to be left to free crossing with its neighbors so that no observations of real value as to its constancy could be made. Nevertheless this was pretty considerable, for, from the seeds of my *fistulosa* I raised 43 plants in 1895 of which 25, that is to say more than half, had the characters of the new variety.

The origin of *Chelidonium laciniatum* from *C. majus* was described in detail in the first volume (p. 189, Figs. 36 and 37); where a series of other cases will also be found. VERLOT (*loc. cit.,* p. 34) describes *Ageratum coeruleum nanum* as a novelty which is sometimes sterile, but sometimes occurs as a fertile and constant variety. *Verbena hybrida,* "à fleur couronnée" arose about 1889 from the variety "à fleur d'auricule," it immediately proved constant and after only two years was put on the market by E. FOURGEOT of Paris.[1] *Robinia Pseud-Acacia rosea* was found by DECAISNE in a crop of ordinary Acacias; and *Gleditschia sinensis inermis* arose in the same way, as also did *Sophora japonica pendula* which appeared in M. JOLY's nursery garden in Paris about 1800.[2] In 1860 a new strawberry "*Reus van Zuidwyk*" appeared in Boskoop. Its leaves and fruits were larger

[1] See his Catalogue for 1891.
[2] VERLOT, *loc. cit.,* pp. 59, 92, 93.

and altogether better than any varieties then known; it was constant from the first and spread rapidly.

I shall conclude this summary with a reference to the new species of Tomatoes which BAILEY has recently described.[1] He describes the origin of two new forms which he has called *Upright* and *Mikado* and which arose in his cultures. They differ from one another and from the parent species by more definite and more numerous characters than many among the older forms which are recognized as good species in the genus *Lycopersicum*. They arose suddenly as usual and were propagated by seed.

The observations recorded in this and the two preceding sections, which are far from constituting complete lists, show that the origin of varieties and of elementary species both in the garden and in the field is amenable to experimental investigation, for the phenomenon is by no means so rare as is generally believed. The botanist will investigate the indifferent and useless forms with just the same result as the profitable ones, to which alone, of course, the practical man pays attention. The cultures need not be very extensive to afford novelties from time to time, though these must not be expected the first or every year. Once obtained, all that there is to be done is to isolate them as soon as they appear and pollinate them artificially. But it is far more important to go back to their ancestors, partly not to lose the historical evidence, but mainly in order to sow the seeds of these ancestors again and to find out if the novelty will be again produced, and if possible to discover the conditions which determine its appearance. Unfortunately there are many plants which do not lend themselves to such experiments,

[1] L. H. BAILEY, *Survival of the Unlike.*

either because they produce no seed or yield too small a harvest when self-fertilized or because they cannot be artificially fertilized on a sufficiently large scale or because the number of seeds produced, even under normal conditions, is too small. Moreover one is almost absolutely confined to annual or biennial species or to such perennial ones as flower freely in the first year.

But in spite of these difficulties and of the incompleteness of the observations made hitherto, we may safely conclude from them the possibility of an experimental study of the origin of horticultural varieties.[1]

[1] I shall describe an experiment of this kind with L*inaria vulgaris peloria* in § 20.

V. ATAVISM.

§ 12. ATAVISM BY SEEDS AND BUDS.

HOFMEISTER in his *Allgemeine Morphologie* defines atavism in these words: "The occurrence of reversions, the offspring of a variety of known origin resembling the parent type" (p. 559).

According to the meaning of the word "known" in the above definition the term atavism may embrace quite a series of phenomena of the most diverse importance. It may mean either that this origin must have been actually observed, or that it can be inferred with sufficient certainty from comparative and systematic studies. If we are merely dealing with morphological questions this distinction may appear unessential, but as soon as our object is to test by experiment the results obtained, it becomes of the highest importance. For to obtain true experimental proof of atavism it is obvious that the origin of the forms should be known directly by observation.

The origin of a whole series of varieties and elementary species from their parent forms, however, is sufficiently established by the historical evidence relating to their first appearance. It seems therefore feasible to confine our attention to such cases and to draw a distinction between physiological and phylogenetic atavism. The former is reversion to actually known ancestors, the latter to systematic ancestors.

Atavism by Seeds and Buds.

But before I proceed to examine these two forms of atavism more closely, I think it desirable to state that I here use the word "atavism" in its narrower sense, for in its wider sense it embraces so large a group of phenomena that it would not be possible to deal with them all within the limits at my disposal. It seems worth while to indicate the more important of these types because they are often confused with one another and because results obtained with one form are often taken to apply to another, simply because they both go by the same name.

We must first of all draw a sharp distinction between atavism as applied to variability and as applied to mutability. In the first case we are dealing with the phenomena presented by a single heritable character; in the latter, with the conflict of two or more. In the improvement of races the offspring do not resemble the selected parents, they always revert partly towards the mean of their ancestors. We are of course dealing in such cases with the phenomenon of regression which was fully discussed in the first volume (pp. 82 and 120); and it would be better to refer to all those individuals of less value which are eliminated in selection as *regressive* and those which exceed the level attained by their parents as *progressive*. But it is customary to call the former atavists; and, as a matter of fact, they exhibit the degree of development of the characters in question as it was manifested by their grandparents and more remote ancestors, and not as in their parents. They could perhaps be called "curve atavists," since this term does not suggest a reversion beyond the curves in question.[1]

The most fascinating section of the subject of atavism is that which deals with so-called "youth" forms

[1] See the pedigree of the many-rowed maize, Vol. I, p. 73, Fig. 18.

and with related phenomena. GOEBEL's admirable investigations have demonstrated the wide distribution of these phenomena and their great importance to the theory of descent.[1] It is now a matter of common knowledge that many plants, and indeed whole groups of species, exhibit characters when young which they either lack in the adult state, or which in later life appear only under definite circumstances. BEISSNER's discovery[2] that whole genera of cultivated *Coniferae*, such as *Retinospora*, are only youth-forms of other known types such as *Thuya*; and REINKE's investigations[3] into the earlier stages of Leguminosae, as well as the work of many others, have resulted in the accumulation of a mass of information relating to this subject. *Sium* and *Berula* in their early stages have the doubly pinnate and finely slit leaves of their close allies; the thorns of *Berberis* on the so-called suckers revert to the foliate form. These phenomena, however, fall mostly within the sphere of systematic botany, and only concern the study of variability in so far as they are dependent on external influences.

We must further exclude from our considerations the effects of crossing. The so-called reversions of the horticulturists which are brought about either by accidental crosses with the parent or by unconsciously using hybrid seed, certainly occupy a very prominent place in the practice of horticultural selection, but they should be rigidly excluded from scientific speculations. And

[1] K. GOEBEL, *Ueber Jugendformen von Pflanzen und deren künstliche Wiederhervorrufung.* Sitzungsber. d. k. bayr. Akad. d. Wiss., Vol. 26, 1896, Part III. For further references see GOEBEL's *Organographie der Pflanzen*, Part I, 1898.

[2] L. BEISSNER, *Handbuch der Nadelholzkunde*, 1891.

[3] J. REINKE, *Untersuchungen über die Assimilationsorgane der Leguminosen*, I-III and IV-VII. Jahrbücher für wissensch. Botan., Vol. XXX, Parts 1 and 4, pp. 1 and 71, 1897.

this is true not only of those cases in which the cause of the reversion is perfectly plain, but still more of those in which the facts observed may lead us to suspect a cross either in the previous generation or in more remote years. By excluding such cases, however, the apparent abundance of data relating to experimental atavism is very much reduced; but it is obviously better to build on a few reliable facts than on the highly insecure basis formed by the numerous data which have hitherto been collected.

With these reservations I shall now turn to the distinction between physiological and phylogenetic atavism. Each has its own sphere. The object of the study of the former is to discover the laws to which this form of variation conforms. That of the latter is to discover the ancestors of the species in question either by the observation of chance deviations, or by cultures and selection.

HEINRICHER'S extensive studies in the genus *Iris* show how fruitful may be the application of selection in the study of phylogenetic atavism.[1] The cultivated plants of this group are well known to be highly variable, and the favorite *Iris Kaempferi* with its large flowers affords numerous opportunities for the study of tetramerous and pentamerous flowers and of other variations. HEINRICHER, starting from occasional anomalies presented by *Iris pallida,* and working on a methodical system of selection, has raised an atavistic race which he calls *Iris pallida abavia.*[2] The individual anomalies could not, it is true, be fixed although they were selected for three generations, but a series of new types gradually

---

[1] CARRIÈRE, *Production et fixation des variétés,* 1865, p. 65.

[2] E. HEINRICHER, *Versuche über die Vererbung von Rückschlagserscheinungen.* Jahrb. f. wiss. Bot., Vol. 24, Part I, 1892, and *Iris pallida abavia* in Biolog. Centralbl., Vol. XVI, No. 1, p. 13, 1896.

appeared and threw a definite light on the probable nature of their common ancestor. This is regarded as being an extinct form, with an hexamerous perigon of equal petals, and six stamens. A still living form, *Iris falcifolia*, possesses such a perigon but has only three stamens.

The reader who is interested in this branch of inquiry and in the highly important results which it has afforded, is referred to the works of this author for further information.

I now return to the main question, viz., that of physiological atavism. Here we are concerned not with the production of new forms but with an inquiry into the processes which underlie the reappearance of preexisting characters. The character in question is, therefore, one that is still retained in that species from which the one under investigation is descended. Atavism is in this case to be regarded as an oscillation between two empirically known extremes. The field of oscillation can obviously not be very considerable, for only in cases of very close relationship is the common origin of two forms historically known to us.

In this restricted province also, atavism may be brought about by fluctuating variation as well as by mutation. In the case of the former it is merely a transitory phenomenon and dependent on external conditions; but in the second case it leads to the origin of a race which externally resembles the ancestors of its parent form. *V*ariational atavism seems to be a phenomenon which plays a large part in the sphere of semi-latent characters. As an example of this I cite the case, described above, of the five leaved clover (§ 5, p. 36) which always bears a certain number of trifoliate leaves especially under unfavorable conditions. These trifoliate leaves obviously

Atavism by Seeds and Buds.

constitute a reversion to the normal clover leaf but, on the other hand, they are merely the extreme variants in the curve of the five-leaved race (Fig. 6, p. 48). A similar state of affairs prevails in numerous cases of semilatency where the range of variation of a character is occasioned by the antagonism of two characters.

Mutational atavism must obviously be as rare as mutation itself. The reversion of striped flowers to self-colored ones, the heritable atavism of *Plantago lanceolata ramosa,* and the inconstancy of the peloric *Linaria,* are facts which we shall have to consider below.

Physiological atavism can be manifested by plants propagated by seeds or by buds. In the case of the former definite proof is only possible under exceptionally favorable circumstances; in the case of the latter it is at once evident (Fig. 16 at A). The published records of atavism in crops of seedlings are always subject to the sus-

Fig. 16. *Cephalotaxus pedunculata fastigiata.* The main stem bears the upright branches with leaves inserted on all sides, characteristic of the variety; but has produced at A, where a branch has been cut off close, several branches with flat spreading biserial leaves such as are characteristic of the parent species.

picions indicated above. I mean that they occur so rarely and in so few individuals that the possibility of a previous cross, by means of insects, with the pollen of allied forms, even if growing a long way off, can never be quite excluded. It is only in cases in which, as in that of *Ocnothera scintillans* (Vol. I, pp. 245 and 377), a species produces a large number of atavistic individuals every year, that the phenomenon easily lends itself to experimental study.

On account of the circumstances indicated, it is not possible to say whether atavism in plants propagated by seed is a common or a rare phenomenon. It is certainly much rarer than the practical gardener usually imagines. I have observed in my cultures a number of cases which might have been called atavistic with more or less certainty, but only the cases of regularly inconstant races, such as those of *Plantago* and *Linaria,* and the phenomena presented by striped flowers, to be described shortly, seem to me to be sufficiently well established to be adduced as instances of atavism.

Atavism by bud-variation, on the other hand, is a well-known phenomenon. One of the best instances is shown in Fig. 16. It represents a vertical branch of a bush of *Cephalotaxus pedunculata fastigiata* (*Podocarpus Koraiana* Hort). Below the middle of the figure can be seen the place where a branch has been cut off, and from the side of its base some lateral branches have arisen with flat spreading leaves (Fig. 16 A).[1] The variety *Fastigiata* has erect branches only and their leaves are inserted on all sides; but the branches at A have the structure of the parent species. *C. pedunculata*; their

---

[1] For a series of interesting experiments relating to this subject see *Mutations et traumatismes* by L. BLARINGHEM (Note of 1909).

leaves project to right and left, and their side branches are horizontal, making the whole shoot flat with definite dorsal and ventral surfaces. The bush which grows in our garden and bears several branches with similar bud-variations, I owe to the kindness of MESSRS. ZOCHER & Co., nurserymen in Haarlem. The variety can only be propagated by cuttings, as it never flowers,[1] and these produce reversions of this kind pretty regularly, both in the nursery of Messrs. ZOCHER & Co. and elsewhere. It appears to have been first observed in 1863 by CARRIÈRE in Paris,[2] and since that time by many others. This remarkable case is well worthy of a closer study. The perfectly analogous *Taxus baccata fastigiata* never exhibits atavism by bud-variations, so far as I know.[3]

The phenomena of bud-variation have hitherto not received from botanists the attention they deserve. In a few cases we know that the phenomenon is preceded by a sectorial segregation, as for instance in striped flowers (§ 13) and variegated leaves (§ 24); but as a rule there is no available information even on this point. Another point which awaits investigation is the nature of the offspring of self-pollinated bud-variants.[4] It seems certain that new types sometimes arise in this way, but much of the proof in favor of this will not bear scrutiny. Under these circumstances it seems desirable to direct more general attention to this phenomenon[5] by means of some

---

[1] BEISSNER, *Handbuch*, loc. cit., p. 181.

[2] CARRIÈRE, loc. cit., p. 44, with Figs. 1 and 2; see also CARRIÈRE, *Traité général des Conifères*, p. 717; and JAMES VEITCH & SONS, *A Manual of the Coniferae*, 1881, p. 308.

[3] See CARRIÈRE, loc. cit., and BEISSNER, *Handbuch*, loc. cit., p. 169.

[4] In the older records attention is seldom paid to pollination; see the literature in CARRIÈRE, loc. cit., p. 59, and DARWIN, *Animals and Plants*, I, 525; II, 442, etc.

[5] CARRIÈRE gives a very complete list; loc. cit., pp. 42-56; see also

further examples. They are taken mainly from woody plants because herbaceous and especially annual plants, with the exception of the instances named and of hybrids, very seldom exhibit bud-variations.

Green branches on red-leaved bushes and trees are not rare and are for instance often seen in the variety *atropurpurea* of *Corylus Avellana, C. tubulosa, Betula alba,* and in the copper beech. The red bananas with their red fruits have given rise to a green variety with yellow fruit in spite of the fact that they are sterile.[1] BRAUN mentions an example of *Kerria japonica plena* which produced some branches with single flowers.[2] On a garden *Hortensia* producing only large sterile flowers, FOCKE observed a branch bearing inflorescences with little fertile flowers in the middle of a circle of large ornamental ones as in the wild form.[3]

Trees with laciniate leaves habitually give rise to reversions on solitary branches, as for instance *Fagus sylvatica aspleniifolia, Carpinus Betulus heterophylla, Sambucus nigra laciniata, Cytisus Laburnum quercifolia, Vitis* and others. (BRAUN, *loc. cit.*) The same is true of *Salix babylonica crispa,* of the parsley grape, of nectarines, and especially of roses and bulbs (*Hyacinthus, Gladiolus,* etc.) although the possibility of previous crosses makes the latter cases still doubtful.

In conclusion, this list shows that the series of cases which are amenable to experimental study is by no means small. On the other hand the number of examples is sufficient to demonstrate the pretty general occurrence

HOFFMANN, *Bot. Zeitung,* 1881, p. 395; DARWIN, *loc. cit.,* I, pp. 476-530; HOFMEISTER, *Allgemeine Morphologie,* p. 560, etc.

[1] FR. MÜLLER, *Flora,* Vol. 84, 1897, pp. 96-99.
[2] *Abh. d. k. Akad. Berlin,* 1859, p. 219.
[3] *Abh. d. Naturf. Vereins Bremen,* Vol. 14, 1897, p. 276.

of reversion of varieties to their parent species, and therefore to suggest that the characters of the latter were not lost when the variety originated, but only became latent.

§ 13. VILMORIN'S SUGGESTION AS TO THE ORIGIN OF STRIPED FLOWERS.

One of the oldest and best-known instances both of bud-variations and of sectorial splitting is afforded by certain so-called variegated garden flowers and particularly by the annual Larkspurs, *Delphinium Ajacis* and *D. Consolida*. All phases of the phenomenon can be followed in this case with great ease, for from time immemorial these varieties have borne flowers which show the most varied striping on a background of a different color; and they also produce flowers a half or a third or some other fraction of which uniformly bears the color which commonly only appears in stripes (Fig. 19). Flowers of this kind may be scattered over the whole plant, but are oftener distributed in such

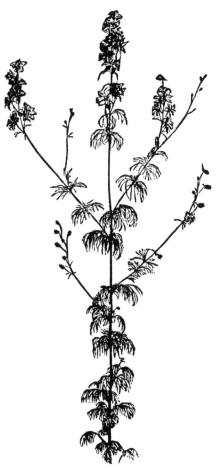

Fig. 17. *Delphinium Consolida striatum plenum*. A plant in flower.

114   *Atavism*.

a way that those on one side of a spike are uniform and those on the other striped.[1] Flowers which are inserted at the boundaries of the two regions exhibit on one side the color of one sector and on the other half, the stripes of the other. A diagram of such a branch is shown in Fig. 18 in which the flowers Nos. 1, 4, 6, 9, and 11 are dark blue, Nos. 2, 5, 7, 10, 12, and 13 pale red with scat-

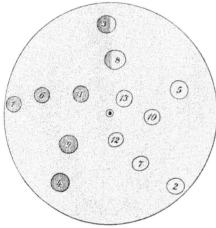

Fig. 18. *Delphinium Consolida striatum plenum*. Diagram of a branch of which the left half was blue, and of which the right bore flowers with fine blue stripes on a pale red background. 1899.

Fig. 19. A sectorial flower of the same variety. The whole right half was dark blue; the left, pale red with scattered blue stripes.

tered blue stripes, and Nos. 3 and 8 half blue and half striped. I obtained this branch in my culture of 1899; similar cases are not at all rare. Branches with nothing but blue flowers also occur, but the seeds obtained from the self-fertilization of such flowers gave rise in my garden to the striped variety and not to a pure blue progeny.

[1] Exactly the same phenomenon is seen in the seedcoats of *Pisum*. The minute purple spotting characteristic of some green-skinned varieties sometimes takes the form of a deep uniform purple. These uniformly purple seeds produce the ordinary form with small purple spots and no more full purples than are usually produced. (Translator's Note.)

On the other hand a certain percentage (often 6% and more) of the plants raised from the seeds of striped flowers and especially of sectorial branches are usually uniform blues.[1]

The phenomena of segregation which we have been describing are quite common in striped flowers, and any one can observe them in *Dahlia variabilis striata* (Vol. I, Fig. 14, p. 54), *Mirabilis Jalapa, Verbena* and many other favorite garden flowers. Sectorially colored flowers appear to manifest a tendency towards a simple proportion between the two parts. Frequently exactly half of the flower is atavistic, sometimes a quarter or three quarters. I observed the proportion ⅜ in white and red striped tulips and in partially dark blue and partially pale blue flowers of *Iris xiphioides,* etc. In these cases the various types frequently occur on the same plant, or in the case of plants grown from bulbs, on examples raised through vegetative propagation from a single original bulb; for instance on the tulips and Iris just mentioned there were also flowers of which one-half of each was atavistic.

Sectorial variability often occurred in my cultures, as for instance in the flowerheads of *Helichrysum bracteatum* and the flowers of *Papaver nudicaule* (Fig. 20), in both of which cases stripes or sectors of the color belonging to the parent species were superimposed on the paler background of the variety. A common balsam (*Impatiens Balsamina*) whose flowers were usually white with fine red stripes bore a branch with red flowers only in my garden. The whole breadth of the fasciated stem

---

[1] A point of great interest to investigate would be the relation between sectorial variability and cell division in the vegetation cone; clues which might lead to the solution of many important questions would probably be afforded by such an inquiry.

of the striped cockscomb or *Celosia variegata cristata,* is traversed by longitudinal stripes of different colors, yellow and red according to the variety. Dahlias however exhibit the most prodigal wealth of color of all variegated flowers, especially those varieties known as Fancy-flowers.[1] In this case the color is in some way connected with the amount of doubling, which often exhibits sectorial variations and bud-variations at the same time.[2] Striped Dahlias give rise to these partial variants sometimes very rarely, but sometimes in such abundance that a good variety is often exposed to the danger of being lost thereby. In most of the cases we have to do with two types which are manifested in various degrees of association and separation. Cases in which more than two forms are combined and which therefore may produce by bud-variation two or more types on the same plant, besides the normal one, have been described; but they were probably hybrids. Central dissociation seems to be a very rare phenomenon, but of *Mad. H. Vourchy,* a variety which usually has

Fig. 20. *Papaver nudicaule.* Yellow variety with dark orange stripes.

[1] See GROOMBRIDGE's *Treatises on Florist's Flowers; The Dahlia,* 1853, and the extensive literature which has appeared since.

[2] VILMORIN-ANDRIEUX, *Les fleurs de pleine terre,* first edition, p. 340.

white flowers with red stripes, I have seen a head whose outer ray florets were dark red whilst the inner ones formed a disc of pure white with only very occasional red stripes. In the center the unmodified fertile yellow disc florets were seen. I have observed the same phenomenon in a few other cases.

The striped varieties of *Cyclamen persicum* are said to bear in some instances only variegated flowers one year and from the same bulb uniformly colored atavistic flowers the next year.

Centaurea Cyanus, the blue corn flower or blue bottle, has a brown variety with double flowerheads which is highly variable in color; it is far from being fixed yet, as a plantbreeder in Erfurt expressed it to me. I cultivated it for five years, always selecting the purest and darkest brown specimens in small numbers as seed-parents. The race produced reversions to the blue form every year. Some plants bore blue flowers exclusively, in others the blue color appeared in segments or in stripes on some of the heads. No advance was brought about by this selection.

The examples given must suffice to show the importance of the striped flowers of horticulture. A *Var. striata* of a number of species is advertised in the catalognes; it is open to any one, therefore, to cultivate them. The *Var. alba* of many other species often reveals on closer inspection scattered stripes of the color of the parent species; these stripes can easily be intensified by isolation and selection as I shall show in one of the following sections (§ 16).

Striped flowers[1] are also of great importance in the

---

[1] Spotted flowers may possibly behave differently; but up to the present time I have not grown them.

science of variability and mutability, and especially in that of atavism of which they perhaps afford the most beautiful examples. As such they have been dealt with especially by Louis Vilmorin whose theory we will now proceed to examine.[1]

Vilmorin starts from the observation that striped flowers only occur on those species which are themselves colored, but which also possess a white variety; or if the color of the flower is composed of red and yellow the uniform yellow variety may behave like the white (*Mirabilis, Antirrhinum*). The first variety to arise is the white (or yellow) from which later on the striped form originates and Vilmorin explains this as a partial reversion to the parent species.

White varieties of a large number of decorative plants have arisen in cultivation, and in fact many favorite ones in M. Vilmorin's own nurseries. They can usually be easily "fixed" in the course of a few years; that is to say, they are generally constant from the very beginning but have to be purged of the consequences of unavoidable crosses, and this takes a few years, as a rule. The striped sorts do not appear in this period, the hybrids resulting from the crosses are like the parent species and segregate into this and the pure white variety. The striping is not the result of crossing therefore; moreover in such cases deliberate crossing has only resulted in the production of self-colored and not of variegated flowers. Also, when such hybrids exhibit sectorial variation, the color is in large patches and not in fine stripes.

It is not until the white varieties have attained complete purity and have proved constant for a considerable

[1] *Société Philomatique de Paris, Séance du 17 janvier, 1852, Procès-verbaux*, p. 9; *Notices sur l'amélioration des plantes par le semis*, 1886, p. 39; and B. Verlot, *Sur la fixation des variétés*, 1865, pp.62-66.

number of generations that the striping appears. It affects almost necessarily, so it seems, every cultivated white or yellow variety. Some are worth putting on the market; others are not. Amongst the latter VILMORIN (1852) has mentioned as an example *Clarkia pulchella,* from bought seeds of the white variety of which the striped form has also appeared in my cultures (see § 16). The same thing happened with *Browallia erecta* and *Commelina tuberosa. Geranium pratense* is only to be bought in two forms, white and blue. I obtained seeds from two plants which were bought as *Var. alba* and raised from them, besides pure whites, plants with all grades of color arrangement from striping and sectorial variations to complete blue (Fig. 21).

If it is thought desirable to put the striped variety on the market it must be purified by selection. The striping first appears as single fine streaks on occasional flowers. If these plants are isolated and their seeds sown separately the majority of the plants raised are pure white, but occasional ones are produced with broader and more numerous stripes. The seeds of these are saved, and so on. The object is to isolate the striped race from the white, and this can be attained in the course of a few years. On the other hand the breeder has to fight against the tendency of the striped form to return to the full blue either by buds or through seeds. It is to guard against this that VILMORIN recommends

Fig. 21. *Geranium pratense album* with piebald blue and white flowers. The dark parts of the petals were blue; the others white.

the selection of seeds from the palest examples of the striped forms.

Convolvulus tricolor was the first species in which this mode of origin of the striped form was observed (1840). It was followed by *Gomphrena globosa, Antirrhinum majus album* and *luteum, Nemophila insignis, Portulacca grandiflora*, and others. Of recent years a large number of blotched varieties have been obtained in various nurseries; and always, so far as is known, in the same way, by so-called partial reversion of a white or yellow variety to the red or blue color of the parent species.

In the following sections we will therefore examine in detail some cases of striped flowers as instances of physiological atavism.

§ 14. ANTIRRHINUM MAJUS STRIATUM.
(With Plate I.)

Amongst the numerous cultivated varieties of the Snapdragon one group is distinguished by the possession of striped flowers. A bed of these produces a fine and varied show of color. On the other hand the horticulturist's handbooks state that, whilst the remaining sorts are practically constant, the striped ones leave much to be desired in this respect.[1] Such a statement naturally invites the investigator to inquire into the mode of inheritance of this character.

The striped varieties owe their character to the fact that the normal red color of the wild snapdragon is confined to broader or narrower longitudinal stripes. Where the red is absent the pure color of the background becomes visible. This may be either white, rose, yellow

[1] VILMORIN'S *Blumengärtnerei*, 3d Ger. ed., Vol. I, 1896, p. 756.

or sulphur as in the corresponding self-colored varieties.[1] It must further be mentioned that each of these types may exist in a tall, medium or dwarf form. In the experiment to be described the form I have used was *Antirrhinum majus luteum rubro-striatum* of medium height.

The richness of types of marking in these striped varieties is very great. The stripes may be sparse and very fine so that the flowers appear at first glance to be pure yellow or white; or the stripes may be bold and broad and very numerous in such a way that the yellow (or the white) appears in about equal parts with the red. Often half of a flower is entirely red whilst the other half is striped, and so on.[2]

If we buy seeds of the striped sorts and sow them, the crop raised is considerably less true than is usual in sowings of bought seeds. In 1899 I sowed samples of different varieties of *Antirrhinum majus* and obtained 26% unstriped individuals from *A. m. album rubro-striatum,* and 19% from *A. m. luteum rubro-striatum.* In other cases a far higher degree of purity is usually obtained, e. g., in *A. m. luteum* I found only 2% impurities.

The admixtures in the striped varieties were in the vast majority of cases uniform reds and therefore closely allied to them. Other deviations were not more numerous in the striped forms than in any other variety. The reason for the abundance of the red flowered individuals has been disclosed by subsequent culture; it is to be sought in the incomplete inheritance of the striped char-

---

[1] *A. m. album rubro-striatum, A. m. sulphureum rubro-venosum, A. m. pumilum roseum rubro-striatum,* etc.

[2] VILMORIN, *Fleurs de pleine terre,* p. 723.

acter. For if the seeds of striped individuals which have been artificially self-fertilized are harvested and sown, we usually obtain some plants with uniformly red flowers.

The striped varieties therefore give rise to red plants from time to time, and in my cultures, which extend over about eight years, *A. m. luteum rubro-striatum* has done so almost every year in spite of being self-fertilized. As the original wild form is uniformly colored (that is, not striped, for the color itself is composed of white, red and yellow) the loss of the striping may be regarded as a case of atavism.

Moreover this phenomenon of atavism was exhibited by my cultures in two other forms (Plate I): on the one hand as a bud-variation in which whole branches of a plant with striped flowers revert to the red type; on the other hand as a lateral or sectorial variation, to adopt HEINSIUS's term,[1] in which one side of the spike bears uniform flowers, whilst the other bears striped ones. Let us examine these two cases more closely.

In the case of bud-variation a striped plant bears a branch all of whose flowers are red, without striping. If, as is usually the case, the plant flowers on 6-8 or more lateral branches the abnormality is very striking. A single plant very seldom bears two branches with red flowers, and it scarcely ever happens, if indeed it ever does, that the terminal portion of the main stem has red, and the branches striped flowers. As a rule it is one of the lower stronger branches which is atavistic and seldom one of the higher weaker ones. I occasionally found a tertiary branch with red flowers, i. e., a lateral twig of a striped branch. As might be expected, the coarsely

---

[1] H. W. HEINSIUS, *Over bonte bladeren*, Genootschap v. Natuur-, Genees- en Heelkunde, Biologische Sectie, May, 7, 1898, p. 2.

striped plants exhibit a stronger tendency to produce bud-variations than the finely striped ones.

Sectorial variation is very diverse in the manner of its manifestation. I found it as a rule on the main stem, but also on the branches. If the inflorescence is looked at from above, i. e., in projection, one sector is red whilst the rest is white. This red sector often consists of a narrow red stripe only, or of one-half or three-quarters of the whole. As a rule the abnormality extends from the base to the top of the spike; but it may also be confined to part of it, especially when it consists of a narrow line only. A single red flower on an otherwise striped spike is by no means a rare occurrence. On the borders of the two sectors the flowers are often striped on one side and red on the other. As in the case of bud-variations it is the coarsely striped individuals which are most prone to the sectorial dissociation of color.

The red color occurs not only on the corolla but also on the stamens. In finely striped flowers the stamens are, as a rule, yellow; in flowers with broad stripes they are striped or red. The individual stamens in the same flower are usually dissimilar in respect to this character; yet it is difficult to find a strong contrast within a single flower, e. g., a single stamen which is almost red, and another nearly yellow. I have spent much trouble in the attempt to find such flowers, especially in those that had one longitudinal half almost or entirely without stripes. But I did not discover any definite relation between the striping on the stamens and that on the corresponding parts of the corolla.

As a matter of fact pure yellow flowers never occur in this race. To a superficial observer it may seem as if they were not rare and even that the red stripes may be lacking on whole spikes and sometimes on entire plants. But such absence is only apparent; closer inspection will reveal the existence of very fine red stripes. I never found a branch on which they were quite lacking, nor a plant, nor even a twig which had reverted to the variety, *A. m. luteum*. On inflorescences on which the striping is very meager it may sometimes occur that on a single flower no stripes can be found; but this is merely an extreme case of that partial variability which all organisms exhibit.

This negative result based on eight years' experience is important because it shows us that we are not dealing here with a segregation into two components, e. g., *A. majus rubrum* and *A. majus luteum*. If we want to speak of a segregation the two units would be the red striped and the uniformly red form.

A glance at a bed of these plants is sufficient to reveal the fact that the breadth of the red stripes exhibits individual variability; moreover that, as might be expected, plants with very fine and those with very coarse red stripes are the rarest. In 1897 I tried to find out if it were possible to express this variability in the form of a curve. At first it seemed impossible to obtain an accurate measure of the striping, for it seemed practically unfeasible to determine the sum of the breadths of all the stripes in a flower and to express this sum in proportion to the circumference of the corolla. I succeeded, however, in attaining my object in the following way: I had the average flower on the main stem of every plant in a bed picked by an assistant, and then I endeavored to arrange these in a series according to their color, ascending from the almost yellow to the completely red. With a group of between one and two hundred flowers this succeeded better than I had anticipated; for at the end there turned up a certain number of groups which corresponded sufficiently closely to equal subdivisions of a scale to warrant their selection as ordinates. I admit of course that this method is not free from the personal factor; but for the case under consideration it sufficed, since, when the same group of flowers was sorted again, the result agreed sufficiently well with the first trial.

I plotted three curves in this way in 1897; each was based on one typical flower of the terminal spikes of all the plants flowering on a bed. The three beds contained the offspring of three individual striped plants of the 1894 harvest, seeds of which had been saved and sown separately; but whose flowers had been left to be pollinated by insects in the midst of a larger culture. Moreover the seed-parents were selected without reference to the degree of their striping, and so the curves give an idea of the average composition of the commercial race.

I thus obtained the following table:

STRIPES	COLOR-EFFECT	A	B	C
Almost absent	Lemon yellow (*g*)	0	6	4
Very fine	Yellow	0	9	18
Narrow	Dark yellow	2	12	30
1-2 mm broad	Reddish yellow	5	15	53
1-3 mm broad	Narrowly striped (*s*)	18	22	84
1-5 mm broad	Coarsely striped	28	22	31
1-6 mm broad	Broadly striped (*b*)	42	21	16
Broad fields	Half yellow, half red	26	12	10
Uniform red	Red (*R*)	37	9	15
	Number of individuals	158	128	261

These figures are exhibited in the form of a curve in Fig. 22; in the case of the figures under C the scale or unit of the ordinates is half of that selected for A and B.

The result of this inquiry shows that the first eight groups merge continuously into one another; but that between the striped and red flowers a broad gulf is fixed. The red are not connected with the striped by a series of transitional forms as the lemon yellow are with the broad striped; red flowers with small yellow patches may occur, but they are at most very rare.

The shape of the curves is far more regular than I had anticipated; but the reds obviously have no place in it; I mean, they are far too numerous in proportion. They are therefore obviously not the extreme variants of the series but constitute a group which is perfectly distinct from the striped although the size of this group varies directly with the amount of striping in the other.

After the composition of the commercial race had been determined in this way, my next task was to discover the nature of the offspring resulting from the self-fertilization of the individual components of this diverse assemblage. I have confined the solution of this problem to the three chief types: finely striped, coarsely striped, and uniformly red. Let us begin with the two former groups.

The offspring of the parent plant A (Fig. 22 and table on page 125) contained many coarsely striped individuals (Fig. 22b); when they were in flower I transplanted some very coarsely striped ones to a special bed, picked off all their flowers and young fruits and enclosed all the buds which subsequently opened to insure self-fertilization. In the same way I treated some plants from the bed B (Fig. 22B) with almost yellow flowers. I harvested and sowed the seeds of each plant separately.

In August, 1898, when the beds were in full flower, I determined the amount of striping by the method already employed, taking care that the boundaries between the individual groups corresponded as closely as possible with those of the previous year. I succeeded in

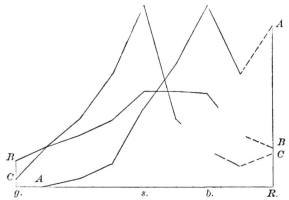

Fig. 22. *Antirrhinum majus luteum rubro-striatum.* A, B, C, curves showing the degree of striping amongst the offspring of three insect-fertilized plants, 1897. g, lemon yellow, almost without red stripes; s, narrowly striped; b, broadly striped; R, uniform red. See table, page 125.

recording the rather scanty offspring of four coarsely striped parents. The result is given below. (The individual seed-parents of 1897 are denoted as A_1—A_4.)

OFFSPRING OF THE COARSELY STRIPED SEED-PARENTS.

Stripes	A_1	A_2	A_3	A_4	Totals
Less than 4 mm broad	0	0	0	0	0
1–5 mm broad	3	2	6	8	19
1–6 mm broad	5	4	6	9	24
Broad fields	7	8	5	6	26
Uniform red	2	0	2	5	9
Totals	17	14	19	28	78

These figures are exhibited graphically in Fig. 23B.

As the extent of this experiment was relatively small and especially as the proportion of self-colored plants appeared to me very small, I repeated it in the following year. I chose from the broad striped bed of this culture a beautiful typical plant with broad stripes but without any broad patches on the corolla, and fertilized it with its own pollen in a bag. In 1899 I raised from its seeds about 250 plants, which covered a bed of about four square meters, and nearly all of which flowered on the main stem and on several lateral branches. There were only a few finely striped individuals amongst them, whereas the majority were very coarsely marked. But the proportion of uniformly red plants was considerable:

Striped individuals	160	64%
Red individuals	91	36%
Total	251	

That is to say, about one-third of the plants had reverted to a uniform red color.

The offspring of the almost yellow parents showed the following distribution of the various types of coloration (B_1—B_4 refer to the individual seed-parents and to the groups of offspring arising from them):

OFFSPRING OF THE YELLOW PARENTS.

Stripes	B_1	B_2	B_3	B_4	Totals
Nearly absent	6	5	12	1	24
Very fine	3	7	18	2	30
Narrow	3	6	12	2	23
1–2 mm broad	9	7	18	3	37
1–3 mm broad	7	4	22	2	35
1–5 mm broad	0	0	3	1	4
1–6 mm broad	0	0	0	0	0
Broad fields	0	0	0	0	0
Uniform red	0	0	0	0	0
Totals	28	29	85	11	153

See Fig. 23A.

These tables, and Fig. 23 which has been constructed from them, show that two races have been produced by the selection and self-fertilization of the extreme variants.

One of them, A, consists almost solely of finely striped individuals and contains no red ones. The other, B, consists almost entirely of broadly striped ones together with 11-36% of uniformly red ones. But the separation is not nearly so sharp as between the striped on the one hand and the red on the other, inasmuch as the two curves overlap.

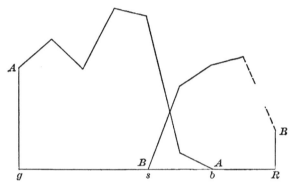

Fig. 23. *Antirrhinum majus luteum rubro-striatum*. Curves to illustrate the distribution of color amongst the offspring of self-fertilized individuals from the culture on which Fig. 22 is based. Experiment in selection with broadly and narrowly striped flowers. Curves representing the offspring: A, of the finely striped seed-parents B_1-B_4; B, of the broadly striped seed-parents A_1-A_4. See tables on pp. 127 and 128. For the signification of g, s, b, R, see previous figure.

We now come to the most important part of the experiment, the question of the inheritance of the red character. On account of this greater importance I had already given it previously much attention.

Here we are concerned not merely with the inheritance of the red flowers in general, but with the study of the special cases already distinguished. First we have to consider the red seed variants, then the bud-variants and lastly the single red flowers on striped racemes. Finally it should be possible to test the red stamens of striped flowers but I have not yet come across suitable material for the investigation of this point.

In 1892 I had raised from bought seed of *A. majus luteum rubro-striatum* a large bed of plants the flowers of which were all striped. I gathered the seed of one individual for the next year's crop (1893). I obtained about 40 flowering plants in this way; the majority bore flowers with fine stripes, and here and there flowers occurred of which one-half was a uniform red. There were four plants which only bore pure red flowers. Of these I selected the strongest, enclosed their spikes in bags and fertilized their flowers with their own pollen. Besides these I dealt in the same way with two striped plants, with few and fine stripes.

As soon as the seeds germinated in the following spring a difference became visible: the seedlings from the seed of striped plants had green foliage, those from the red, however, were reddish brown. This difference was particularly striking on the under surface of the later leaves of the young plants. On the former bed 152 plants flowered, on the latter 71. Both groups consisted of plants with striped flowers and plants with red ones, but as I had expected, in very different proportions. The proportions in the offspring from the two types of parents were as follows:

	STRIPED	RED
Finely striped parents	98%	2%
Red flowered parents	24%	76%

Most of the striped flowers were finely striped; coarsely striped plants only occurred in the proportions of 6 and 7%.

The characters of both races are therefore heritable but, so to speak, incompletely so. We may describe the production of individuals of the opposite race in both cases as atavism. The striped offspring of the red parents resemble their grandparents. The red offspring of the striped parents resemble the wild species, that is, their very remote ancestors. Thus the difference in the intensity of inheritance could be expressed in the statement that the influence of the nearer ancestors is greater than that of the remoter ones. But this is merely a restatement of the facts in conventional terminology. It affords no clue to the solution of the problem.

Amongst the finely striped individuals in the culture under consideration there were thirteen plants which had, besides the striped terminal portion of the main stem and the several striped lateral branches, one or two twigs with red flowers exclusively. A good opportunity was thus offered of studying inheritance in bud-variants. I owed it to the fact that the seeds had been sown early, the plants had been grown far apart and the ground well manured; circumstances which together brought about a profuse branching in all the plants. I transplanted these individuals to a separate spot, picked off all the open flowers and young fruits and superfluous twigs, and enclosed 1-2 striped and 1-2 red spikes in bags to insure pure self-fertilization.

	Plant No.	Red	Striped	Totals	PERCENTAGES Red	Striped
From the red spikes	1	73	27	100	73	27
	2	21	12	33	63	37
	3	25	5	30	77	23
From the striped spikes	1	3	93	96	4	96
	2	0	75	75	0	100
	3	1	36	37	3	97

I obtained a sufficient harvest of the striped and red spikes of the same plant from three individuals only.

132   *Atavism.*

These produced in the summer of 1895 the result shown in the table on page 131.

In other words, the average intensity of inheritance for striped spikes was 98% and for red ones 71%.

If we compare these figures with those derived from the previous generation we do not observe any appreciable difference between them. In other words, the intensity of inheritance exhibited by the red bud-variants is essentially the same as that of the red seed-variants.

In the following year I continued this experiment through one more generation by self-fertilizing some striped and some red individuals amongst the offspring of the bud-variants. The seeds of three striped parents gave rise to 67 offspring that flowered, only 5% of which were red; the seeds of the five red seed-parents, however, gave rise to 127 offspring of which 84% were red. (The percentages in the five individual groups were 71-78-84-88 and 100.) Thus the proportions were similar to those of the previous year.

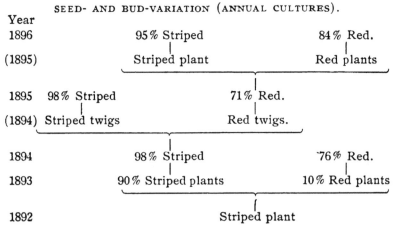

ANTIRRHINUM MAJUS LUTEUM RUBRO-STRIATUM.
SEED- AND BUD-VARIATION (ANNUAL CULTURES).

I have exhibited on the opposite page the whole experiment in the form of a pedigree.

The result of our experiment can be given in yet another form. The intensity of inheritance in the finely striped spikes in successive generations produced by self-fertilization was always about 95-98%. The intensity of the inheritance of the red character in the various subdivisions of the experiment was as follows:

    1. For seed variants 76%
    2. For bud variants 71%
    3. For the offspring of bud variants . . 84%
                                     Average 77%

Finally I have endeavored to investigate the mode of inheritance in the case of sectorial variation; that is, of spikes which on one lateral part bear striped flowers and on the others red ones. It is obvious that this phenomenon may be due to two entirely different causes. First the red flowers may be genuine bud-variants and, in such cases, they will presumably exhibit an intensity of inheritance which corresponds with that found for the bud-variants dealt with above. But it may also happen that on a very coarsely striped spike some of the flowers may possess this striping in so extreme a degree that they appear uniformly red. In this case their mode of inheritance will presumably not differ from that of the remaining flowers on the same spike.

The latter was the case in the only experiment which I have so far had the opportunity of making. In the summer of 1898 I employed for this purpose a broadly striped plant from the crop referred to on page 128. One side of its terminal spike bore red and the other striped flowers. There were 8 of the former and 7 of the latter.

I enclosed the whole branch, before it flowered, in

a bag, fertilized each flower with its own pollen, and gathered the seeds separately. Five fruits of each color ripened, though some of them contained little seed. I sowed the seed in 1899 on ten separate beds; they flowered in July. On each bed one saw at a glance that about half the plants bore exclusively red whereas the other, slightly larger half, bore striped flowers. I recorded the numbers separately for the ten groups; but do not consider it necessary to give the separate numbers. There flowered:

From the seeds of:	Plants	Reds	Average
1. Red flowers	67	33%	42%
2. Striped flowers	137	46%	

* *

The result of all the experiments described above may be summarized in the following theses:

1. *Antirrhinum majus luteum rubro-striatum* (Plate I) is an inconstant race consisting of striped and of red flowered plants.

2. The striping of the commercial race varies continuously, but the continuity does not include the red ones; these are separated by a gulf from the striped (Fig. 22).

3. The intensity of inheritance of the finely striped plants is about 95-98%. They pass into the red type either when propagated by seeds or by buds.

4. In the same way the broadly striped individuals produce many more reds; the mean of three experiments (11-36-42) was about 30%.

5. The red plants resemble the wild ancestral form externally but are not constant as this is. The intensity of inheritance of their character is only about 70-85%; and the remainder of their offspring revert to the striped

type. I have not yet observed this to happen by means of bud-variation.

6. *Antirrhinum majus luteum* does not arise from these striped and red races.

7. If we compare the forms which we have been considering,[1] with the half races and middle races which we distinguished in § 3, p. 18, we find that between the two constant elementary species (the systematic species, *A. majus* and the systematic variety, *A. majus luteum*) there exist two intermediate forms which are perfectly distinct from these two, but not from one another. We can distinguish,

a. The eversporting variety, *A. majus luteum striatum*, with striped flowers and a high degree of fluctuating variability, from which a faintly striped and a broadly striped race can be raised by selection. These three races however merge continuously into one another.

b. The atavistic type in this race is uniformly red, but with incomplete inheritance and gives rise, when self-fertilized, in each generation to about 25% striped individuals besides the red ones.

In contrast with the previously described cases, the transition from the atavistic type to the eversporting variety and the reverse process here occur every year but always with a slight gap. The red type arises from the striped race by seeds and by buds, but the striped race has, hitherto, arisen from the atavistic type only by

---

[1] The mode of inheritance in the coarsely striped individuals will have to be more closely investigated; so also must sectorial variation. Moreover the experiment should be repeated with other striped varieties, and the spotted forms investigated to see if they behave in the same way. But it is most important that pure cultures of the various types should be made by breeding for several generations. For this purpose the tall varieties should be chosen preferably, since they promise a much better harvest than the half-dwarf ones which I employed in my experiment.

seeds. The transition from the red to the striped oscillates round 25%, the transition from the striped to the red is largely dependent on the degree of striping, which points to the existence of factors as yet incompletely understood.

It may perhaps be mentioned here in anticipation that the varieties of *Hesperis* and *Clarkia* (§ 15 and § 16) with striped flowers behave in the same way , whilst both in *Plantago* (§ 17) and *Linaria vulgaris peloria* (§ 20) the eversporting variety is inconstant and reverts more or less easily to the atavistic type.

§ 15. HESPERIS MATRONALIS.

The flowers of the dame's violet are violet as the name indicates. There are three varieties on the market: a white flowered, a double, and a dwarf variety, all of which are constant so far as I know. A *Forma lilacina* and a "mixed" sort are offered in the catalogues. The plants are perennial; if the seed is sown in the spring, the majority of the plants will not flower until the following year; but if the seed is sown as soon as it ripens, or is allowed to fall on the ground instead of being harvested, the plants generally flower the next year. I have employed both of these methods at different times.

Fig. 24. *Hesperis matronalis*. A flower of the pale finely striped form, with half of one of its petals dark violet.

I obtained my seeds in 1890 from a mixed group of white and violet flowered plants which were growing in our Botanical Garden. I grew them for two generations and found that the "white" were not pure white but pale lilac. Then I kept only plants of this variety through the winter, and first examined them in 1894, when they were in full flower. They flowered in isolation and partly pollinated themselves with their own pollen, partly were fertilized by insects. In later years also I have not enclosed this species in bags but have either grown them in an isolated position and left them to be pollinated by insects, or have had them flowering in a little greenhouse entirely built of fine metal gauze, where they fertilized themselves.

My object was to test the degree of inheritance of the pale, the lilac, and the violet types separately. I shall first give a summary of my experiment. In this table, W denotes whitish, L lilac, and V violet (that is, the color of the wild species). The numbers in each case are percentages of the particular culture; where the culture was too small I have omitted the numbers.

HESPERIS MATRONALIS.

(WHITISH, LILAC, AND VIOLET IN PERCENTAGES.)

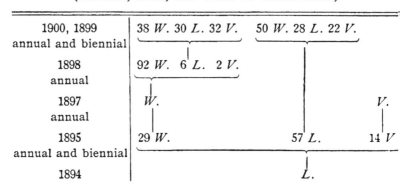

Before I come to the description of this experiment it is necessary to give some more details as to the variability of the color of the flower.

Plants with pure white flowers such as those belonging to the variety *Alba* did not occur in my cultures. I have compared the *Alba* and also the *Alba plena* directly with my plants. Certainly the difference is sometimes very slight, especially as the petals of *Alba* acquire a pale lilac color when they fade. They are all gradations between the whitest examples and those with the full lilac color; the variability in this case is perfectly continuous. But between the lilacs and the violets there is always a gap; the darkest lilacs seem to be about half as dark as the violets; intermediate stages do not occur.

The vast majority of the plants have all their petals of the same color, but mixed conditions also occur. As in other cases there are striped flowers, sectorial and bud-variations. Examples of these three groups appeared in various years in my cultures but only sparingly. On the striped petals the stripes were fine and rare, but they exhibited the dark violet hue of the original species. The instances of sectorial variation have so far been occasional dark flowers on pale clusters, and on the other hand flowers of which one-half of a petal was whitish and the other violet (Fig. 24). Bud-variations occurred on plants with very pale flowers, especially when they were richly branched and flowered on into the autumn. They were always stray twigs on the lower part of the main stems; their flowers were all of the normal violet color. But so far I have not been able to obtain seed from them.

A glance at a large bed reveals the general distribution of color. At once the pale flowers are seen to be in the majority, whereas the whitish on the one hand and the lilac on the other are obviously rarer. The violet stand out conspicuously because they are not connected with the rest by any gradations. Except for this the variation is so continuous that it is almost impossible to express it in numbers. I have tried to arrange the plants in groups and to count the numbers of each group. And I give the numbers obtained in this way, only with the object of conveying to the reader the general impression which a bed makes on the observer, for it is inevitable that the limits between the groups should be somewhat arbitrary. Nevertheless I trust that I have succeeded in keeping fairly well the same limits between the groups during the successive years of my experiment, and this is the most important point.

For the purposes of these color valuations I picked a flowering cluster, if possible the terminal one, from each of the plants on a bed, brought them to my house and sorted them there. I made out the following more or less clearly defined groups:

W. Whitish, always without stripes.
- W_1. Almost white; buds and withering petals almost white.
- W_2. White suffused with lilac, not darker when withering.
- W_3. Very pale lilac; buds lilac; only slightly darker when withered.

L. Lilac, sometimes striped or spotted.
- L_1. Definitely lilac, although pale; darker than W_3.
- L_2. Lilac; half as dark as V.

V. Violet, the color of the typical species.

I shall now give the composition of the culture of 1898 which was raised from the seeds of plants with whitish flowers. On July 14, I sorted 250 individuals by the method described and found:

HESPERIS FROM THE SEED OF WHITISH FLOWERED PLANTS.

W_1 5% ⎫
W_2 57% ⎬ 92% W.
W_3 30% ⎭
L_1 4% ⎫
L_2 2% ⎬ 6% L.
V 2%   2% V.

I determined the composition of the cultures of the next year, 1899, in the same way; they were both raised from the seeds of lilac plants. One of them (5th generation) flowered partly in 1899 and partly in 1900; but the other only in 1899. The result was as follows:

HESPERIS FROM THE SEEDS OF LILAC FLOWERED PLANTS.

Color	1st Experiment (5th Gen.)		2d Experiment (3d Gen.)	
W_1	3%	⎫	4%	⎫
W_2	15%	⎬ 38% W	22%	⎬ 50% W.
W_3	20%	⎭	24%	⎭
L_1	17%	⎫ 30% L	22%	⎫ 28% L.
L_2	13%	⎭	6%	⎭
V	32%	= 32% V	22%	= 22% V.

The first list is based on 155 flowering plants, and the second on 219.

The seeds of the whitish *Hesperis,* therefore, in this experiment, produce their like with a small percentage of lilacs and violets. The seeds of lilacs, on the other hand, give rise to the three types in about equal numbers, though it must be remembered that the limit between W_3 and L_1 is to a certain extent arbitrary.

I have not yet made a sufficient number of observations on the inheritance of the violet color in this race. In the only experiment which I have carried out, only five plants flowered and they had the same color as their parents.

Let us now pass on to a detailed description of the

experiment. It began in 1894 with seven plants which had already flowered in 1893 and had been noted as lilac flowered. Many of their flowers were more or less striped, some of them produced in August the violet bud-variations mentioned above, when the rest of the flowers had been through blooming for a long time. Seed was only saved from the lilac flowered branches; a part of it was sowed in August, the rest as soon as it was ripe. Most of it germinated in the following February and March; more than half of these plants produced stems and flowered in August. I obtained altogether 234 plants in flower of which 29% were pale, 57% were lilac and 14% normal violet. I selected the strongest plants from among the most typical of each group and transplanted them in the autumn to three as isolated spots as I could find in my garden. Here they grew freely, branched abundantly and flowered in the following year (1895) for a second time.

There were three violet plants which however set very little seed. This was sown and the offspring flowered in the summer of 1897 in a conservatory. I took precautions to prevent their being visited by insects in order to render impossible the transference of their pollen to the other plants. As soon as the color of the flowers could be determined with certainty for any plant, this was pulled up. There were, as I have already stated, only five plants and their flowers were violet.

I did not allow the lilac flowered plants to flower in this year but kept them for the next. Of the plants with pale flowers which had been planted out separately in the autumn of 1895, only one plant flowered in 1896. Its seeds were sown immediately and gave rise to 12 plants which flowered in the summer of 1897; they were all pale with no more than the faintest indication of the lilac color. The seeds were sown in pans in the autumn, the seedlings were pricked out in November and planted out in April 1898 on a large bed. In June 250 individuals flowered, and the percentage composition of the color, as given above, was determined. Then the four lilac individuals falling into the group L_2 were taken up and transplanted with all possible care to the metal gauze greenhouse. Before doing so all open flowers and young fruits were of course removed. It may be noted that in this experiment the lilac flowered individuals began to flower conspicuously later than the pale and violet ones.

The seeds of these four plants were sown partly in October and partly in November, separately for each parent. Only one of the four resultant groups flowered in the following year (1899); the rest remained in the rosette stage and flowered in 1900. The proportions of the various colors were very much the same in the four groups. I recorded them separately but did not find any significant differences. The numbers in the first column (1st Experiment, 5th Gen.) on the table on page 140 give the composition of the whole culture.

I transplanted some lilac plants of the first crop (1895), but only kept one of them which caught my eye with its beautifully striped flowers. It grew up into a sturdy plant, flowered in 1898 in an isolated spot and set an abundance of seed. From this 219 flowering plants were raised in 1899, and their colors are recorded in the last column of the table on page 140 (2d Experiment, 3d Gen.).

If we consider the results of these experiments, extending over seven years, in their relation to other known facts we find that we can distinguish the following races:

1. *Hesperis matronalis alba,* the constant commercial variety.

2a. A whitish, pale lilac, seldom or never striped sort (W_1-W_3), which can reproduce the violet color by sectorial, bud- and seed-variation; violet seed-variation about 2%; lilac offspring about 6%.

2b. A lilac, often striped or spotted, race which gives rise to an inconstant but mostly considerable number of whitish and violet offspring. Its color merges continuously into that of No. 2a, but is sharply separated from No. 3.

3. A violet variety which has arisen from 2a and 2b and is presumably inconstant, on the analogy of *Antirrhinum majus*.

4. *Hesperis matronalis,* the original, constant, violet species.

The analogy with the corresponding races of *Antirrhinum majus* seems to me to be obvious and can be expressed as follows:

1. The systematic variety which is perfectly constant (*H. m. alba, A. maj. luteum*).

2. The eversporting variety with lilac or striped flowers (*H. m. lilacina, A. maj. luteum striatum*). It can be split by selection in a plus and in a minus direction; into a pale lilac, or finely striped race on the one hand, and on the other into the dark lilac and frequently striped dame's violet and the broadly striped snapdragon.

3. The self-colored but inconstant atavistic type which has the color but not the constancy of the original species.

4. The original violet, or red, perfectly constant species (*Hesperis matronalis, Antirrhinum majus*).

§ 16. CLARKIA PULCHELLA.

A white variety of this pretty red species is offered by seedsmen.[1] Besides this a striped race sometimes occurs which has more or less numerous red bands of varying breadth on the petals.[2] The red in these cases has the same intensity as that of the species. Moreover the white flowers are not pure white; a very delicate but distinctly visible red flush can be seen on any bed of them in full flower. Sometimes occasional plants or individual flowers are somewhat richer in pigment, so that it is at once obvious that they are not pure white.

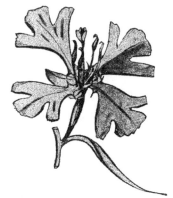

Fig. 25. *Clarkia pulchella.* A white flower of which one petal and a half are dark red, while there are dark red stripes here and there on the other two petals.

I have only made an incomplete series of experiments with this plant because it does not lend itself easily to artificial fertilization and, as a rule, does not stand transplanting while in flower. But the results obtained suffice to demonstrate their essential correspondence with those obtained with *Antirrhinum* and *Hesperis*.

We can distinguish in this as in the other two cases between a pale race poor in stripes and a richly striped one; moreover these two races possess the characters of the corresponding ones in the two species named. But in *Clarkia* the broad stripes appear chiefly as sectors, as

---

[1] There is also a variety, Carnea, which is constant so far as my experience goes.

[2] See p. 119. It was referred to by VILMORIN and by B. VERLOT, *Production et fixation des variétés*, 1865, p. 64.

for instance, whole or half petals; I shall therefore call such flowers and plants sectorial.

In 1896 I had a bed of about 50 plants all of the flowers of which were whitish. The majority bore no red stripes, or only such fine ones and so rarely that they were overlooked, which is always possible since the plants produce very many and rapidly fading flowers. Only one plant stood out amongst the rest; at the end of July it bore a flower with two red petals and at the beginning of August a petal the middle third of which was also colored red. Otherwise, the bed was practically white throughout the summer. Some of the seed of the whites was saved.

From the seeds of a white flowered specimen I obtained in 1897 a culture of about 100 plants. Amongst these again there was only one sectorial example; I saved its seeds separately although it had been fertilized by insects in the midst of the others. In the majority of these others I had not seen red stripes, but on a few of them there had been some insignificant ones.

The seeds of the pale flowered plants gave rise to a generation equally poor in stripes; in 1898 I only saw one striped one amongst 30. This race therefore remained poor in red sectors as a result of a continued selection of almost white plants.

From the seeds of the sectorial plant I at once obtained a race which was rich in red petals and red sections of petals, and often produced whole red flowers and twigs with red flowers only (Bud-variation). I grew it for two generations (1898 and 1899). The seeds for the first were gathered in 1897 from a seed-parent which had not been isolated; in 1898, however, I pulled up all of the non-sectorial plants whilst they were in flower and on the remaining seed-parents only harvested the seeds from those flowers which opened after that operation.

The single sectorial plant of 1897 bore one flower with one, and another with two red petals. Their seeds were harvested separately and sown. The other flowers were pale; I also harvested their seed separately. The first named seeds, naturally few in number, gave rise in 1898 to about 40 plants which flowered; the latter to 200. In both groups the red stripes and sectors were remarkably numerous in comparison with the previous year. At the end of July I found amongst the former about 25%, and amongst the latter 23% sectorial plants. Besides these, a plant bearing red flowers exclusively, occurred in the former group. If I had repeated these observations from time to time the two percentages would of course have been considerably increased. But in order to isolate the sectorial plants I pulled up all those which up to that time had exhibited only few and narrow stripes. As already mentioned, I harvested seed only from the fruits of those flowers which had opened after this operation. I saved two kinds of seed: one was from a number of sectorial flowers which I had marked on a large group of individuals; the other was from a particularly striking plant which I had also marked, and which had a fair number of sectorial and occasional perfectly red flowers, exhibiting also red bud-variations on its lower branches. I harvested seeds only from the narrow striped flowers of this plant.

I have one more case of sectorial variation to mention before I proceed to give the results obtained from this harvest. A green lateral branch in an inflorescence on an otherwise white or finely striped plant had a narrow red longitudinal line on it which was not much broader than a flower stalk and extended over four internodes. The upper, lower, and middle flowers of the tract stood on this line; the two former were completely red, the middle one only partly so. The two flowers occupying intermediate positions but on the green side of the branch were almost white.

The culture of 1899 was richer in sectorial plants than that of 1898, as the isolation of the seed-parents would have led us to expect. From the mixed seeds referred to above, I had about 300 plants of which five were wholly red whilst the proportion of sectorial ones was 40%. The single selected seed-parent, however, gave rise to only 50 offspring which flowered, of which one was red, whilst the proportion of sectorials mounted to 70%. The average number of reds in the two cultures was 1-2%; and that of sectorial plants 45%.

These experiments show that the pale flowered plants, selected as seed-parents, give rise to a fairly constant progeny amongst which the proportion of sectorial plants is quite small.

The progeny of sectorial plants, on the other hand, consists of about 45% broadly striped and 1-2% red plants, the remainder being pale tinged with red, or at any rate very poor in stripes.

The cultures of the pale flowered plants are ordinarily in flower some weeks before the first stripes appear; but in the beds of sectorial plants the red may be seen among the very first flowers. Here also the white flowered ones are always in a large majority; among a thousand flowers of this race I counted 34 striped and 8 sectorial ones, that is to say only 4% altogether.

§ 17. PLANTAGO LANCEOLATA RAMOSA.

Plantago lanceolata is one of of those plants which are remarkably rich in anomalies. PENZIG mentions a considerable number of them such as leafy stalks, ears the tops of which bear tufts of foliage leaves,[1] forked spikes with two or more tips, torsions etc. These and many other malformations such as split leaves, pitchers consisting of one or more leaves, occur commonly in this neighborhood and also in my cultures. It is worth mentioning that all or nearly all of these abnormalities can occur in the same race, and sometimes indeed in a single stout individual. Evidently every plant must contain a number of latent or semi-latent characters which lie outside its proper range of form; these characters constitute, as I have already said, the outer range of the forms of the species (p. 27).

A form also frequently mentioned[2] in the literature of the subject is one with branched ears (*Plantago lanceolata ramosa*).[3] In this variety sessile secondary spikes are produced in the axils of the bracts at the base of the main ear. They are often small, but sometimes nearly attain the size of the central ear. Their number is highly variable. Under good conditions of cultivation each head may have from 2-7 lateral ears, but on single ears the number may rise to 20 and more (Figs. 26, 27).

I have been carrying out experiments on the inheritance of this *ramosa*-character since 1887. It proved to

---

[1] I have often picked these tufts and made cuttings of them; they take most quickly and grow to strong rosettes of radical leaves, the ears arising from which may repeat the phenomenon of the tufting to a certain extent (*Plantago lanceolata coronata*).

[2] PENZIG, *Teratologie*, II, p. 252.

[3] *Kruidkundig Jaarboek*, Gent, 1897, pp. 76 and 91.

be only partial. In spite of the most careful selection and isolation during the time of flowering this race every year produces plants not one of whose spikes, even when there are a hundred to the plant, exhibits the smallest trace of branching. They are obviously to be regarded as atavists.

The proportion in which these atavists occur seems to be fairly constant, fluctuating however from year to year. It can be slightly increased or diminished by the choice of favorable or unfavorable seed-parents; but it does not seem possible to effect an essential and permanent improvement by continued selection, at least not to a degree that would open a chance of altogether eliminating the atavism.

In the first years of my cultures I did not pay particular attention to this phenomenon; moreover my experiments were on too small a scale to afford numerical data of any value. But I found atavists as well as *ramosa*-plants every year, although I always collected my seeds from the former. I did not determine the proportion until the fifth generation (1892) was reached. I should state that I have isolated my seed-parents every year, cutting off as many as possible of their unbranched ears before they flowered. Pollination which had to be left to the wind was therefore confined to the group of selected seed-parents, whose number scarcely ever exceeded 10. It was as pure as it was possible to have it.

I obtained the following figures:

GENERATION	PERCENTAGE OF ATAVISTS
5.—1892	46%
6.—1894	50%, 58%, 59%
7.—1897	47%
8.—1898	45%, 56%, 59%
8.—1900	52%

Plantago lanceolata ramosa, therefore, produces a proportion of about one-half atavistic individuals every year.

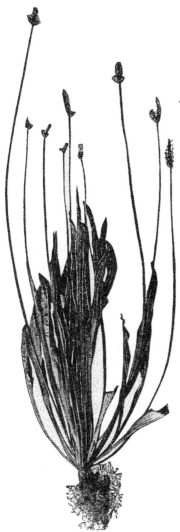

Fig. 26. *Plantago lanceolata ramosa.* A whole plant.

The variability in the figures given is at least in part dependent on external influences (nutrition and selection). Closer examination of the individual years proves the truth of this. In 1892 I had 48 plants in flower; nine of these plants produced split leaves and pitchers at the time when they were being transplanted, about three weeks after the seed had been sown, and seemed especially desirable on this account. In the summer they turned out to be all *ramosa-*plants with richly branched spikes. They were cultivated the following year also; and the sixth generation was raised partly from their seed and partly from the seed harvested in 1892 from two other seed-parents. From the latter were raised 103 plants which flowered, of which 50% were atavists, this proportion being nearly the same for the two seed-parents. In order to investigate this, the seeds of the individual seed-parents were, as usual, sown separately. The higher figures 58% and 59% were derived from the offspring of a plant which had been divided into two in 1893, after which one-half of it was grown on sand and the other on ordinary garden soil. I shall have to revert to the effect of this treatment on the plant itself; but it will be observed that the differential treatment had no visible effect on the offspring of the two halves. (The numbers of individuals dealt with in the two cultures were 57 and 60 respectively.)

The seeds of the typical individuals of my race of 1894 I sowed in 1897 under normal conditions, as usual (seed sown in the greenhouse; seedlings pricked out into pots, and afterwards transplanted to the beds). The seed had been saved from two plants with richly branched ears. It produced a culture of 70 plants which flowered and contained 47% atavists. Whilst flowering was proceeding I transplanted all the *ramosa*-plants whose ears were only slightly branched, and marked among the remainder a specimen which seemed to be the most profusely branched. I harvested seed from those flowers only which protruded their stigma after this separation had taken place and after the atavists had been weeded out. Seed was harvested separately from each plant. In the following summer (1898, 8th generation) it was seen that the seed of the best seed-parent had only produced 45% atavists (among 100 plants that flowered). The seeds of the average seed-parents gave 56%, and those of the worst, 59%. Selection had therefore a distinct, although not a very great effect. It should be remarked that the number of average seed-parents was 8, and that of the worst ones 10. The composition of the progeny was determined separately for each seed-parent, but the differences were not greater than the extent of the experiment would lead us to expect. There were 1033 offspring from the average seed-parents and 732 from the ten worst plants. The two separate cultures which deviated most from the mean contained 37% and 65% atavists respectively. The value of 52% given

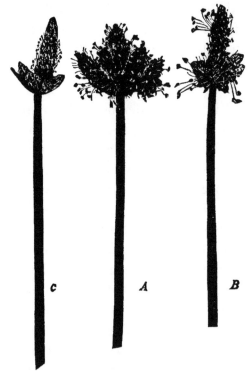

Fig. 27. *Plantago lanceolata ramosa.* A, B, C, three branched ears.

above for the same generation but grown in 1900, will be dealt with below.

Bud-variations occur in this as in the inconstant races of other species, although very rarely. In such cases it is one or several lateral rosettes which vary. The structure of our plant is a very simple one. The stem of the seedling grows out into a short, somewhat oblique, rhizom which produces a rosette of radical leaves. Ears are formed in the axils of the higher leaves but rosettes of the second order grow out from the axils of the lower ones. In the second summer the primary and secondary rosettes behave in the same way, again producing ears above and secondary rosettes below. If the plant grows very robustly it may consist of as many as 10-20 single rosettes; if it is a *ramosa* every rosette produces branched ears, at least on some stalks. Sometimes all the ears of the whole plant are branched, in which case it is perfectly easy to see that there is no bud-variation. In its second year a single plant may often produce more than 50 branched ears.

The culture of 1897 contained a plant which exhibited a bud-variation. The seeds of its branched ears, harvested in the first year, had produced 89 individuals that flowered, of which 36 (40%) were atavists. The plant in question consisted, in the autumn of its second year, of more than 25 single rosettes which were carefully isolated, and planted separately. Only the seven strongest ones survived this operation. I kept them all in their pots until a sufficient number of ears were visible and then planted them out on two distant beds. On the one I planted four rosettes with unbranched ears, on the other, three with branched ears. The four former produced, together, over 200 strong ears, all unbranched with the exception of a single one which bore a small lateral branch at its base. The three latter formed both unbranched and more or less richly branched inflorescences, but during the whole summer the unbranched ears were all cut off before they flowered. The harvest from the two beds, gathered and sown separately, gave rise to two cultures in 1900. They had the following composition:

Ears of seed-parent	Extent of culture	Atavists	With branched ears
Branched	44 individuals	52 %	48 %
Unbranched	206 individuals	92 %	8 %

The rosettes with branched ears gave rise to rather more atavists than the seed of the branched inflorescences of the same plant in the first year (52% as against 40%), which was probably due to the fact that it had a less sunny position in 1899 than in 1897. But the rosettes with unbranched ears, although they were in a good position in 1899 and grew very healthily, gave a progeny dissimilar to that hitherto produced by any of the branched plants of this race (see Table on page 149 which gives the results of more than 25 individual sowings from separate seed-parents).

The four lateral rosettes with unbranched ears, therefore, formed in this case a clear instance of bud-variation, producing a race poor in branched ears.

The question of the constancy of the atavists in my race is a point of considerable interest. Hitherto I have found them completely constant. With a view to testing this I did not weed out the atavists in the fifth generation in 1894, but simply cut off all their ears before the branched plants flowered, and repeated this operation from time to time when new ears appeared before they could protrude their stamens. After the harvest I weeded out all the branched individuals; most of the atavists survived the winter and flowered luxuriantly in 1895 in isolation. The majority of them produced over one hundred ears per individual. I harvested the seeds separately for each seed-parent.

The sowings took place in 1896 and in 1897. They

gave rise respectively to three and six cultures derived from the nine seed-parents. Each culture consisted of from 35 to 100 plants, making together 600 flowering individuals bearing 4000 inflorescences. These were unbranched without exception.

The question suggests itself, whether the seed-atavists and the bud-atavists belong to the same type. On the one hand it is possible that the constancy of the former is not always so absolute as it appeared in my experiment. On the other hand, branched bud-variants might occasionally appear in the race derived from the atavistic bud-variants, and such might have been the cause of the occurrence of branched individuals (8%) in my culture of 1900. But further investigations are necessary to provide a satisfactory answer to this question.

Plantago lanceolata ramosa, therefore, *gives rise to atavistic individuals, either by seed (about 50%) or by buds (very seldom) which are either absolutely, or at least in a high degree, constant from seed.*

It still remains to describe briefly the fluctuating variability of our race of plantains. This is considerable, and conforms to the common laws; especially is it dependent to a large extent on external conditions and, within certain limits, capable of being altered by selection. The observations, which I now shall give, refer to true *ramosa*-plants, and not to atavists and bud-variations.

The variability of this race corresponds with that of other monstrous races inasmuch as the curve describing it is dimorphic.[1] During July and August 1893 I

---

[1] *Sur les courbes galtoniennes des monstruosités,* Bull. Scientif. de la France et de la Belgique, publié par A. GIARD, XXVII, 1896, p. 397.

picked all the ears of a small group of plants, and obtained the following figures:

Ears without any branching	191
Ears with one lateral ear	80
Ears with two lateral ears	136
Ears with three lateral ears	93
Ears with four lateral ears	33
Ears with five lateral ears	12
Sum of ears	545

The degree of branching in this group was fairly low; nevertheless the apex of the curve of the atavistic ears is distinct from that of the branched ones. This phenomenon could indeed be easily observed, even without any counting, on account of the relative scarcity of heads bearing a single lateral ear, a fact which I have also observed repeatedly since. This is a character of the eversporting variety and suggests the possibility that the one-branched ears which are so common in nature (where the *ramosa*-form, as is well known, is not at all rare) presumably constitute the half race; but I have not investigated this point.

The number of compound ears per plant, and the degree of branching in each, are to a great extent dependent on the conditions of life. The stronger the growth of the whole plant, and the richer the foliage, the more pronounced will the anomaly be. Therefore, a more profuse branching of the individual ears usually goes hand in hand with a richness of branched inflorescences. The branching also manifests a certain periodicity. The young plants almost always begin with unbranched ears; it is not until later that the monstrosity appears, gradually increasing in strength. Then towards the end of the summer I often observed a diminution in the amount of branching and often the formation of more numerous unbranched ears. In the second summer often almost all the ears on healthy individuals are branched even when their number reaches 50-60 per plant. In the first year I found that as a rule there were 10-20 branched ears, and sometimes as many as 30 or even more occurred on each plant. In fact we may assume that, on the average, and with ordinary methods of cultivation, about one-third of the ears will be branched during the first summer; for instance, in 1898 I found amongst 439 ears on 30 individuals 136 or 31% which were branched. It goes without saying that the atavistic individuals were excluded from these countings.

I have also made some direct experiments to determine the influence of individual vigor on the development of the anomaly. In the first place I have grown very weak plants and have then got them to grow stronger gradually. For this purpose I made use of the plantain's well-known property of producing buds from its roots. As the roots are all very thin, the plantlets obtained in this way are very weak at first, nor do they grow up as quickly as seedlings.

For the purpose of this experiment I selected (March 1893) ten plants which had had 10-25 branched ears each in the previous year. I pulled them out of the ground, cut off the mass of their roots and planted these, throwing away the rosettes and any leaf-buds that might be present. I put the roots of each individual straight into the ground without separating them. Radical buds were produced in hundreds, often so many from one bundle of roots that there was not room for all of them to develop. In the middle of June, that is, after about three months, they began to flower. At first there were only 40% branched ears, with only one or two lateral ears (on the 46 first flowerstalks). In the next 100 the proportion mounted to 60%, and 3-4 partite inflorescences also occurred. Later on, about the middle of July, the first stalks with five lateral ears appeared, and the number of branched ears gradually increased to 70%, and in August the strongest rosette had 67 ears of which 52 were branched, i. e., about 78%.

A question at one time much discussed was whether adventitious buds had the power to reproduce the variations and anomalies of the parent plant. At that time malformations were not regarded as heritable, but since the inheritance of monstrosities has become generally recognized,[1] it must be considered evident that adventitious buds will behave like normal ones; and the only question that can arise is whether they are more liable to produce bud-variations or not. If they are weak the abnormal character will be less pronounced; but if their strength is equal to that of ordinary buds the abnormal character must be developed to the same extent. It is therefore almost superfluous to lay much stress on the reproduction of the branched ears from the radical shoots of our plantain.

The rest of my experiments deal with divided plants. In the spring of 1893 I selected for this purpose two fine rosettes that had survived the winter and which had proved to be particularly rich in branched ears in the previous year. Both plants were divided as equally as possible into halves. Of the first plants one-half was planted in sand and of the other one-half was put in the shadow of a tree, the control halves of both plants being cultivated under ordinary conditions for the purpose of

[1] *Erfelyke Monstrositeiten,* Kruidkundig Jaarboek, Gent, 1897, p. 62.

comparison. At the beginning of the period of flowering no difference was discernible in either experiment between the two halves, but it gradually became visible during the course of the summer. I picked off all the ears from the culture in sand at the end of July and at the end of August; here is a record of them:

		\multicolumn{6}{c}{Number of lateral ears per primary ear}	Totals
		0	1	2	3	4	5	
July 28th	Sand	3	3	4	6	3	1	20
	Control	9	7	9	6	0	0	31
Aug. 31st	Sand	14	10	12	8	3	1	48
	Control	12	2	10	7	6	2	39

The difference though slight is distinct. It is more clearly brought out if the mean number of lateral ears per primary ear is calculated. In August in the plants on sand this was 1.5, in the control half 2.

A similar effect was produced by shade which exerted a most deleterious effect on the whole growth of my experimental plants as will be seen from the small number of ears produced. I obtained the following figures in the same way as in the previous experiment.

		\multicolumn{7}{c}{Number of lateral ears per primary ear}	Totals
		0	1	2	3	4	5	6	
July 28th	Shade	7	6	2	7	5	2	0	29
	Control	1	1	2	8	19	20	1	52
Aug. 31st	Shade	15	1	1	2	0	0	0	19
	Control	21	9	20	16	10	3	0	79

The mean number of subsidiary ears per primary ear in August in the shadow half was 0.5 and in the control half 2.0.

In conclusion, the results of the whole series of experiments which has lasted over more than ten years may be summarized as follows: The *Plantago lanceolata ramosa* of my experiment constitutes an "inconstant" *middle race or eversporting variety; that is to say, a race which produces in every generation a fairly constant proportion of atavists.* This proportion is about 50%.

The segregation of atavists occurs regularly in generations grown from seed, but sometimes also in those grown by means of bud-variation. The atavists are perfectly, or at least very nearly, constant.

The true representatives of the race (i. e., all other than atavists) produce both unbranched and more or less profusely branched ears, and are largely dependent, in regard to this character, on their environment and their individual vigor (fluctuating variability). The stronger the plant and the more favorable the conditions the more pronounced is the anomaly.[1]

[1] Compare the behavior of *Papaver somniferum polycephalum* in Vol. I, Part I, § 16, p. 138; and also the end of this part.

VI. EXPERIMENTAL OBSERVATION OF THE ORIGIN OF VARIETIES.

§ 18. THE ORIGIN OF CHRYSANTHEMUM SEGETUM. PLENUM.

(See Plate II.)

The double corn marigold constitutes a new variety which has recently arisen in my cultures. It has never occurred before. *Chrysanthemum segetum* is, of course, a favorite annual garden plant, and so is a variety of it called *C. segetum grandiflorum*. A form called *C. segetum Gloria* is announced amongst this year's novelties;[1] its flowers are said to attain a diameter of 10 centimeters, but it is not double. If a double form ever had appeared, it would without any doubt have been put on the market as a noteworthy improvement, even as the double varieties of *Chrysanthemum inodorum* and other composites are so widely grown.

My "conquest," as the breeders of hyacinths in Haarlem call their novelties, is the counterpart of the well-known *Chrysanthemum inodorum plenissimum*. It is inferior to it in the matter of color, inasmuch as white flowers are always in greater favor than yellow ones. The doubling of the heads of composites is never so perfect that tubular florets are completely absent from all inflorescences. Nevertheless it frequently looks as if this were so (Fig. 28); but if we look a little closer we

[1] Seed-catalogue of HAAGE and SCHMIDT in Erfurt, 1900.

will always find between the tongue florets, more or less numerous tubular florets which are hidden from view by the others. Moreover, the degree of doubling is to a considerable extent subject to fluctuating variability; one plant has more and another fewer, transformed florets.

Fig. 28. *Chrysanthemum inodorum plenissimum*. A plant with a high degree of doubling in the inflorescences, and, consequently, perfectly sterile.

This is an important point, for the white tongue florets of *C. inodorum plenissimum* are female and inflorescences such as those shown in Fig. 28 and Fig. 34C on page 184 set absolutely no seed. The variety is therefore maintained by saving seed from plants such as those figured in Fig. 34 A and B, p. 184.

These remarks also apply to my new *Chrysanthemum segetum plenum*. Many specimens set absolutely no seed because the doubling has gone too far. For the same reason others afford only a meagre harvest. Too drastic a selection at the beginning of the flowering period would destroy any prospect of a harvest and might even result in the extinction of the variety.

Moreover plants with a high degree of "doubling" produce no pollen for the fertilization of the others, because they are almost exclusively female; so that they can take no part in the perpetuation of the race in this way either.

My novelty is probably the first horticultural variety which has arisen in an experimental culture. By this I mean that pure fertilization has been insured since the beginning of the culture and that exact and detailed records of the course of the experiment have been kept every year. Moreover the selection of the seed-parents has constantly been carried out from the very beginning, with a view to the same ideal. Selection began in 1897, the "double" race was obtained in 1900. The selection occupied, therefore, a period of three years.

The corn marigold, being a composite, is admirably adapted to form the material for a statistical investigation of its variability. The number of ray florets fluctuate in accordance with the well-known law of LUDWIG based on BRAUN and SCHIMPER's series. By this means the exact composition of a culture can be expressed in figures and plotted graphically by recording a sufficient number of inflorescences. The course of the selective process can in this way be displayed in all its details. Although an explanation of BRAUN-SCHIMPER's series is still wanting, each of the numbers in it (e. g., 13, 21, etc.) may figure as a specific character; that is, it may be the constant mean for a particular species. On the other hand they may constitute stages of variation or characterize races whose nature is still unknown to us. We must therefore limit ourselves to a purely empirical description.

Fig. 29. *Chrysanthemum segetum plenum.* An almost completely double inflorescence. See also Plate II.

It seems desirable to give a general outline of the significance of my experiment before I proceed to describe the details.

The corn marigold is very common in cornfields over the greater part of Europe, as also its German names *"Saatwucherblume"* and *"gelbe Kornblume"* imply. It has thirteen ray florets in the inflorescence and fluctuates around this number according to QUETELET's law. A commercial variety, called *Chrysanthemum segetum grandiflorum*, whose origin is not known, is distinguished by the possession of larger and more numerous tongue florets.[1] So far as my experience goes, bought seeds of this variety, give rise to a mixture of this and of ordinary *C. segetum*, no doubt on account of the fact that in the nurseries both are grown close together, for practical

[1] RÜMPLER, *Vilmorin's Blumengärtnerei*, 1896, II, p. 507.

reasons. In botanical gardens, too, both sorts are often grown together; and, frequently, simply under the name of *C. segetum*.

This mixed assemblage gives rise to a dimorphic curve;[1] but the two groups of individuals which compose it can easily be isolated by selection. Then the *C. segetum grandiflorum* proves to have a mean of 21 ligulate florets, around which variation practically takes place in the same way as in the 13-rayed race (i. e., the wild species), except that it has a tendency to multiply the number of rays beyond the limits of a normal QUETELET's curve; a fact which indicates discontinuous variation.[2]

This slight indication was the starting point for my experiment. In 1897 I chose a seed-parent with 34 rays for the 1898 crop, and reached 49 rays.[3] Proceeding in the same way I reached 67 in 1899 and about 90 in 1900 in the best inflorescences. Up till 1899 the ligulate florets only appeared in the circumference, the disc consisting solely of tubular florets. In this year, however, there appeared 2 or 3 ligulate florets in the midst of the disc of a few flowerheads on a single plant. This was the first indication of the double race. Therefore I only sowed the seeds of this one plant in 1900, and from that the race was fully developed (Plate II). Apart of course from eliminating the possible effects of crossing, it needed no further selection; a too rigid selection was moreover

---

[1] *Eine zweigipfelige Variationscurve*, Archiv für Entwickelungsmechanik der Organismen, Leipsic, 1895, p. 52.

[2] Compare the half curves (p. 26) and the note on page 29. See also *Ueber halbe Galton-Curven als Zeichen discontinuirlicher Variation*. Berichte d. deutschen bot. Gesellschaft, Vol. XII, p. 197.

[3] *Ueber Curvenselection bei Chrysanthemum segetum*. Same journal, 1899, Vol. XVII, p. 84.

to be avoided on account of the sterility of the most highly modified individuals.

My cultures embraced, as a rule, a hundred individuals each, but sometimes a few hundreds. There can hardly be a doubt that if I had carried out more extensive sowings I should have attained my object at least one year earlier. But the more stringent the selection is, the smaller are both the harvest and consequently the next year's crop.

Of course the reader will ask, has this transition been a gradual or a sudden one? I consider it sudden; but much depends on the meaning that we attach to the words. At any rate the change did not occupy centuries, as is commonly supposed by the current theory of selection; it did not even require one decade. Three years were sufficient in a culture of no more than a few square meters in extent.

I now come to the details of the experiment and shall first give a short description of the original wild species.

The species does not grow around Amsterdam. The herbarium material collected by me in various parts of the Netherlands points to the general occurrence of a mean number of 13 rays. HEINSIUS plotted curves from plants from two localities in the province of North Brabant, and obtained the following numbers. The first row relates to plants which were collected near *V*ucht, the second to a collection from Hintham.[1]

NUMBER OF LIGULATE FLORETS (L. F.) IN THE NETHERLANDS.

L. F.	6	7	8	9	10	11	12	13	14	15	16	17	18	19	20	21
Vucht	0	1	13	5	3	8	18	78	37	22	11	17	2	3	3	0
Hintham	1	0	9	9	8	15	14	33	9	4	1	0	1	0	0	0

[1] *Ber. d. d. bot. Ges.*, Vol. XVII, p. 87. I have already exhibited the variation in both localities united into a single curve in Vol. I (See p. 152, Fig. 32).

In all 221 and 104 flowers were examined. The curves are monomorphic and symmetrical.

The same is true of this species in Thuringia. LUDWIG gives the following data derived from 1000 plants collected at Brotterode.[1]

DATA FROM THURINGIA.

L. F.	7	8	9	10	11	12	13	14	15	16	17	18	19	20	21
Inflorescences	1	6	3	25	46	141	529	129	47	30	15	12	8	6	2

We may therefore assume that the mean number of rays for the wild corn-marigold is 13.

I investigated the mixed race occurring in botanical gardens for the first time in 1892. The result proved to

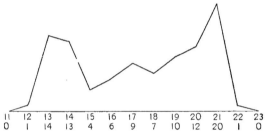

Fig. 30. *Chrysanthemum segetum.* Mixed crop. Curve of the ray-florets in the primary inflorescences of 97 individuals in 1892. The upper series of figures gives the number of rays, the lower series the number of those individuals possessing the scale character written above it.[2]

be a dimorphic curve (Fig. 30), which at the time was the first compound curve to appear in botanical literature.[3] I had obtained the seed for the experiment by exchange from a number of botanical gardens. I mixed it thoroughly and sowed it on a single bed, where 97 plants

---

[1] F. LUDWIG, *Ueber Variationscurven und Variationsflächen*, Bot. Centralbl., Vol. LXIV, 1895, p. 5. Also F. LUDWIG, *Die pflanzlichen Variationscurven und die Gauss'sche Wahrscheinlichkeitscurve*; same journal, Vol. LXXIII, 1898, p. 71 (p. 16 of the offprint).

[2] From the *Archiv f. Entwickelungsmechanik, loc. cit.,* p. 58.

[3] *Archiv für Entwickelungsmechanik*, 1895, *loc. cit.* See also LUDWIG in *Botan. Centralbl.*, Vol. LXIV, 1895, p. 71.

168 *Observation of the Origin of Varieties.*

flowered altogether. I picked off and recorded a head from each of these during the course of the summer. On every plant I selected the terminal inflorescence of the main stem as soon as it opened; plants in which this failed were pulled up before they flowered. Only primary inflorescences were, therefore, employed, and the curve obtained was an index of individual variability, that is to say each unit in it represented a whole plant. The figures obtained are represented in the following series. The upper row gives the number of ligulate florets (L. F.) per inflorescence; the lower, the number of individuals which possessed these numbers.

VARIATION IN NUMBER OF RAYS IN C. SEGETUM, 1892.

L. F.	12	13	14	15	16	17	18	19	20	21	22
Individuals	1	14	13	4	6	9	7	10	12	20	1

The curve based on this series of figures is given in Fig. 30. One of its two apices corresponds to that of the wild species, the other to that of the curve for *Chrysanthemum Leucanthemum* and *C. inodorum*.

My next task was to separate the components from this mixture and to do this in such a way as to place their existence in the mixture beyond doubt. On account of the inevitable interference of insects in pollination it seemed to me impossible to do this for both supposed races at the same time, so I determined to isolate the 13-rayed form first, and the 21-rayed later on from a new mixed crop. I devoted the two years 1893 and 1894 to the former inquiry.

With this object in view, I eradicated every individual of the mixed crop of 1892 which had more than 13 rays, as soon as I had counted the rays on its terminal flowerhead. In this way only 15 plants were saved, of which one had 12 and the rest 13 ligulate florets; the rest were removed so early that there was no danger of these 15 being fertilized by them. These plants flowered abundantly from their lateral shoots but exhibited no tendency to form a curve with an apex at 21. They were therefore sufficiently pure representatives of the supposed race.

In September I harvested the seeds of the 13-rayed plants which I had spared, and sowed half of them in the

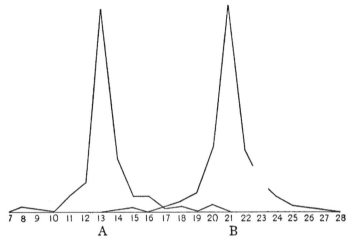

Fig. 31. A. *Chrysanthemum segetum*. B. *Chrysanthemum segetum grandiflorum* (after purification). Curves of the races after isolation. A, Curve of the 13-rayed race in 1894. B, Curve of the 21-rayed race in 1897. The ordinates give the number of individuals with like number of ray-florets in the primary inflorescences of the individual plants. The numbers of ray-florets themselves are given below the abscissa.

following spring (1893). I raised 162 flowering individuals, and recorded the numbers of rays on their terminal heads. The curve representing this generation was steep, monomorphic and symmetrical (see Fig. 31A for 1894), and agrees satisfactorily with the curves, given above, for the plants from the wild locality (p. 167 and Fig. 32, Vol. I, p. 152). Therefore there can be no doubt that the wild form exists in the mixtures grown in botanical gardens. But in order to strengthen this proof I have cultivated the isolated race for one more generation. For this purpose I selected three vigorous plants from amongst the 1893 crop whose terminal inflorescences had 12 ray-florets, and left them to be fertilized by themselves and by their like after all plants with 13 or more rays had been eradicated. From these three seed-parents I harvested the seed separately and raised three families, in 1894, on different beds. The rays of the terminal inflorescences were recorded, and the experiment brought to an end.

I shall now give the results of these three counts made in 1894 together with that of 1893. It will be seen that the series of figures correspond with one another exactly; at any rate as nearly as is necessary for the object of this experiment. The composition of the four cultures in the two generations was as much the same as we should expect four samples of an ordinary species to be.

THE 13-RAYED RACE.

VARIATION IN NUMBER OF RAYS, IN TWO GENERATIONS.

Ray-florets	8	9	10	11	12	13	14	15	16	17	18	19	20	21
1893	2	1	0	7	13	94	25	7	7	1	2	0	3	0
1894. First family	0	0	0	1	10	59	18	2	3	4	1	0	2	1
" Second "	0	0	1	4	11	89	11	5	0	0	2	1	0	0
" Third "	0	1	2	3	10	73	21	1	2	0	0	0	0	0
Total, 1894	0	1	3	8	31	221	50	8	5	4	3	1	2	1

The total for 1894 is given in the form of a curve in Fig. 31A. The whole number of individuals dealt with in this year was 338.[1]

In order to isolate the 21-rayed race out of the same

[1] For a detailed comparison of the curves of the two years see *Archiv für Entwickelungsmech.*, II, 1895, *loc. cit.*, p. 62.

The Origin of Chrysanthemum Segetum Plenum. 171

mixture, I had to provide more seed because the previous stock had been completely exhausted. I procured it in the same way, by exchange from botanical gardens, and from a similar number of them (about 20). It was not to be expected that the identical form of curve would be obtained again, because the relative height of the two apices obviously depends on the proportion in which the two constituent races are mixed; and this must be left to chance. I was therefore curious to find out whether the 13-rayed race alone was cultivated in some gardens and the 21-rayed exclusively in others. With this object I sowed the various samples separately, and on a sufficient space to bring as many specimens to flower as possible. I then recorded the terminal inflorescence of each plant. From no single garden had a pure race been sent, neither of the 13-rayed nor of the 21-rayed form. In every case both forms were found mixed and in the most diverse proportions. The mixed race was therefore the only one generally cultivated at that time.

The variation in the number of ray-florets in the terminal inflorescences of the 589 individuals of the whole culture of this mixed race from the botanical gardens in 1895 was as follows:

L. F.	8	9	10	11	12	13	14	15	16	17	18	19	20	21	22	23	24	25	26
Individuals	7	3	3	5	14	153	77	60	55	31	33	39	41	56	10	1	0	0	1

That is to say, the same apices as in 1892; but in this series the 13-rayed race is obviously more strongly represented than the other.

With a view to discovering also the character of the race which is sold as *Chrysanthemum segetum grandiflorum,* I sowed a quantity of its seed. When the plants flowered in July an extraordinary variety of forms was exhibited by the ligulate florets. These were in some cases very short, in others very long; in some cases so narrow that they did not touch one another, in others more than twice as broad as those of the wild form. The color varied between golden and straw yellow, the tips of the florets were entire or indented, and so forth. This was sufficient to indicate the presence of several races. With regard to the number of ray-florets the differences were not so great, as in the mixtures we have already dealt with. There was only one perfectly distinct apex, that at 21. The other at 13 was more or less obscured. It was obvious that the commercial race was the 21-rayed one, and that it had been adulterated by admixture with the other only as much as is unavoidable and therefore admitted in all cultivation on a large scale.

The terminal inflorescences of the 282 plants of this culture of *C. segetum grandiflorum* were recorded with the following result:

L. F.	8	9	10	11	12	13	14	15	16	17	18	19	20	21	22	23	24
Individuals	1	0	0	1	1	27	27	24	31	30	21	29	24	58	7	0	1

These figures confirm what I have said and show that the 21-rayed race of *C. s. grandiflorum* contains an admixture of a relatively small number (which probably fluctuates from year to year) of the 13-rayed race.

My next task was to isolate from this mixture the 21-rayed race, whose existence had so far been merely inferred. I devoted the two following years to this inquiry, and in the summer of 1895 selected the necessary seed-parents from the mixed crop.

We here encounter an obstacle in the shape of transgressive variability, to which we have already referred,[1] and which has often raised difficulties in the earlier in-

[1] See Vol. I, Part I, p. 56; and Part II, § 25, p. 430.

vestigations. In order to bring this phenomenon into bolder relief let us imagine that the isolation has already taken place and the new race isolated. In other words let us examine Fig. 31 (p. 169) and the data from which this is derived (pp. 170 and 176). Let us first fix our attention on the ordinate at 21. It contains only individuals of the 21-rayed race. But in 1894 a single extreme variant appeared, which, although it belonged to the 13-rayed race, nevertheless had as many as 21 rays (p. 170). If the cultures of 1893 and 1894 had been more extensive the number of these extreme variants would obviously have been greater. For the ordinates 20, 19, 18, etc., it is still more evident that individuals of both races can occur.

If we choose plants which have 21 or more ligulate florets in their terminal inflorescences we cannot at all be certain that they all belong to the race which is being sought. And if they are left to pollinate one another, or if their seeds are mixed in the harvest, there is small likelihood of the strain being pure. Amongst the majority of pure seed-parents a number of individuals of inferior value may exist and it is necessary to remove these as soon as possible, at any rate before the harvest.

The possibility of doing this is afforded by the later flowers. By means of them a curve can be determined for each plant, and in this way values can be obtained which are independent of the chances inseparably connected with small numbers. The curves describing the separate parts of one individual are called its "part-curves." I have therefore plotted such curves of all the individuals selected at the beginning of the flowering period as having 21 and more rays in their terminal inflorescences. The result proved my view to be correct and showed the necessity of the correction which it had suggested. For there were 22 plants which, although their terminal inflorescences were 21-rayed, had a part-curve with an apex at 13-14. The following are the data as obtained at the end of August:

L. F.	12	13	14	15	16	17	18	19	20	21
Lateral flowerheads of 22 indiv.	2	54	58	51	28	19	19	12	2	2

These plants therefore belonged to the 13-rayed race, and were consequently eradicated.

Besides these, there were five plants with doubtful curves; they were also not retained. All that was left was a group of 6 individuals whose curves seemed to me sufficiently distinct and certain to justify the harvesting of their seed. The following line gives the sum of their data:

L. F.	12	13	14	15	16	17	18	19	20	21	22
Lateral flowerheads of 6 indiv.	0	1	3	5	4	6	11	21	30	29	1

All in all there were 111 inflorescences.[1] If the terminal inflorescences of these plants (5 with 21, 1 with 26 rays) had been included, the maximum would have been exactly at 21. Seed was saved only from these six plants for the 1896 crop. It was harvested separately from each parent.

The fertilization of these plants had not been wholly pure, because the rejected plants referred to above could not be recognized nor removed before the latter part of August, and because flowers which bloom in September set hardly any seed with us. Each of the six crops actually gave a curve which had a distinct maximum at 21, but only one of them (No. 1) wholly lacked

[1] The curve is figured in *Ber. d. d. bot. Ges.*, Vol. XVII, Plate VII. Fig. 2B.

The Origin of Chrysanthemum Segetum Plenum. 175

the other maximum, without however being symmetrical. Even in this group the race was therefore still far from being pure. Below I give the curve for the offspring of the single best seed-parent together with the sum of the curves representing the offspring of the five remaining seed-parents (Nos. 2-6). These curves therefore refer to the initial culture of the 21-rayed race (1896).

L. F.	9	10	11	12	13	14	15	16	17	18	19	20	21	22	23	24
No. 1	0	1	0	3	9	15	15	22	30	33	36	64	123	15	5	0
Nos. 2–6	1	5	11	12	70	84	69	92	79	77	114	150	416	46	3	1

All in all 370 plants were recorded for No. 1 and 1220 for Nos. 2-6.

Only the first named group, that is to say the offspring of the plant numbered 1 in 1895, was used for the continuation of the experiment, and from it the best seed-parents for the purification of the race were selected on the basis of an examination of their lateral branches. These were two plants the lateral flowerheads of which gave the following curves (1896):

L. F.	12	13	14	15	16	17	18	19	20	21	22
No. 1 a	0	1	1	4	3	2	2	0	3	3	0
No. 1 b	0	0	0	0	0	0	2	0	3	14	0

Of the two, No. 1b most obviously belongs to the race I was looking for.

I harvested only the seeds of these two plants, and sowed them separately in the following year. In harvesting this seed I confined myself to flowers which had bloomed after the other plants had been removed and had therefore been pollinated with their own or similar pollen. The result corresponded with my expectation, for in the following summer the race was pure on both beds.

This is seen at a glance from the two series that follow and from Fig. 31B which relates to the second group. The data were obtained in the same way as in previous years, only the terminal inflorescence of the main stem of each plant being recorded. The character of the second generation of the 21-rayed race in 1897 was, therefore, as follows:

L. F.	14	15	16	17	18	19	20	21	22	23	24	25	26	27	28	32
No. 1a	0	0	1	2	0	2	3	41	4	1	2	0	0	0	0	0
No. 1b	1	3	0	3	7	14	43	142	43	21	11	5	3	1	0	1

Both groups are very symmetrical, a fact which can be immediately seen in Fig. 31B which is even more regular than the corresponding figure of the 13-rayed race (Fig. 31A). There were only 56 flowering offspring of No. 1a but 298 of No. 1b.

If I had not limited myself in the previous year to such a small number of seeds, I should have had to sow the seed either of less suitable individuals or from flowers on the same plants which had opened earlier, i. e., which had been pollinated with inferior pollen. In that case my race would have been just as incompletely pure in 1897 as it had been in 1896. I have convinced myself, by special experiments with such seed, of the correctness of this view, but do not consider the details worth printing.[1]

By this result the isolation of the races supposed to exist in the mixture, was accomplished. Let us therefore once more examine Fig. 30 on page 167 and Fig. 31 on page 169. The first thing that we see is that the maxima are the same in both figures; they lie at 13 and 21. The

---

[1] Races differing in their number of ray-florets can be mixed by crossing (*Ber. d. deutschen bot. Ges.*, Vol. XVII, p. 92). This mixture is an extremely interesting phenomenon in many respects, but needs a closer investigation.

The Origin of Chrysanthemum Segetum Plenum. 177

explanation suggested by the double curve has thus been fully substantiated by the result of selection. On the other hand it is perfectly plain that the dimorphic curve is not simply the sum of the two monomorphic ones. The mixed assemblage does not simply contain the two mixed races, either in equal parts or in any other proportion. It cannot be synthesized from its components. This is proved by two circumstances: on the one hand by those parts of the curve that lie outside the maximum ordinates, on the other by the middle part. The two component curves begin at 7 and end at 28 (32) and their sum should do so too. But the curve of the mixed race is limited by 11 and 23. This is seen more clearly by looking at the ordinates 12 and 22, since there are far too few individuals in these in Fig. 30. Thus we see that the limits of the curves are, so to speak, "drawn in" in the mixture. On the contrary the individuals are heaped up between the two apices. Moreover in this part there is a secondary maximum. This is seen at 17, but in the commercial mixture of 1895 falls on 16[1] according to the figures given above (see p. 172).

We come now to the double race. It is a well-known saying amongst horticulturists, that any one who wishes to obtain novelties must be eagerly on the lookout for small differences (See *V*ol. I. Part I, p. 185, and this volume, § 2, p. 9). If these deviations are not cases of fluctuating variability but strike the eye by the fact that they are much rarer than these, it is probable that they are the external manifestations of semi-latent characters. If this is actually the case it is further probable that the character can be brought by isolation and selection to

[1] 16 (= 3 + 5 + 8) is one of the subsidiary numbers in LUDWIG's law. The question arises whether by the crossing of two pure races these subsidiary numbers may arise elsewhere also.

partial if not to complete predominance. The success of the experiment of course depends on factors still unknown to us, for it is by no means always successful.

My belief in these principles, which DARWIN himself often refers to, led me to pay special attention, from the very outset of my experiment, to part-curves, i. e., to curves derived from the lateral flowers of the single plants (see p. 173). It is useless to give the numerous cases which afforded no indication of a latent character, and so I will proceed at once to that plant which was the first to do so. It was a specimen of the 21-rayed race of 1896, which had 21 ray-florets in its terminal inflorescences and gave the following part-curve on the 12th of August:

L. F.	14	15	16	17	18	19	20	21	22
No. 1c	1	1	2	2	2	3	0	3	4

I refer to this plant as No. $1c^1$ in order to indicate that it belonged to the same culture as Nos. 1a and 1b whose part-curves were given on page 175. It agrees with those two plants in the fact that there is not a trace of a maximum at 13; but it differs from them and from all the other plants that were examined on the same bed, by the possession of four flowers with 22 rays. On no other plant was there a single lateral flower with more than 21 rays.

This indication was no doubt pretty small. It would not have been discovered but for the counting of the ligulate florets. Without this statistical method of investigation it would certainly never have been grasped, for the plant 1c grew in a culture of about 1500 specimens. It was noted first, along with 500 others, as hav-

[1] *Berichte d. d. bot. Ges.*, Vol. XVII, p. 91, where No 1c is given as No. 12 in the series.

ing 21 rays in the terminal inflorescence, and as thus complying with the main condition for the new race. By means of the grouping of the figures for the offspring, that of one seed-parent (1895, No. 1 of page 175) was first proved to be far better than that of the five other parents. Then amongst this chosen group the individuals with the largest number of florets in their terminal heads were selected and amongst the best of these was found the one which gave the faint indication already described.

CHRYSANTHEMUM SEGETUM GRANDIFLORUM.

ANCESTORS OF THE DOUBLE RACE.

YEAR	PLANT	NUMBER OF RAYS IN TERMINAL INFLORESCENCE	ADVANCE
1895	No. 1	21	—
1896	No. 1c	21	—
1897	—	34	13
1898	—	48	14
1899	—	66	18
1900	Maximum	101	35

But small though this indication was, it sufficed to bring the latent character to light. All that was still necessary was to carry the process of selection on through three years in the same direction and on similar principles.

I chose only one seed-parent each year for the continuation of the experiment, isolated it together with some of the next best as early as possible, and harvested its seeds separately from those of its neighbors. Completely isolated plants of *Chrysanthemum segetum* usually set so little seed that it is impossible to rely on them, and therefore fertilization has to be effected to a certain extent by inferior individuals. If this were not the case my object would most certainly have been reached earlier.

CHRYSANTHEMUM SEGETUM PLENUM.

(See page 182.)

CURVES OF RAYS IN THE ANCESTRAL GENERATIONS.[1]

(Only the terminal inflorescence of each individual was employed in plotting these curves.)

T. F.	14	15	16	17	18	19	20	21	22	23	24	25	26	27	28	29	30
1896[2]	15	15	22	30	33	36	64	123	15	5	0						
1897	2	1	2	1	0	12	10	169	102	45	30	19	21	3	1	2	1
1898						1	2	10	17	17	20	21	30	17	13	10	11
1899						1	1	1	0	0	3	2	9	6	6	7	3

(CONTINUED.)

T. F.	31	32	33	34	35	36	37	38	39	40	41	42	43	44	45	46
1897	1	0	0	1												
1898	6	9	13	21	6	3	5	3	1	2	0	0	0	0	0	2
1899	8	12	13	12	14	10	10	8	6	7	4	8	6	1	10	6
1900				1	0	0	1	1	1	0	0	1	0	1	2	1	0

(CONTINUED.)

T. F.	47	48	49	50	51	52	53	54	55	56	57	58	59	60	61	62
1898	0	1														
1899	4	4	2	2	2	0	2	0	1	0	2	0	0	0	0	1
1900	4	0	0	2	0	1	1	2	0	1	0	2	0	0	0	1

(CONTINUED.)

T. F.	63	64	65	66	67	68	69	70	71	72	73	74	75	99	101
1899	0	0	0	1	1										
1900	1	0	1	0	0	1	0	0	0	1	1	1	1	1	1

[1] These series of figures, with the exception of those for 1896 are exhibited in the form of curves in Fig. 32.

[2] For the complete curve of 1896 see page 175. The individuals with 10-13 rays are left out here.

The Origin of Chrysanthemum Segetum Plenum. 181

Chance may also be unfavorable in another respect. It often happens that the best plant is not sufficiently vigorous to be chosen as seed-parent, but fortunately this

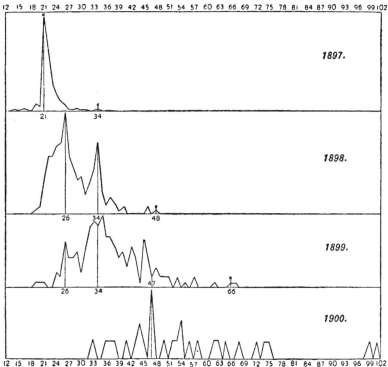

Fig. 32. Ancestral generations of *Chrysanthemum segetum plenum*. Curves of the number of rays in the terminal inflorescences in the several individuals of the generations of 1897-1900. For the numbers themselves see page 180. The seed-parent of 1896 was 21-rayed ($\times$ at the top of the 1897 curve); the other chosen seed-parents are indicated in the various curves by a $\times$ over the appropriate ordinate. The original plant, from which the culture was derived, was the individual grown in 1895 as No. 1 (p. 175) from which No. 1c arose in 1896, and from the seeds of this came the culture of 1897.

difficulty did not present itself in the experiment under consideration, partly because of the favorable conditions of culture.

The progress was uniform and regular and the simplest index of it is the series of successive seed-parents. The number of ray-florets in the terminal inflorescences of my selected plants in successive years was as shown in the table on page 179.

The progress can be seen still better from the curves which I have plotted of the terminal inflorescences in the various generations. The reader is referred to the table on page 180 and Fig. 32 on page 181. The original plant of 1895 referred to as No. 1 arose from a seed which, as already stated, was obtained by exchange from a botanical garden. Indeed this particular lot of seed came from Groningen but produced a mixture the curve of which would obviously not have any special interest. The cultures of the subsequent years were every time the offspring of a single individual whose fertilization by its like had been insured as much as possible.

The following considerations arise out of an inspection of Fig. 32.

The curve of 1897 was monomorphic like that of the typical examples of the 21-rayed race (Fig. 31 B, p. 169); but it was markedly asymmetrical, a fact which afforded a pretty strong indication that the race could be improved by selection in the plus direction. It confirmed the expectation based on the part-curve of the parent of this culture.

The curve of 1898 relates to the offspring of the 34-rayed plant of 1897. In it new maxima appear. These conform with Ludwig's law, for they lie on the figures of the well-known Braun-Schimper or Fibonacci series. One of them is at 34 ($= 13 + 21$) which belongs to the primary series; the other is at 26 ($= 5 + 8 + 13$) which is one of the subsidiary numbers. The maximum in this year was offered by a plant with 48 rays which was healthy enough to be chosen as a seed-parent. But this figure lies very close to the next figure in the series $(13 + 34 = 47)$. The maximum at 21 has disappeared, but the form of the curve clearly indicates its participation in the composition of the whole.

In the following year the advance was much less considerable. The maxima at 26 and 34 and that near 47 became more distinct, but the maximum number of rays increased to 67. At the same time another still more important difference appeared since now for the first time ligulate florets appeared between the tubular florets of the disc. This only occurred on a single plant and not till the beginning of September. This plant had 66 rays in its terminal inflorescence, and was one of those which had been selected as seed-parents, and accordingly isolated at the beginning of the flowering period. On account of its possession of this first sign of real doubling it was chosen for the continuation of the experiment in 1900, to the exclusion of all the rest.

Fig. 33. *Chrysanthemum segetum plenum.* One of the six inflorescences which in 1899 first exhibited true "doubling." The figure represents the parent plant of the "double" variety.

It is well known that in other species of this genus (e. g., *Chrysanthemum indicum* and *C. inodorum*) the doubling consists in exactly the same phenomenon. In the midst of the tubular florets (Fig. 34A) ligulate florets are developed (Fig. 34B). If the "doubling" is carried very far the former are completely covered by the latter (Fig. 34C), and can only be seen by pulling out the ligulate florets or by turning them aside. If this is done a large number (and not merely a few scattered ones, as might perhaps be expected) of tubular yellow corollas

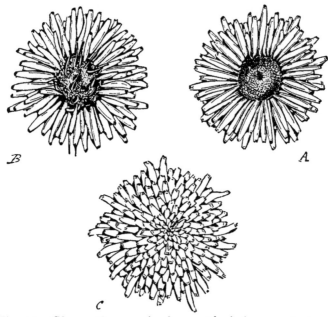

Fig. 34. *Chrysanthemum inodorum plenissimum.* A, inflorescence with central disc of tube florets (fertile); B, with scattered tongue-florets in the disc (half fertile); C, highest degree of "doubling" (sterile).

are ordinarily found; and the less the amount of doubling the more conspicuous they are. Moreover we often find, in both species, inflorescences with a broad yellow disc over which occasional white ligulate florets are scattered (Fig. 34 B). Such flowerheads look like anomalies, though, as a matter of fact, they are less anomalous than the apparently completely "double" forms.

The Origin of Chrysanthemum Segetum Plenum.

This 66-rayed plant was the first of my race to betray the fact that it contained the much desired double character. From this moment the attainment of my object was assured.

The six first "double" inflorescences referred to, had about 40-50 ray-florets around their circumference and moreover one to three in the disc. But as they flowered too late to ripen seeds, I have photographed and preserved them (Fig. 33).

Unfortunately this plant gave but a poor harvest, producing only 31 plants with terminal flowerheads. A curve representing these heads is given in Fig. 32 under 1900. The number of observations is of course much too small to furnish a proper curve or to justify the drawing of conclusions as to its maxima. On the whole, however, the figure indicates a definite advance over the earlier years, and this advance is especially expressed in the fact that amongst this small number there were two plants which far outstripped all previous ones in the number of their ray-florets. Their terminal inflorescences contained respectively 99 and 101 rays, whereas the next maximum expected would be $34 + 55 = 89$.

"Doubling" now appeared quite suddenly in full development in this culture (Plate II). For convenience of reference I shall call the white ligulate florets situated amongst the little yellow tube florets, "disc-tongues." These disc-tongues were now quite common. From no single plant were they completely absent if attention was paid to both the terminal and lateral flowerheads. But their number was subject to a high degree of fluctuating variability. As a rule flowers with less than 40 rays had no disc-tongues, and the number of these increased with the total number of the outer rays. For instance, a terminal flowerhead with 56 rays had 53 on the periphery and 3 inside; while one with 74 rays had 58 on the circumference and 16 in the disc. In the records on which were based the table on page 180 and Fig. 32, both kinds of ligulate florets were counted together. The two flowers with 99 and 101 ligulate florets respectively were to all appearance almost entirely double.

The "doubling" was also exhibited on the lateral branches. When these were in full flower, I selected the twelve best "double" plants and pulled up the rest. The lateral inflorescences of the rejected plants gave a curve whose maximum was at 47 ($= 13 + 34$) in accordance with the indications referred to above and apparent in Fig. 32. The worst flower had only 28, the best one 94 rays. The average of the whole lot was 47; but the curve, in spite of the coincidence of the mean and the maximum was not symmetrical. Altogether the rays of 378 inflorescences were counted.

As was to be expected, the selected seed-parents exhibited great differences in the degree of "doubling" in the lateral inflorescences. On some this was inconsiderable. On others the mean was from 2-5 disc-tongues per inflorescence whilst on two a mean of 11 was counted. One plant bore nothing but wholly double flowers. It had seven flowers on which 279 disc-tongues were counted, giving an average of 40. In consequence of this the plant was absolutely sterile; it bloomed well afterwards, but in spite of every care I could not get a single seed from it. But the finest specimens of *C. inodorum plenissimum* are also known to set no seed. Likewise the two plants whose lateral flowers had on the average 11 disc-tongues, set no seed.

We thus see that the limit has been reached. Any further improvement of the race will only increase the number of doubles and consequently of sterile individuals. Seed-parents therefore must always be chosen amongst the plants with the same degree of "doubling" as in this year (1900). In this respect my new race behaved, immediately after its origin, exactly like the old-established *Chrysanthemum inodorum plenissimum*.[1]

It still remains to give some account of the general conditions of fertilization of the seed-parents in the various years. No doubt the experiment would have been purer and more demonstrative if the corn marigold were capable of self-fertilization. But this either does not occur at all, or only to a totally inadequate extent. Each year I have therefore left a group of a few selected plants to flower together after the eradication of the rest; and have been obliged to content myself with collecting the seed of each in a separate package. Future experiments will have to determine what the effect of this crossing may have been on the progress of the race. Meanwhile it may be of interest to place on record the number of plants which have flowered together each year, showing the stringency of selection to which they were subjected.

In the summer of 1895 the original parent of the whole race, which was raised from seeds obtained by exchange (1895, No. 1), could not be isolated until late and then incompletely, but as the plants flowering at the same time also belonged to the 21-rayed race the curve of the offspring was very "pure" in this respect (p. 176). In the next year the number of seed-bearers was reduced, about the middle of August, to three very vigorous indi-

---

[1] The *Matricaria flore toto albo plenissimo*, described by MUNTING in 1871, the best specimens of which also set no seed, was probably the same variety (*Waare Oeffeninge der Planten*, p. 527).

viduals which had 21, 21, and 22 rays respectively, in their terminal inflorescences. One of those with 21 served for the continuation of the experiment but all three had exhibited correspondingly high numbers in their lateral flowers. The fertilization in this year was therefore sufficiently pure.

This was not the case in 1897. The 34-rayed seed-parent of that year was pollinated at first amongst all the other plants, and later amongst the rest of the selected ones which were however as many as 25 in number. It set so little seed that it was impossible to rely solely on the seed due to the later pollinations (i. e., the purest seed) for next year's crop. The two maxima of the curve of 1898 are therefore, at least to some extent, due to mixed pollination (Fig. 32 under 1898).

In 1898 I selected the seeds for the continuance of the race in two periods on the chosen seed-parent after having marked the flowers separately for them. The first harvest was from flowers which had bloomed before the removal of the rest, the second from those which had bloomed later. The latter must therefore have been fertilized by the seven remaining seed-parents whose terminal inflorescences, however, all had had more than 34 rays (the numbers were 35-36-37-38-39-40 and 46). The two specimens were sown separately and their curves determined; but no essential difference between them could be detected, either in their limits, or in their means, or in their general shape. The seed-parent of 1899 with 66 rays and with the first 1-3 ligulate florets on its discs (Fig. 33), belonged to the first series, the 67-rayed plant shown in the table on page 180 for 1899, however, to the second.

In the summer of 1899, towards the end of July, I

saved 17 seed-parents with 48-67 rays in their terminal inflorescence. From these I collected the seed from the flowers which bloomed in July separately from those whose flowers had opened after the selection had taken place. But from the former specimen I raised only three plants that flowered (with 41-44-47 rays in their terminal head), which evidently could have no effect on the shape of the curve and were soon removed. Fertilization in 1899 therefore was again very pure.

Having arrived at the end of the account of our experiment, all that remains is to compare the course of the selective process in this case with the ordinary process of selection carried out in the improvement of agricultural plants. I refer the reader to FRITZ MUELLER's experiment with the many-rowed maize already described.[1]

That the difference is essential will be clear from the description given. In the case of the maize the object was to intensify the racial character (12-14 rows) as much as possible by selection; in the case of the *Chrysanthemum* the object was to uncover a latent character and to bring this to its full development. In the first case a visible character that had been known for ages had to be increased as much as possible; in the second, according to current conceptions at least, a new character had to be called forth. The 26-28 rowed ears fall within the range of fluctuation of the 12-14 rowed race; and they would without doubt have appeared within it without any selection, if cultures of sufficient extent, which could be calculated beforehand, had been grown (Vol. I, p. 162).

Without any doubt my crop of 1897 would have

[1] See the pedigree in Vol. I, Fig. 18, p. 73.

given rise immediately to flowerheads with central ligulate florets if it had been on a sufficiently large scale. But it would not have produced them in a proportion which could be predicted by QUETELET's law, but according to the principles of discontinuous variation which are still unknown to us.

The course of the improvement is different in the two cases. The results obtained with the maize conform to the law of regression, the increase in the number of rows in the ears becoming slower and more difficult to secure, the further we get from the starting-point. Exactly the reverse is the case in the *Chrysanthemum*. The progress was continuous and did not materially change until 1899, when the first central ligulate florets appeared. Then it took a leap, all the offspring of this plant having more or less double flowers. More strictly speaking, the leap had already taken place, the plant with the first central ligulate florets (Fig. 33) having already crossed the threshold. Its offspring behaved like the offspring of a pure race, such as for instance *C. inodorum plenissimum*.

A break therefore occurred, and obviously before 1899; either in the origin of the seeds of 1898 from which the plant in question arose, or even earlier.

And since *C. inodorum plenissimum* has maintained itself for many years without selection, it is probable that the new *C. segetum plenum* will do the same. But the reverse was the case with the maize which reverted to the old form within a few years after the cessation of selection (Vol. I, p. 125).

Hitherto I have taken the number of ray-florets in the terminal inflorescence almost exclusively as a character of the whole plant, and the curves have been plotted from the figures obtained in this way. But there is, as we have already seen, another method of estimating the individual value of a plant, namely that based on a determination of the mean character of as many flowers as possible on a single individual. This raises two points for consideration: first the mode of branching of the corn marigold, and secondly the influence of the deviation of the individual from the mean of its race.

The mode of branching in *Chrysanthemum segetum* is as follows. The main stem which arises from the plumule bears two groups of branches; strong ones at its base from the axils of the radical leaves; and, higher up, weaker ones whose vigor first gradually increases and then decreases, as they succeed one another from below upwards. This applies both to their length and to the number and strength of their secondary branches. These secondary branches are, therefore, of the third order; they often bear branchlets of the fourth and even of the fifth order. The flowers that bloom in July, with us, are mostly of the second order, those blooming in August and September of the third and fourth.

In the course of the summer, and with the flowers on the successively higher orders of branches variability is seen to manifest a general decrease. The sides of the curve are, so to speak, drawn in; the curve itself becomes narrower. The amount of deviation of the various individuals from the mean of the race decreases, and the mean consequently comes to stand out more boldly. This is of especial importance in cases in which a curve has been shifted laterally by stringent selection (such as Fig. 32, p. 181) where it might remain doubtful what the shape of the curves would have been if selection had effected nothing more than the isolation of the individuals of the new race.

We have therefore to examine the "late summer" curves of the 13-rayed, 21-rayed and double races. Let us begin with the first.

The extreme limits of the curve of this race at the beginning of August were 11 and 21 ray-florets. These numbers gradually decreased until September, when only heads with 13 and 14 rays were formed. In the next year at the end of July the limits were 10-19, but in August 12-14.

I examined the 21-rayed race, with reference to this character, in the summer of 1898, dealing with the individuals which had been saved for seed. The data for three of the plants[1] are summarized in the following table:

PLANT	FLOWERS	NUMBER	MIN.	MED.	MAX.
A.	Terminal	1	—	48	—
	September 1st	32	29	33	45
	October 10th	42	18	27	36
	November 1st	28	19	26	31
B.	Terminal	1	—	35	—
	September 1st	36	24	28	36
	October 10th	33	16	22	27
	November 1st	23	15	21	25
C.	Terminal	1	—	46	—
	September 1st	14	26	28	35
	October 10th	18	18	26	30
	November 1st	8	21	23	28

We see that the numbers gradually shift in the direction of the maximum at 21 (in the case of one plant actually reaching it), without any indication of the maximum at 13 of the other race. The plants dealt with, therefore, clearly belonged to the 21-rayed race.

[1] *Over het periodisch optreden der anomalien op monstreuze planten.* Kruidkundig Jaarboek, Gent, XI, 1899, pp. 57-58.

The Origin of Chrysanthemum Segetum Plenum. 193

This was apparently no longer the case in the following summer. The following are the records of five plants taken in late summer:

	TERMINAL INFL.	MIN.	MED.	MAX.
A	67	33	39	50
B	55	31	42	50
C	51	37	47	54
D	50	33	51	60
D	—	33	40	51
E	66	28	47	62
E	—	32	43	52

D' and E' were counted 6 weeks later on the same plants as D and E. The plant E is the seed-parent in Fig. 32, p. 181, under 1899, marked with a ×. The figures may be regarded as the expression of a tendency to fall back on the secondary maximum at 47 ($= 13 + 34$), and the same result was reached by the other countings, which it is not worth while to reproduce here.

In the following year (1900) the maximum of the lateral inflorescences was still higher. I give the data derived from three plants which were "double" and consequently sterile, and of the four next best which were chosen as seedparents.

	MIN.	MED.	MAX.
Sterile I.	72	87	100
II.	48	62	94
III.	46	56	79
Seed-parent I.	47	63	76
II.	51	62	91
III.	44	60	94
IV.	46	56	86

The curve of the "double" race thus seems to have its maximum at about 55 ($= 21 + 34$). The possibility of attaining higher mean numbers seems to be excluded by the sterility of the more perfectly double plants.

Let us now briefly summarize the results of this experiment. There is, on the market, a 21-rayed race of the normally 13-rayed *Chrysanthemum segetum*. It is not strictly pure, but can easily be made so; it bears the name *C. segetum grandiflorum*. From a plant which, in 1895, caught my attention by a few 22-rayed lateral flowers, I succeeded in raising, by a process of selection, a hitherto unknown race with double flowerheads, the new *C. segetum plenum* (Plate II). The course of this process is exhibited in Fig. 32, p. 181, in which the × × × × indicate the individuals selected as seed-parents for each succeeding generation. *C. segetum plenum* behaves with regard to its double character, exactly like the double commercial varieties of other species of the same genus (*C. inodorum, C. indicum* etc.).

The new variety was therefore obtained by bringing to light a character latent in *C. segetum grandiflorum*.

§ 19. DOUBLE FLOWERS AND FLOWERHEADS.

The experiment described in the foregoing section (§ 18) justifies an attempt to form some conception of the manner in which this phenomenon of "doubling," which is widely distributed among cultivated composites, may have arisen in other cases. If we examine the facts closely we shall discover in the majority of cases an extraordinarily close agreement with our own specimen, at least so far as the absence of scientific observations admits the possibility of a comparison.

There are, it is true, certain abnormal types of "doubling," such as the development of secondary flowerheads

(*Cineraria*), the transformation of the little yellow discflorets into long white tubes (*Pyrethrum*, see Fig. 36) etc. We shall however leave such cases out of consideration; they may be regarded provisionally as cases of spurious doubling.

The genuine "doubling," on the other hand, as exhibited by the most diverse species, presents a very marked agreement with the conditions found in *Chrysanthemum segetum*. Indications of a tendency to "doubling" occur both in forms of which a double variety is not offered by seedsmen and in those of which such are already on the market. For instance in 1892 I observed occasional tube-florets more or less completely transformed into ligulate florets in a culture of *Bidens grandiflora* in my garden. In other cases the variation is only seen when curves are plotted. For example I obtained the following very asymmetrical curve from a culture of the single variety of *Chrysanthemum coronarium*, a favorite garden plant whose double form has long been known (Fig. 35). 130 flowers on 25 plants of a single crop were recorded, the flower at the top of the main stem and those on the primary branches alone being taken into account. I found:

Ligulate florets	7	8	9	10	11	12	13	14	15	16	17	18	19	20	21
Inflorescences	1	0	1	2	2	12	25	19	21	15	14	6	7	2	3

That is to say, 18 on one side of the mean and 87 on the other, with a faint indication of a second maximum at the next figure in the BRAUN-SCHIMPER series, 21. It is clear that the double variety of this species could probably be obtained from these plants, exactly in the same way as in *C. segetum*.

These considerations evidently lead to the hypothesis that the secondary maxima on the positive side of LUD-

WIG's ray-curves[1] may also indicate the existence of a latent character, which, if it could be made active, might perhaps give rise to the double variety of the species.

If we now examine the double varieties of the composites, we see that the structure of their inflorescences corresponds in every detail with that of *Chrysanthemum segetum plenum*. The amount of doubling is always

Fig. 35. *Chrysanthemum coronarium*.

highly variable. The best cases reveal no tube florets as in the case of *Chrysanthemum inodorum* in Fig. 34, p. 184. But if one looks between the ligulate florets small yellow tubes will be found in no inconsiderable quantity. This is true of *Calendula officinalis* and many other species. Such flowerheads are usually sterile, for

[1] May not the secondary maxima on the negative side similarly indicate the variety without ray-florets, the *Var. discoidea* (See § 8, pp. 78-79 and Fig. 9, p. 80).

the ligulate florets are female, and inasmuch as very often all the flowerheads on a single plant attain to this degree of doubling the best variants cannot serve as seed-parents. But two further types are always found with yellow discs which are either uniform (Fig. 34A) or contain scattered ligulate florets amongst the tubular ones, as is so often seen in *Chrysanthemum indicum* and *Zinnia elegans.* The double *Bellis perennis* also, if grown from seeds, is highly variable in this respect. These two types are fertile and therefore constitute the seed-parents of the variety; if the plants with central ligulate florets (see p. 185) furnish sufficient seed the harvest is saved exclusively from them; but they often set little or hardly any seed.

This unavoidable restriction in the choice of the seed-parents and the frequent difficulties of selection dependent on it account for the fact that bought samples of the seed of double composites often give rise to only a relatively small proportion of the desired type, as has long been known[1] to be, and still is, the case (*Chrysanthemum coronarium* sometimes only 50%, *Centaurea Cyanus* 40-50%, *Tagetes africana* with rare exceptions double etc.).[2]

Many double varieties of composites seem to be almost as old as horticulture itself (See Vol. I, p. 183). According to the oldest accounts the degree of doubling and the range of its variation were formerly the same as now.

Finally I have to mention the fact that bud- and sectorial variations are found in this case as well as in others.

[1] E. g., *Pyrethrum roseum, Dahlia, Chrysanthemum indicum,* according to VERLOT, *Production et fixation des variétés,* 1865, p. 83.

[2] See the catalogues of BENARY, and HAAGE & SCHMIDT of Erfurt, VEITCH & SONS of London and SUTTON & Co.

I refer to a very beautiful instance of the latter (Fig. 36) which I owe to the kindness of MR. ERNST H. KRELAGE in Haarlem.

The origin of double flowers in other groups of flowering plants has probably occurred on similar lines with that of double inflorescences. I restrict myself to a consideration of doubling by the transformation of stamens into petals, that is, the petalody of the stamens, referring the reader for an account of the other types of doubling to GOEBEL's well-known monograph.[1]

Occasional petaloid stamens occur fairly commonly both in culture and in nature; they are so well known that it is not necessary to cite special instances. The curve which represents this variation is unilateral, indicating thereby the existence of a latent or semi-latent character.[2] The attempt to render this active may be made, and if it succeeds[3] the origin of a double variety may be expected.

Double varieties of this kind tend to vary in the same way as those of composites. If, for instance, the commercial *Varietates plenae* of *Clarkia pulchella, Clarkia elegans, Phlox Drummondi* and others are examined, almost all the intermediate stages between nearly hemispherical double flowers and flowers with normal stamens are met with. In such cases it is usually obvious that favorable conditions tend to increase "doubling," a fact which has been known for a long time in the case of *Anthemis nobilis,* of some species of *Narcissus,*

---

[1] K. GOEBEL, *Beiträge zur Kenntniss gefüllter Blüthen,* in PRINGSHEIM'S Jahrb. f. wiss. Bot., Vol. 17, 1886, p. 207.

[2] *Ueber halbe Galton-Curven als Zeichen discontinuirlicher Variation.* Ber. d. d. bot. Ges., Vol. XII, 1894, p. 197.

[3] Which is, however, by no means always the case. See the experiment with *Ranunculus bulbosus* in § 23 of this part.

Double Flowers and Flowerheads. 199

and other bulbous plants.[1] There is a certain periodicity in this case too; for sometimes the first, but more usually the later, flowers are less double than those which bloom in the height of the flowering period. This fact is well known to breeders,[2] especially in the case of certain double varieties of Begonia in which seeds can only be saved from the autumn flowers.

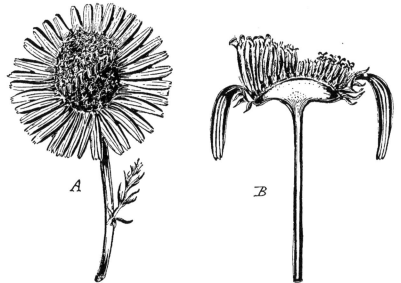

Fig. 36. *Pyrethrum roseum*, from the nursery of Messrs. KRELAGE & SON in Haarlem (1899). In one half (the rear half in A, the left in B) the inflorescence is made "double" by the elongation of the tube-florets; in the other half it is "single." A, oblique view; B, section.

The majority of double varieties are constant from seed, even in the case of trees and shrubs (varieties of the peach and the apple for instance),[3] others appear to be only slightly so, and others not at all (*Prunus spi-*

---

[1] LINDLEY, *Theory of Horticulture*, p. 333.

[2] CARRIÈRE, *Production et fixation des variétés*, 1865, pp. 66 and 67 (*Camellia alba plena, incarnata, Fuchsia*, etc.).

[3] VERLOT, *loc. cit.*, p. 83.

nosa).[1] For instance, 80% is the figure given for *Dianthus Caryophyllus*,[2] and double varieties of *Campanula* are said always to produce a certain number of single plants. In the case of double stocks one may reckon on between 50-60% double offspring according to the treatment and selection of the seed. Pot culture favors "doubling."[3]

The transformation of stamens into petals often goes so far that no pollen is formed. When this occurs the stigma of the double flower must be fertilized with the pollen of a single flower or left to be pollinated by insects. The result is that the race gives rise to both forms every year. For instance *Papaver nudicaule aurantiacum plenum*, the seeds of which give rise to between 40 and 60% of double-flowered specimens every year. It is the other way round with the double *Petunia* whose capsules are usually malformed; but they develop a few stamens, with the pollen from which the stigmas of single flowers are dusted,

Fig. 37. *Anemone coronaria*, "The Bride." Double on one side, single on the other. From the cultures of Messrs. E. H. KRELAGE & SON of Haarlem.

[1] *Ibid.*

[2] Seed catalogue of D. SACHS, Quedlinburg, 1890-91. (*Dianthus Caryophyllus c. fl. Margaritae*, novelty 1889).

[3] CHATÉ, *Culture pratique des Giroflées*. NOBBE, Botan. Centralblatt, Vol. XXXII, 1887, p. 253.

preferably after castration. The seeds collected after this operation are said to give from 25 to 40% double plants the number varying directly with the care with which the castration was carried out.

Double flowers are also subject to sectorial and bud-variation. A chestnut tree (*Aesculus Hippocastanum*) at Geneva, a single branch of which has borne double flowers for many years,[1] is perhaps the best known example of the latter, whilst our Fig. 37 gives an interesting case of the former. It is a flower of the pure white *Anemone coronaria*, "The Bride," which, like the *Pyrethrum*, I owe to the kindness of MR. KRELAGE. It grew in a bed of the single variety; the plant which bore it had exclusively single flowers with the exception of this one. On the one half there were stamens only, as is shown by the figure; in the other half, however, the vast majority of stamens were transformed into narrow petals, just as happens all round the stigma in the double form. The single variety frequently exhibits more or less definite traces of doubling, and from these MESSRS. KRELAGE have succeeded in producing a double sort and putting it on the market. But a sectorial variation like that figured has only been observed once in the course of many years.

§ 20. THE ORIGIN OF LINARIA VULGARIS PELORIA.

About ten years after the appearance of the first edition of DARWIN's *Origin of Species* (1859) HOF-

[1] A. P. DE CANDOLLE, *Physiologie végétale*, 1832, II, p. 479, and ALPH. DE CANDOLLE, *Géographie botanique*, 1855, II, p. 1080. This tree stood in the garden of M. SALADIN DE BUDÉ near Geneva. Many cuttings made from the double-flowered branch have been distributed.

MEISTER wrote the following words at the end of his account of pelorias.[1]

"One of the most remarkable features of the variations of plants is, without question, the sharpness and suddenness of the origin of profound deviations from the normal form of structures such as we see it in the phenomena just considered, in many analogous cases, and especially in the formation of monstrosities. The new form does not come into existence by the gradual summation of small deviations in one direction, during succeeding generations; it appears all at once, perfectly distinct from the original form."

This highly important and perfectly correct statement rests even now simply on the absence of transitional forms, and does not rest on direct observation. If the peloria had originated by a gradual process it would be reasonable to suppose that at least in some of the relatively numerous instances the intermediate steps would have been found; but as this was not the case it was concluded that they did not exist and therefore that the origin of the variety had been immediate.[2]

But it is hardly necessary to point out that nothing short of direct observation can furnish the final proof. Direct observation will moreover inaugurate a new stage in the study of this remarkable phenomenon, by making

---

[1] W. HOFMEISTER, *Allgemeine Morphologie der Gewächse*, 1868, p. 564.

[2] On the pelorias of L*inaria*, especially of L. *spuria*, see H. VÖCHTING, *Ueber Blüthenanomalien*, Jahrb. für wiss. Botan., Vol. XXXI, No. 3, 1893, and L. JOST, *Blüthenanomalien bei Linaria spuria*, Biolog. Centralblatt, Vol. XIX, 1899, p. 145. Also J. H. WAKKER, *Over pelorien*, Ned. Kruidk. Archief, Vol. V, p. 1, July 1889, with Plate X. P. VUILLEMIN, *Monstruosités chez le Linaria vulgaris*, Bull. Soc. Sc., Nancy, Dec. 1893, with one plate (Vol. XIII, 1894, p. 33). W. and A BATESON, *On Variations in the Floral Symmetry*, Journ. Linn. Soc. Bot., Vol. 28, 1871, p. 381.

accessible to investigation the mode of its appearance and the external causes to which it is due.

For these reasons I have endeavored to induce the occurrence of the *Peloria* from the ordinary form in my experimental garden. It is obvious that the success of such an experiment, at least at first, is dependent on chance. This chance however can be favored by making the cultures as extensive as possible, and by widely variable conditions of life. Fortune has favored me, and after seven years' work my object has been attained. The *Peloria* appeared quite suddenly in the fifth and sixth generation of my culture.

The signification of my observations will be more properly understood if I premise my account of them with a short general and historical account of the subject, referring the reader for the literature to the following section (§ 21) and to Penzig's *Teratologie*.[1]

Fig. 38. A, B, *Linaria vulgaris.* C, D, Peloric flowers.

Peloric flowers in *Linaria vulgaris*[2] were first discovered, as is well known, in 1742 by Zioberg on an island near Upsala and described by Linnæus in the

---

[1] O. Penzig, *Pflanzen-Teratologie*, Vol. II, p. 195.

[2] The Pelorias have five spurs: *Peloria nectaria*. But there is also a *Peloria anectaria* in which the flowers are regular but without spurs. See Penzig, *loc. cit.*, and Verlot, *Production des variétés*, p. 90. This variety is nearly sterile, setting very little seed, but it breeds true.

204 *Observation of the Origin of Varieties.*

Amoenitates academicae.[1] The plant grew there together with the ordinary *Linaria* and formed a "constant" race

Fig. 39. *Linaria vulgaris peloria.* A richly branched stem of a plant of the second generation. Raised in 1898 from seed of the first generation of 1897 and photographed in August 1900. All flowers are peloric.

through propagation by the buds on its roots. All the flowers of this plant were peloric (as in Fig. 39). LIN-

[1] *Amoen. acad.*, I, p. 55, p. 280 (1744). See MOQUIN-TANDON, *Pflanzen-Teratologie*, 1842, p. 170, and HOFMEISTER, *loc. cit.*, p. 563.

The Origin of Linaria Vulgaris Peloria. 205

NÆUS described this form, which was new then, under the name of *Peloria,* derived from the Greek πέλωρ, a monster.

It was not till later that the occasional occurrence of isolated peloric flowers on the ordinary *Linaria vulgaris* was noticed. Moreover in the course of time further specimens of the true *Peloria* were found scattered over most of Europe.

Such plants have been brought into cultivation by many investigators. They have remained constant and could be propagated by means of their numerous radical buds. In the occasional cases in which the plants apparently reverted to the one-spurred form it is possible that some roots of the ordinary *L. vulgaris* were accidentally transplanted amongst the roots of the peloric specimen. So many descriptions of the flowers exist that I think it is hardly necessary to repeat them. In Fig. 39, however, will be seen a freely branched specimen of our plant. I have also given a figure of a spike of the ordinary *Linaria vulgaris* in Fig. 40, for the sake of comparison.

Fig. 40. *Linaria vulgaris.* A normal flowering stem.

The common opinion of those who have worked with the *Peloria* is that it is in a high degree sterile. The pollen is poorly developed and the capsule is practically atrophied; but not to such an extent that fertile seeds are never produced, as some investigators seem to think,[1] for some attempts to harvest seed have been successful. WILLDENOW records an experiment in which such seed has given rise almost exclusively to peloric plants.[2]

The *Peloria,* or *Linaria vulgaris peloria* is characterized by the fact that all its flowers are peloric. This character is, it is true, subject to considerable fluctuating variability, especially in the number and degree of development of the spurs. But I never found normal one-spurred flowers amongst them, although since 1894 I was able to observe in my cultures several hundreds of peloric flowers every year, and in favorable years even many thousands of them.

Besides this *Peloria,* as already stated, there are sometimes found on the ordinary *Linaria vulgaris* isolated peloric structures, which are subject to a high degree of fluctuating variability (Fig. 41). The most usual case is a single flower on a plant which does not bear another afterwards during the whole course of the summer. Sometimes I found 2 or even 3 peloric flowers on the same plant, both in the wild and in the cultivated state, but seldom a larger number. It often happens that an individual which has produced the abnormality in its first year will not produce a single one in the second, although it branches more freely and bears many more flowers; on the other hand the abnormality sometimes reappears. Such isolated pelorias are not limited to any particular position;[3] although in my garden they usually occurred

---

[1] VERLOT, *Production et fixation des variétés,* p. 90.

[2] DE CANDOLLE, *Physiologie végétale,* II, p. 692. My experience is in full agreement with that of WILLDENOW. (See p. 216.)

[3] See PENZIG, *loc. cit.,* p. 195.

on the highest lateral twig below the main flower-spike.

The question suggests itself, Is the power of producing isolated peloric flowers inherent in all plants of *Linaria vulgaris*? Or are there two races, one with and one without this faculty? This question seems not to have been investigated as yet. From the observations already described it must be concluded that this point can never be determined in the field, for the absence of the abnormality on particular days, or even in particular years proves nothing in itself. Personally I think it likely that both kinds exist and that there are localities for *Linaria vulgaris* in which these abnormalities are never found.

Holland however is not one of these. By paying attention to them when out on an expedition, one will find isolated peloric specimens fairly frequently and in the most diverse localities. When I wanted a specimen to

Fig. 41. *Linaria vulgaris hemipeloria.* Branch of a normal flowered plant with a single peloric flower. Zandpoort, Aug. 1900. *a,* normal one-spurred flower. *b,* a Peloria.

photograph for an illustration (Fig. 41), I asked my wife to look for one in the neighborhood, and it was not long before I had one. The power to produce them is, therefore, widely distributed in this country; and also obviously heritable although in a latent state as a rule. Whether or no there are localities in which this character does not occur, I cannot tell.

So long as it is not certain whether a *Linaria vulgaris apeloria* exists, I propose to call the plants with this power provisionally *L. vulgaris hemipeloria* (Fig. 41). This name of course refers both to those plants on which isolated peloric flowers have been observed, and to their offspring.

Cases of true *Peloria* (Fig. 39) are also occasionally seen in this country in the wild condition. A few localities for it are recorded in the Floras. I myself had some plants from a spot near Zandvoort in 1874, but since then it has not been found there again. Only one new locality has since become known to me, and this was near Oldenzaal (1896). It is of course not known whether the *Peloria* occurred spontaneously in these various localities and had not been introduced from elsewhere, but its high degree of infertility makes the likelihood of such an introduction very remote.

For the purposes of my experiment I transplanted some plants from the country into my garden in the summer of 1886. I selected plants with occasional peloric flowers and freed their roots as carefully as possible of fragments of roots whose connection with the hemi-peloric plants was not absolutely certain. The plants came from Gooiland. I also collected, at the same time, the *Linaria vulgaris* with *Catacorolla*,[1] and obtained the three-spurred variety (see § 8, p. 87) from DR. WAKKER. These three forms flowered together in the following summer in my garden.

In 1888 I sowed the seeds which I had collected in 1887, to produce the second generation, but the plants did not flower till 1889 and again in 1890. In the first year a single peloric flower was produced amongst in-

[1] See Chapter II of this part, § 4, p. 31.

numerable flowers with a single spur; in the second year, however, two appeared. I collected the seeds of these plants in 1889.

From this I raised the third generation in 1890. Here again the plants did not flower till the second year, and again there was one case of a *Peloria* amongst thousands of normal flowers. I harvested the fruits of this peloric plant separately and it furnished me with sufficient seed for the culture of 1892.

This year I adopted the plan of sowing the seeds in pans, containing good garden soil, in the greenhouse of my laboratory. Hitherto I had simply sown the seed in the bed, for which method, however, a much larger quantity of seed is required. The seedlings were planted out singly in pots containing richly manured soil as soon as they began to develop a hypocotylous bud; then they were kept under glass, and were not transferred to the open bed until June. The result was that they not only flowered in the first year, but did so very luxuriantly. There were about twenty individuals in all. On one of these I saw a single peloric flower at the end of August. In the autumn I pulled up all the plants except two, one of which had exhibited the peloria. These two plants flowered in the following year in complete isolation, a profusion of flowers being borne on the freely branched stems, but they did not then develop a single peloric flower. They produced 13cc of seeds, an abundant harvest. I sowed a small proportion of this in the following year, and as it gave rise to the *Linaria vulgaris peloria* I was looking for, I sowed the rest in 1896, and some again in 1899.

Before we proceed to give an account of this main section of the experiment let us briefly summarize the results obtained in the years 1886-1893. They comprise four generations, each of which produced only one or two peloric flowers amongst thousands of normal ones. The anomaly, therefore, seems to recur every year and is obviously due to the existence of some heritable semi-latent potentiality which only very seldom becomes active.

This result of the experiment supports the conclusion based on the repeated occurrence of isolated peloric flowers in nature. *Linaria vulgaris hemipeloria* is thereby shown to be a heritable form. The question whether it is identical with *Linaria vulgaris* itself, or constitutes a variety or a race of this, cannot be answered for the present. From it my *L. vulgaris peloria* arose, as I shall now show.

In order to make this part of my experiment more easily intelligible I shall first describe it in the form of a pedigree. This contains the four generations already dealt with, and two further ones of the *Hemipeloria* (1-6), together with the first, second, and third generations of wholly peloric plants (I-III). The meanings of the abbreviations are:

h and *H:* *Linaria vulgaris hemipeloria.*
p: "   "   *peloria*, 1st generation.
P: "   "   "   2d and 3d generations.

Wherever necessary the number of plants is prefixed to these letters either in absolute numbers or in percentages. For the fifth and sixth generation I have, as will be seen, made repeated sowings in various years. The sign (^{2}) means that the examples in question were the same as in the previous year, and bore seed a second time. Finally I have denoted by *H* the two plants of 1893 which in their second year produced the seed from which the *L. vulgaris peloria* first arose. These *H* plants were therefore the parents of the peloric race.

PEDIGREE OF THE ORIGIN OF LINARIA VULGARIS PELORIA.

GENERATION		H and h = *Hemipeloria;* p and P = *Peloria.*
III	1899 annual	28 P + 4 h
II	1898 annual	75 P + 4 h
I–II	1897 annual and biennial	3 P + 5 h (2)
6	1895, 1897 annual	15 h + 2 p 6 h + 1 p 1895 1897
5	1894, 1896, 1899 annual	57 h + 1 p h + 1% p h + 1% p 1894 1896 1899
4	1892–93 biennial	H
3	1890–91 biennial	h
2	1888–89 biennial	h
1	1886–87 biennial	h

We will begin the further account of the experiment with the parent plants (H) of the peloric race (1893). As I had not of course observed anything extraordinary up to that time I only sowed a little of its seed. This was done in pans in the greenhouse; the young plants were transferred into pots with manured soil until they were planted out in June. As a result of this treatment they all flowered in the first year,—58 plants in all, of which 45 were dicotylous and 13 tricotylous. Amongst the dicotyls there were eleven plants, each of which bore one, two, or three peloric flowers, while in one case a peloric flower replaced a whole raceme. Amongst the tricotyls I did not find any such flowers, partly because the majority of these were removed by the middle of August; but there appeared amongst them one plant which bore peloric flowers exclusively on all of its stems and their branches. It bore no seed in spite of repeated careful pollination, partly with pollen from the neighboring plants; it survived the winter and flowered freely in the following year, again producing exclusively peloric flowers.

This experiment seemed to suggest that the *Peloria* arose from the hemipeloric parent in a proportion of about 1-2%. So in order to obtain closer knowledge of this proportion, I made a larger sowing in 1896 from the same lot of seed, and was able to plant out about 1850 seedlings in pots. By the middle of July some wholly peloric individuals had appeared, which were promptly taken up and transferred to a remote part of the garden. The further examples of *Peloria* which appeared from time to time, were planted beside them. By the middle of August all healthy plants were in flower and were recorded. There were altogether 16 totally peloric plants and 1759 with ordinary flowers, and here and there occasional peloric structures. This gives a total of 1775 plants which flowered, of which 1% (strictly speaking 0.9%) belonged to the new peloric variety.

For the harvest the flowers of the best peloric plants were enclosed in parchment bags and each fertilized with the pollen of another peloric plant. I also selected a beautiful hemipeloric plant which bore a profusion of flowers, one of the earlier of which was peloric. It set a quantity of seed after self-pollination.[1]

I repeated the experiment in 1899 with the rest of the seed of the parent plant H, and obtained the same result, as was to be expected. I raised slightly over 300 flowering plants, of which 3 were wholly peloric; that is to say, a proportion of 1% again. I observed on the rest a certain number of stray peloric flowers during the course of about two months.

These three cultures constituted the fifth generation of my experiment. The sixth generation therefore could be raised from the seeds of the hemipeloric plants in it. I did this partly in 1895 from the plants of 1894, and partly in 1897 from those of 1896. The plants which bore the seed had flowered in bags and had been fertilized partly by their own pollen and partly by pollen which I had transferred from one seed-parent to the other.

In both cases the mutation was repeated. Wholly peloric individuals again arose from hemipeloric ancestors, in spite of the smallness of the crops occasioned by the poorness of the harvest.

In 1895 I raised 17 flowering individuals from seeds of the dicotylous plants mentioned on page 559; two of them were wholly peloric, all their flowers being of this type. In 1897 I sowed the seed of the fine hemipeloric plant of 1896 referred to above, but obtained only 7 flowering individuals, one of which again, however, was wholly peloric.

I come now to the consideration of the question as

---

[1] This frequently fails in *Linaria vulgaris,* but sometimes succeeds more or less completely on very vigorous plants.

to whether the mutants are immediately constant from seed. An almost insurmountable obstacle in the way of providing an answer to this question is the low fertility, or rather the almost complete sterility, of the peloric flowers. Practically no results can be obtained with self-pollination, and when artificially fertilized with one another's pollen the majority of the flowers set no seed. I have pollinated thousands of flowers in the course of several years, only to obtain a little over one hundred fertile seeds. Under these circumstances it is obviously difficult to avoid mistakes; stray pollen grains may happen to reach the stigma from distant groups of normal plants, by the agency of insects, or in the operation of artificial pollination.[1] These circumstances evidently tend to invalidate the conclusion in cases in which the abnormality would seem to be incompletely inherited.

Only three of the wholly peloric plants of 1896 set seed in that year. From this seed only 8 plants were raised; five of them had one-spurred flowers and 3 were wholly peloric. I kept the peloric plants of 1896 through the winter, and took much trouble in 1897 in the attempt to fertilize their flowers. Every other day I pollinated all the open flowers with pollen from two other seed-parents. I obtained a very small quantity of seed most of which was empty (0.2 cc). About 100 seeds germinated, but some of the young plants were so weak that they soon died. 79 plants flowered most of which were very vigorous and branched freely; 75 were wholly peloric, and 4 normal, the latter being removed as soon as possible. The former exhibited great variability in the structure of their flowers, but did not produce a single one-spurred corolla. During July and August they

[1] Such crosses give normal one-spurred individuals.

filled an entire bed of over 3 square meters in extent, with hundreds of vigorous spikes which bore exclusively peloric flowers.

I again obtained only a very small harvest from this bed (0.3 cc); it was the result partly of artificial and partly of insect pollination, the plants flowering in sufficient isolation. Very few of the seeds germinated (1899) and only 32 plants flowered; 28 of them were peloric but 4 were normal.

The progeny of the peloric race was therefore a mixed one, in the three experiments which were continued over two generations. It consisted altogether of $3 + 75 + 28 = 106$ peloric and $5 + 4 + 4 = 13$ normal (including hemipeloric) individuals, a total of 119 with about 10% atavists. WILLDENOW (see p. 206) also found the peloria character inherited, though incompletely. As already stated, however, insufficient isolation may have played some part in bringing about this result, but hardly to such a degree that we might infer from our experiments that the peloria comes true.

If we now look back over this experiment, which occupied 13 years, its result may be summarized as follows:

1. *Linaria vulgaris hemipeloria* is a race with an inherited semi-latent character, which manifests itself from time to time among thousands of flowers, but seldom in more than one instance on a plant. It is widely distributed in the wild state.
2. From it the *Linaria vulgaris peloria* may arise but the conditions under which this happens are not yet understood.
3. This origin is a mutation; it takes place suddenly, and without any visible preparation. Especially

in those individuals from the seeds of which the mutation arises the latent character is not more highly or more often developed than in the rest of the race.

4. The mutation is repeated in successive generations. I observed it for two years, but did not follow it further.
5. The mutation occurred in about 1% of the individuals.
6. The new character was exhibited by the mutants, in a full state of development, in all their flowers; although it was subject to considerable fluctuating variability.
7. The mutants are to a large extent, perhaps even perfectly, constant from seed. The intensity of inheritance observed was about 90%, but it is probably more.

* * *

Let us next see how these results can be applied to the explanation of the occurrence of the *Peloria* in the free state. Wholly peloric plants have been found wild by numerous botanists and in the most diverse localities; but, so far as the published information extends, always as rarities. They maintained themselves during a larger or shorter period of years by means of their radical buds, perhaps produced some scanty seed but could not spread nor reach new localities by this means. They must therefore have originated in each case in the spot where they were found.

I imagine that this origin is determined everywhere by the same general laws, and thence conclude that it occurs in the wild state in the same manner as in the particular case observed by me, i. e., from *Linaria vul-*

garis hemipeloria, and always suddenly. The very general occurrence of this race and the fact that intermediate forms between it and the fully developed *Peloria* have never been mentioned by botanists, give support to this hypothesis

If this view is correct we have here a mutation which is not limited to a period but continues to appear from time to time during the course of the ages. Its appearance in every single case is independent of the others, at least so far as external conditions are concerned. In this sense it is polyphyletic.

A point which favors this view is the fact that it is not a member of a definite group of mutations as are the subspecies of *Draba verna, Viola tricolor* and others. *Linaria vulgaris,* it is true, frequently gives rise to other kinds of variations such as the *Peloria anectaria* and the *Catacorolla,* both of which have occasionally appeared in my own cultures, but nothing is on record concerning the relation between these and the *Peloria nectaria* which I have studied.

If we compare these results with those which we have described above for *Antirrhinum majus striatum* (§ 14, p. 134), we see that *Linaria vulg. hemipeloria* is obviously a half race; and that *L. vulg. peloria,* whose partial constancy seems analogous to that of the striped snapdragon, may perhaps be regarded as parallel to this. These two races fluctuate so as to approach one another, so to speak, occasionally overstepping the common boundary either in single flowers (*L. vulg. hemipeloria*) or in whole plants (*L. vulg. peloria*).

* * *

We now come to the most important point to which our results and conclusions lead us—namely the comparison of this mutation with those of *Oenothera Lamarckiana*. The two processes have several features in common, but possess others which are more or less strongly opposed.

The points of similarity are: the sudden and immediate origin, the repeated appearance, the mutation-coefficient of about 1% (see *V*ol I, Part II, § 14, p. 337), the completeness of the new type, and its high degree of heritability.

These common characters justify the description of the origin of *Linaria vulgaris peloria* as a mutation.[1]

But it is a mutation of a special kind. The structural change does not extend to all parts of the plant, but is confined to the flowers; in their youth the two types cannot be distinguished. In the mutations of *Oenothera Lamarckiana* the new characters are analogous to the specific characters of related species already existing; in the case of *Linaria* no such analogy exists. On the contrary the new character in *Linaria* occurs as a variety in numerous other species, and even in distantly related

---

[1] LINNAEUS, as is well known, expressed the view that the *Peloria* is a hybrid between the common *Linaria vulgaris* and some other unknown plant. Its comparative sterility favored this view, but as the second of the two parents could not be found this view has since been given up. Here, however, I might discuss the possibility that L. *vulg. hemipeloria* might be a cross between L. *vulgaris* (*apeloria*) and L. *vulg. peloria*. If this were so the appearance of the latter from the former would perhaps have to be regarded not as a mutation, but as a segregative process in a hybrid race. If this view were true the *Peloria* should first have arisen from the *Apeloria*, without the mediation of the *Hemipeloria*, a process which has still to be observed. It is, however, no more than a pure assumption that the hybrid *Apeloria* × *Peloria* would be a *Hemipeloria*; in fact our knowledge of other cases would lead us to suppose that it would be like one of the parents, in this case the *Apeloria*, and so long as there is no direct information on any of these points a further discussion of this view seems barren. Moreover it is by no means certain that *Linaria vulgaris apeloria* exists at all, or ever has existed; the variety, in this genus particularly, may well be older than the species.

plants. Lastly the mutation in *Linaria* does not appear along with others in space and time, but occasionally, and scattered perhaps over the whole area of the parent form and probably over the whole period of the life of this race.

The mutations of *Oenothera Lamarckiana* necessitated the assumption of a definite premutation, but the origin of the *Peloria* is obviously a phenomenon of a different kind.

Peloria is often regarded as an instance of atavism.[1] The correctness of this interpretation obviously depends primarily on whether this term is used in a narrow or a broad sense. Atavism is a reversion to ancestral characters; in the narrow sense to the complete type of particular ancestors, in the wider it refers only to single characters. But it is clear that the spurs which form a distinctive character of the genus *Linaria* must be older than the species *L. vulgaris,* which cannot therefore have had ancestors without the spur but with the other characters of the species; so that *L. vulgaris anectaria* can occupy no place in the series of ancestors. The symmetry is ever so much older and *L. vulgaris* with regular flowers has certainly never existed amongst the ancestors of the common toadflax. Moreover the sterility of the peloric plants does not favor such a view.

If the *Peloria* must be regarded as atavistic, this view can mean no more than the assertion that it has arisen by the loss or latency of a character of the common *Linaria*. Therefore we are concerned here with a *retrogressive* mutation, and the question arises, how far the differences between this case and the progressive mutations with which we have become familiar in *Oenothera*

[1] See L. JOST, *Biolog. Centralbl.,* 1899, p. 149.

are thus to be explained. The explanation is so simple that it follows directly from the preceding discussion. It is merely necessary to point out that the most important condition for a character to become latent is its presence; and this explains how it is possible that the *Peloria* so often appears over the whole area of distribution of the species. Neither a premutation nor a period of mutation is necessary for such an occurrence.

If the loss or latency (for the inner potentiality is obviously not lost but only becomes inactive) affects single flowers we have a partial atavism, but if it affects the whole plant we have the complete and heritable *Peloria*.

It is on this basis that the atavistic phenomena of the striped flowers, of the many-spiked *Plantago* (§ 17, p. 148) and of the peloric *Linaria* fall in line. They are retrogressive phenomena, reversions to ancient characters which have externally become lost but are still present in a latent state. Their agreement with one another on the one hand, and their contrast with the progressive mutations of *Oenothera Lamarckiana,* on the other, thus receive a satisfactory explanation.

§ 21. HERITABLE PELORIAS.

Pelorias are very rarely met with in nature as a specific character. As an instance I may quote *Mentha aquatica,* the apical flowers of which according to SCHIMPER'S discovery are always regular and consequently peloric,[1] and the orchid *Uropedium Lindenii,* which is regarded as the peloric form of *Cypripedium caudatum.*[2]

[1] A. BRAUN, *Abh. d. Berliner Akad.,* 1859, p. 112; and DELPINO, *Mém. R. Instit. di Sci.,* Bologna, 5 Ser., Vol. I, 1890, p. 269.

[2] A. BROGNIART, *Ann. Sc. nat.,* 3 Ser.1, Vol. XIII, p. 113 (Plate 2); and J. M. JANSE, *Maandblad voor Natuurwetenschappen,* Vol. XIV, No. 3, 1887, p. 29. *Uropedium Lindenii* appears to be by no

In a state of cultivation peloric races are also very rare, and the common *Gloxinia superba erecta* with its numerous color varieties and hybrids is the best generally known cultivated example.[1]

Our present knowledge of the origin of *Linaria vulgaris peloria* as described in the foregoing pages, justifies us in attempting to form some idea concerning the origin of such forms in these perfectly analogous cases and also to sketch the details of this idea on a basis, or background of facts.

But there are still difficulties in the way. The low fertility and the incomplete constancy of the *Peloria* distinguish it from true species.[2] Most systematists would evidently not consider *L. vulgaris peloria* to be a true species unless the common *L. vulgaris* were extinct.

Besides the examples named, there is a whole series of heritable cases of peloria, which either appear as rare anomalies, or are familiar cultivated races, and repeat the abnormality regularly and in a fairly large number of individuals every year.[3] In both cases, however, the development of the anomaly is, as usual, in a high degree dependent on external conditions.

There are, as we have stated in the foregoing section

means rare in Colombia (South America); it was discovered there by LINDEN in 1843 (LINDEN, *Pescatorea, Iconographie des Orchidées*, 1860, Plate II.

[1] The spurless varieties of certain species of *Viola* and *Tropaeolum* may also be regarded as pelorias: see the following page.

[2] From this point of view it would be very important to know whether the *Mentha* and *Uropedium* cited are perfectly constant, that is, never produce atavists without pelorias.

[3] It is extremely doubtful whether, besides these, there are pelorias, the origin of which is solely due to external influences and does not need the existence of a corresponding internal potentiality.

(§20), various kinds of pelorias according as one or another form of the petals of the parent species has become the one which prevails in the subspecies. In spur-bearing species they are distinguished as *Peloria nectaria* and *anectaria*. Both possess a very low degree of fertility but are, so far as is known, heritable. Peloric flowers without spurs are well known in *Linaria*,[1] *Antirrhinum*,[2] *Viola*,[3] *Tropaeolum*,[4] etc.[5]

There are few heritable peloric races beyond those which have been named. The best known are *Corydalis solida peloria* which in GODRON's experiments[6] was found to transmit the abnormality through a series of generations, and *Digitalis purpurea monstrosa* (Fig. 42). This latter, the peloric foxglove, has been a favorite garden plant for a long time, and has often been the subject of morphological investigations. The oldest descriptions and figures are due to my predecessor G. VROLIK, whose preparations are still to be seen in the collection at Amsterdam.[7] Since his time the variety has been cultivated in our botanical garden more or less regularly, and is still growing there.[8] It is very constant; its peloric flowers

---

[1] C. BILLOT, *Annotations à la Flore de France et d'Allemagne*, quoted in Bot. Zeitung, 1872, p. 278.

[2] J. T. C. RATZEBURG, *Animadversiones ad peloriarum indolem*, 1825, Plate I, Figs. 64-76.

[3] J. C. COSTERUS, *Pélories du Viola tricolor*, Archiv. Néerl., Vol. XXIV, p. 142, Table II; DE CANDOLLE, *Organographie*, Pl. 45.

[4] E. VON FREYHOLD, *Ueber Pelorienbildung bei Tropaeolum aduncum*, Botan. Zeitung, 1872, p. 725 and Plate IX.

[5] D. A. GODRON, *Mém. Acad. Stanislas*, 1865 and 1868 (*Delphinium chinense*, etc).

[6] GODRON, *loc. cit.*, 1868, pp. 3-8, Cultures from 1862-68, with more than fifty peloric plants.

[7] G. VROLIK, *Ueber eine sonderbare Wucherung der Blumen bei Digitalis purpurea*, Flora, 1844, p. 1, Plates I and II; also *Fortgesetzte Beobachtungen über die Prolification von Digitalis purpurea*, Flore, 1846, p. 97, Plates I and II.

[8] The following selection of references may be of use: W. F. R.

are, however, highly variable and only too frequently accompanied by other malformations. The commonest of these are an increase in the number of organs, the formation of catacorollas and the production of a secondary raceme from the axis of the flower. These are the cases which are most commonly described and figured in literature. In order to find more regular and even perfectly pentamerous flowers we must look to the tops of the weak lateral branches of vigorous plants (Fig. 42); these hardly ever proliferate, are often still pleiomerous, but there will also occur amongst them flowers with a perfectly regular corolla with five lips and five erect stamens.

The peloric flowers of *Digitalis purpurea* are always terminal, whether they occur on the main stem or on branches. The same is true of most other *Scrophu-*

Fig. 42. *Digitalis purpurea monstrosa*. A lateral branch with a terminal pentamerous peloria.

SURINGAR, *Plantaardige Monstruositeiten*, K. Akad. v. Wetensch., Amsterdam, 1873, 2d. R., Vol. VII, Plates I-II
   P. MAGNUS, *Digitalis purpurea*, Sitzungsber. Prov. Brandenb., Vol. XXII, 1880, p.
   J. C. COSTERUS, *Teratologische Verschynselen by Digitalis purpurea*, Ned. Kruidk. Archief, 1885, Plate VII.
   ANGEL GALLARDO, *Fasciacion, Proliferacion y Sinantia*, Ann. Mus. Nacion., Buenos Aires, Vol. VI, p. 37, Pl. 3; also *Sobre algunas anomalias de Digitalis purpurea* (with complete bibliography), same journal, Vol. VII, pp. 37-72.

lariaceae,[1] and of many other families, especially orchids.[2] The relation between this position and the regular form of the flower is still without a proper explanation; and the question whether the anomaly is due to high nutrition or to the absence of the factor which determines the bilateral symmetry or both, still awaits a definitive answer. Laterally situated peloric flowers are very rare but sometimes occur as we have seen in *Linaria*

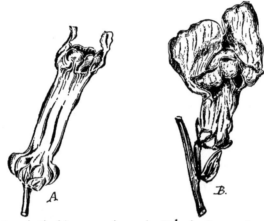

Fig. 43. *Antirrhinum majus.* A, Peloric flower from the middle of an otherwise normal raceme, August 1899. Two slips of the corolla stand erect; the other three are bent downward. B, Normal flower of the same spike.

vulgaris hemipeloria (Fig. 41, p. 207), and as is shown by *Antirrhinum majus* (Fig. 43), etc. Of great importance, also, is the hitherto little noticed fact that in *Digitalis* and one or two other cases, the peloric terminal flower opens first of all, whilst the order of opening of all the other flowers on the stem is normal, i. e., acropetal.

[1] Eichler, *Blüthendiagramme*, I, p. 208.

[2] Pfitzer, in Engler and Prantl's *Natürl. Pflanzen-Familien: Orchid.*, p. 61. For further information on pelorias of Orchids see Penzig, *Mém. Soc. nat. Sc. Cherbourg*, Vol XXIX,, 1894, pp. 79-104.

Peloric flowers occur as chance anomalies in a large number of plants. A speciment of *Scrophularia nodosa* which I have had growing for the last ten years produced them abundantly. On the other hand my cultures of *Antirrhinum majus* although of twelve years duration and carefully guarded gave rise to no more than two peloric flowers, one of which is shown in Fig. 43A. Both sprang from the middle of the racemes, that is, they were lateral. I have also observed occasional cases of peloria on *Aesculus Hippocastanum, Melampyrum pratense, Orobanche Galii,*[1] *Cytisus Laburnum,* etc. In my cultures of 1892 a peloric flower occurred on a plant of *Lupinus luteus.* The tube-shaped peloric flowers of the cultivated *Calceolarias* are also well known. In these and similar cases the mode of inheritance has still to be investigated. In this respect the observations of PEYRITSCH are of great importance He has shown that in the case of peloria in *Leonurus Cardiaca,* an annual Labiate, the anomaly can be reproduced from seed whether this originates from the peloric or the normal flowers of the same plant.

PEYRITSCH's memoir is one of the most valuable of those which deal with peloria, and is indeed an almost complete monograph so far as the Labiates are concerned.[2] He has also investigated the influence of the environment on the anomaly as occurring in a series of Labiates.[3] I select the following observations for notice here:

[1] See also W. F. R. SURINGAR, *Orobanche Galii,* Ned. Kruidk. Archief, 1874, Vol. I, p. 330, Plate 18.

[2] J. PEYRITSCH, *Ueber Pelorien bei Labiaten,* Sitzber. d. k. Akad. d. Wiss., Vienna, Vol. XL, Part I, 1869, p. 343, Plates I-VI; and Vol. XLII, 1st section, 1870, p. 497, Plates I-VIII.

[3] J. PEYRITSCH, *Untersuch. über die Aetiologie pelorischer Blüten-*

Lamium maculatum and *Galeobdolon luteum* commonly produce peloric flowers in the neighborhood of Vienna. They often bear them every year on the same plant, but one or more years are sometimes skipped. A sunny position increases the number of anomalous flowers whilst dense shade diminishes it; consequently one locality often furnishes instances of peloria in several species of Labiates (e. g., *Calamintha* and others), whilst the same species growing together in another locality will not produce a single symmetrical flower or only very few. Whenever the conditions affecting a plant were improved by cutting down timber, peloria occurred in profusion, and the transference of a plant to a sunny spot in a garden often resulted in its appearance. Other authors, and particularly VUILLEMIN,[1] also assert that the conditions of life play an important part in inducing the anomaly, provided that the inherited potentiality for it is present.

bildungen, Denkschr. d. k. Akad., Vienna, Vol. XXXVIII, Part II, 1877, with Plates I-VIII. See also GOEBEL, *Organographie,* I, p. 163.

[1] *Loc. cit.,* 1894, p. 33.

VII. NON-ISOLABLE RACES.

§ 22. TRIFOLIUM INCARNATUM QUADRIFOLIUM.

Few experiences are so well fitted for enabling us to obtain an insight into the nature of specific characters as the failure of an experiment in selection. I am not speaking of practical experiments because in such cases the breeder is often disappointed by the fact that the result is not superior to what he has already, or is not suitable for cultivation on a large scale from other causes. This kind of failure only concerns the practical breeder and does not affect the scientific investigator. The object of the latter is simply to find out whether a race specified beforehand can be obtained or not.

According to the theory of selection almost anything ought to be obtainable. Almost all characters manifest fluctuating variability to the extent requisite for selection. If the range of variation is considerable, selection should proceed rapidly; if it is within narrower limits it should merely require longer series of generations; and if, moreover, the familiar but undemonstrated opinion is assumed that fluctuating variability increases as the result of the selective process, there is no reason why in any given case the attempt to breed a desired race should not succeed.

But this discussion, in my opinion, only applies to ordinary fluctuating variability, and if thus limited, I willingly agree with the prevailing view. In the sphere of mutability, on the other hand, matters are entirely different. Here species, subspecies, varieties, races, etc. arise by mutations which are induced by rendering active a hitherto latent or semi-latent character. The first condition for a desired mutation therefore is the existence of the character in question in a latent or semi-latent state. Without this nothing can be achieved, at least in the present state of science, and it is only in the case of semi-latency that we can have any sort of evidence that the desired character is present. Horticultural breeders are well known to be continually on the lookout for any such indication.[1]

But the presence of a latent character is not of itself sufficient, according to my experience, to insure the success of an experiment in selection. For many an experiment has failed in spite of years of labor.

This proves nothing in itself, because it is often due to lack of sufficient experience, and this experience can only be acquired by carrying out a successful experiment in an analogous case; in other words, by making exactly the same experiment with a related plant, preferably with another species of the same genus.

For this reason I have more than once endeavored to breed a race analogous to one already existing in a closely related species of the same group, which is either on the market, or has appeared in my own cultures. Experience has taught me that the end may often be attained with greater or less ease according to circumstances; but that in many other cases, so it appears, insurmountable obstacles bar the way.

A very definite and simple case is afforded by the

[1] See Vol. I, Part I, § 25, p. 188; and this volume, Part I, § 2, p. 9.

Trifolium Incarnatum Quadrifolium. 229

attempt to breed a five-leafed race of the crimson clover (*Trifolium incarnatum*) analogous to the five-leafed race of the red clover (*Trifolium pratense*) which has already been described (§ 5, p. 36). I started the experiment in 1894; since then I have devoted a great amount of trouble to the task without any result, until in 1900 I gave it up. The attempt simply does not succeed, with my material at any rate.

The object was worth a great effort. At first I believed that I had artificially made the five-leaved red clover, or as it is often expressed, that I had created it. The gradual development of my theory, however, led me to doubt the correctness of this opinion. It seemed possible that I had merely found the race already existing in nature, but in a condition in which it was not recognizable as such. Eight years however had gone by since the beginning of that culture, and it was practically out of the question to go back to it. I resolved therefore to endeavor to raise a new five-leafed clover and selected the crimson clover. This

Fig. 44. *Trifolium incarnatum.* A flowering branch with a single 4-foliate leaf; the result of an experiment in selection lasting six years.

choice was largely determined by the fact that there were no published records of 4- or 5-foliate leaves of this clover,[1] which means that the character, if present in a latent state, is much rarer than in the red clover.

I take this opportunity of calling attention to the inestimable value of PENZIG's "Teratology." This lies perhaps rather on the negative than on the positive side, for it is of course possible to collect the main literature relating to a given question oneself, although not without the expenditure of a great amount of time; but if one is not a teratologist by profession, it seems hardly possible without some such help, to satisfy oneself that

Fig. 45. *Trifolium incarnatum,* 4-foliate leaves, the middle one with incomplete segregation of a lateral leaflet.

absolutely no records relating to a particular phenomenon exist.

The first step in a purely scientific breeding experiment evidently is to find out whether the deviation in question has occurred before, and if so, whether it is rare or common. My belief is that the commoner anomalies are heritable characters with a high index of inheritance (often about 30-40% or more), but that the rarer ones are the occasional expressions of latent or semi-latent characters. These are also inherited in their latent state, and if they turn up here and there this latent condition must probably be widely distributed.

[1] O. PENZIG, *Pflanzenteratologie,* Vol. I, 1890, p. 385, where *T. incarnatum* is not even mentioned.

If *Trifolium incarnatum* with 4-foliate leaves had often been mentioned it would therefore seem probable that a five-leaved race of it occurs in nature, although just as little separated from the ordinary crimson clover as the five-leaved race of the ordinary clover is from this.

Latent characters, in my opinion, are often older than the species which bear them. I regard the division of the leaf into four blades in this case as an atavistic phenomenon, and I believe that this latent potentiality is as old as the whole group of clovers with trifoliate leaves (*Trifolium, Medicago, Melilotus* etc.), that is, older than the individual genera of this group. In many species this power of reproducing quadrifoliate leaves may have been completely lost, for it is mentioned in PENZIG's book only for a relatively small number of them. In others, however, it has persisted to the present day. If the trifoliate leaves of the clovers are derived from Papilionaceae with pinnate ones, the multifoliate leaves which they occasionally produce must evidently be regarded as atavistic phenomena. The correctness of this view is proved by those very rare cases in which, in the races in question, pinnate leaves appear instead of the ordinary multifoliate ones. I have observed this from time to time in my *Trifolium pratense quinquefolium* (Fig. 46) and the same thing has been found by other authors in *Trifolium minus* and *Trifolium repens*.

Fig. 46. *Trifolium pratense.* An atavistic pinnate leaf.

I have myself found 4- and 5-foliate leaves in *Medi-*

cago lupulina, whilst BRAUN has observed them in *M. sativa*. They are well known in *T. pratense* and *T. repens*, and WYDLER has recorded 4-foliate leaves in *Lotus major* and *Tetragonolobus biflorus*. In some successive sowings which I made with *Medicago lupulina* I found the character to be inherited although in a moderate degree only, but I have not continued the experiment.

But let us return to the crimson clover. The question is, what prospects were present at the beginning of the experiment, and what may be expected from such experiments in general? There are three main possibilities to be considered. We may find at the beginning of the experiment (See § 3 p. 20):

1. A race which often exhibits the anomaly in question, and bears it as a heritable character, i. e., an ever-sporting variety;

2. A half-race with a semi-latent anomaly which is only occasionally manifested;

3. An ordinary plant of the species with the character in question in a latent condition.

In the first case the race already exists and all that is necessary is to isolate it; in the second it may possibly be obtained; in the last there is little prospect of doing so.

In order to present a clearer idea of the mutual relations of these three cases let us examine *Trifolium repens* and *T. pratense*. That the anomaly is by no means very rare is testified in both cases by the popular belief in the so-called lucky four-leaved clover as well as by common experience. If looked for in a field of clover, or in a meadow, or along the roadside, a four-leaved clover will be found from time to time. If repeated attempts are made to find them they will certainly prove to be rare but not so rare as we might have imagined. I have found them almost every year, and often quite soon after I had been asked for one. On the other hand there is on the market the 5-foliate *T. repens atropurpureum* which is often cultivated in gardens for its dark brown leaves, and for *T. pratense* I have described the five-leaved form in detail in § 5.

Plants of *T. pratense* are sometimes found in the field with two or more 4- or multifoliate leaves. I found one in 1866 in the Cronesteyn estate near *Leyden*, and another in 1886 near *Loosdrecht*. The first had several 4-foliate leaves, and also some 5-6-foliate ones. I secured the former but did not cultivate it; the latter formed the starting point of my race. In view of my present knowledge I must assume that in both cases the plants already belonged to the race when I found them; and I also consider it as probable that this race had arisen on these very spots, or at least not far from them.[1]

Whether the same race can also be produced from the occasional stray four-leaved clovers I do not, of course, know; but I anticipate that the attempt would sometimes succeed and at other times fail. If this view is confirmed by future experiments we shall have proof of the existence of the two races, the eversporting variety and the half-race, existing simultaneously within the limits of a single species. For the present we must be satisfied with the knowledge that there exists a race rich in anomalous leaves in the red and in the white clover, and one in the crimson clover which bears the character only in the semi-latent state.

I shall now proceed to the description of the latter.

In the winter of 1894-95 I bought a kilo of the seed

[1] A polyphyletic origin, therefore, as in *Linaria vulgaris peloria*.

of the ordinary crimson clover and sowed part of it on a bed of about five square meters. Two of the seedlings were tricotylous and one was tetracotylous, and these were transplanted to a special bed as soon as possible in the hope that they would exhibit the desired abnormality. This hope was based on the principle of the correlation between different kinds of anomalies.[1] If a plant exhibits an anomaly in its early stages it will, according to this principle, be more likely than any other individual in the same culture to give rise to other deviations later on. In this particular case my expectation was fulfilled, for the tetracotylous plant produced one 4-foliate and one 5-foliate leaf in the course of the summer. Such were not found on any other plant, either during the course of the experiment or at the end of July when the plants were in full bloom and were pulled up and minutely examined. There were about a thousand plants.

I left the three selected specimens to flower together and sowed their seeds in April 1896. Over 600 seedlings came up, all of them with only two cotyledons. In all of them the first leaf was single, which is the general rule in clovers (Fig. 47 A). The second and third leaves developed in May; they were quite normally trifoliate, with the exception of one, of which one of the three leaflets was split laterally, although not completely divided. The form of this blade was similar to that figured in Fig. 45 B. About 250 individuals of the whole group were planted out. The seed had been sown in pans; the young plants were transplanted into pots and were planted in the beds in the middle of May. At the end of June,

---

[1] *Eine Methode, Zwangsdrehungen aufzusuchen,* Ber. d. d. bot. Ges., Vol. XII, 1894, p. 25.

at the beginning of the flowering period, several individuals exhibited one or more 4-foliate leaves; the anomaly was therefore a heritable one.

Moreover the multiplication of the blades had also increased considerably as the result of selection, as the following figures prove. These refer to the offspring of that seed-parent which had already exhibited the anomaly in the previous year. There were 90 of them; among the offspring of the tricotylous parents "4-leaves" were not entirely absent, but they were relatively scarce, and

Fig. 47. *Trifolium incarnatum.* A, a seedling with normal primary leaf. B—D, seedlings with 2- and 3-foliate leaves. The former arise from the larger; the latter from the smallest seeds.

the whole group was consequently pulled up at the beginning of the following period. About ⅔ (58 out of 90) of the rest were perfectly normal without any increase of the number of leaflets. On the average they had about 10 stems and 100 leaves per plant. The remaining plants formed a half-curve[1] of the following composition. The first row gives the number of 4- or 5-foliate leaves per plant, and the second the number of

[1] § 4, p. 26; and *Ueber halbe Galton-Curven,* Ber. d. d. bot. Ges., 1894, Vol. XII, p. 197.

individuals on which these numbers were observed (culture of 1896):

Abnormal leaves	0	1	2	3	4	5	6	7	8	9
Individuals	58	10	12	4	2	2	1	0	0	1

The 58 normal plants were pulled up. Of the rest four were weak and died; there remained 28 which all flowered together. Their seed was harvested separately after the number of 4- and 5-foliate leaves on each parent had been recorded.

In March 1897 I sowed a part of this seed in pans, separately for each seed-parent. The object of this was to find out whether there was any difference between the individual seed-parents with regard to the number of anomalous offspring which they produced. From an examination of the pans it was easily seen that the abnormality had already appeared in the primary leaves of some of the seedlings. In the great majority of cases these were perfectly normal, consisting of one leaflet as in the whole of the previous generation. In some cases however this primary leaf consisted of two or three leaflets (Fig. 47 B-C). Such occurred in the crops raised from 6 of the 21 plants whose seeds had been sown. Each seed-parent had given a crop of about 300 seedlings. Five of the crops contained not more than 2 abnormal seedlings, but the remaining one had a very large number, namely 14 amongst 335 seedlings or about 4%. It is worthy of notice that the parent of this crop had only had two 4-foliate leaves itself and thus had not given the least sign that it would produce offspring with so much higher a degree of the abnormality. Moreover I could not find any relation between the number of abnormal leaves on the other seed-parents and the proportion of abnormal offspring raised from their seeds.

The plant with nine 4- or 5-foliate leaves did not give rise to a single anomaly amongst 300 seedlings.

Amongst breeders of animals it is generally recognized that the visible characters of an animal are of very little use as an indication of its value for breeding. The offspring which the animal has already produced afford a much more reliable indication.

On the basis of the choice of the seedlings, the 14 abnormal offspring of the seed-parent with 4% were planted out in the beds in June 1897, together with the seven next best plants. The latter produced very few 4- to 5-foliate leaves per plant, the first 0 and 1 in eleven cases, but 9, 9 and 4 in three cases. There was therefore no marked advance on the previous year in this respect.

The progress was just as inconsiderable in the harvest of that year. The percentage of abnormal individuals amongst the seedlings ranged in 1898 between 1 and 4% and in one case reached 6%. On the other hand all the (19) seed-parents investigated had at least one and usually two or more seedlings with a divided primary leaf. But here again no relation was manifested between the number of abnormal seedlings and the number of 4- or 5-foliate leaves on the seed-parents which produced them.

227 seedlings were planted out, most of which were perfectly normal at the time of flowering. I obtained the following half-curve (1898):

Number of multifoliate leaves per plant:	0	1	2	3	4	5
Individuals:	188	29	7	1	1	1

That is to say, about 20% of individuals with the inherited anomaly in from 1 to 5 of the whole number of leaves counted on the plant (about 100). The numbers were therefore smaller in this than in the previous year.

For this culture I had planted out the normal and abnormal seedlings of the most abnormal seed-parents and some abnormal seedlings of the remaining seed-parents. No essential difference between these three groups could be detected when they were recorded at the time of flowering.

Pitcher formation was observed both amongst the seedlings and during the later stages; this is another indication of correlation amongst the various characters.

In the summer of 1898, 41 of the selected plants furnished a sufficient quantity of seed. In the following spring I determined the proportions of seedlings with compound primary leaves in the crops from each of these parents and reckoned them in percentages. The composition of the 1898 harvest with respect to this character was:

Percentage of abnormal offspring 0 1 2 3 4 5 8 11 15 16 20 24 27
Parents 3 12 7 5 4 2 1 2 1 1 1 1 1

That is to say, a considerable advance which at once becomes evident if this series of figures is compared with that given above for the 1897 harvest (1-4 and 6%). This advance has moreover taken place in spite of the falling off in the number of 4-foliate leaves in the seed-parents.

In the spring of 1899 I only selected seedlings with trifoliate primary leaves for transplanting (see Fig. 47C), and only from amongst the offspring of the four seed-parents with from 15-24% abnormal offspring. At the time of flowering, however, my hopes were disappointed. In the middle of July there were amongst 120 richly branched flowering plants 45% without the anomaly,

27% with a single abnormal leaf each, and 28% with two to four 4- to 5-foliate leaves each. That is to say, 55% abnormals as against 20% in the previous year—which indicated a marked advance.

But my hope of obtaining a leaf with more than five leaflets was not fulfilled. In spite of repeated search I never found one. Nor did I obtain plants rich in four-bladed leaves; for there were none with more than four of them.

Therefore I have since abandoned the hope of breeding a race of four-leaved clover, corresponding to my *Trifolium pratense quinquefolium,* from this material.

A striking feature of this experiment is the apparent absence of a relation between the degree of abnormality of the adult plants and that of the seedlings. For the paucity of four-bladed leaves in the grown plants seems incompatible with the abundance of multifoliate primary leaves in the seedlings from which they grew.

The failing of this relation has led me to the discovery of a most remarkable connection between this variability and the size of the seeds, for the smallest seeds are those which give rise in the largest number to compound primary leaves.

Small seeds germinate somewhat later than larger ones and also give rise to weaker plants. It had often struck me that the selection of the most abnormal of the seedlings was frustrated by the fact that many of the individuals with compound primary leaves were too weak to be planted out, or died soon after the process. It also struck me that all the seedlings in a pan could not be recorded at the same time. At first view the plants appear to germinate very regularly, and hundreds in the same pan seem to unfold their leaves at the same moment.

At this point they were recorded and, if the first leaf was single, were usually pulled up. Those which were saved were usually weaker, more stunted and backward in growth. Several had not yet unfolded their first leaves, and amongst them a great number of the anomalies were found when the examination was repeated a few days afterwards.

I then convinced myself by a very simple experiment of the correctness of these conclusions. All that was necessary was to isolate the large and the small seeds in a sample and to sow them separately. But as there is no absolute limit between the two it was necessary to know how many seeds should be separated out, as the smallest. And this can only be done by the number of anomalies, i. e., compound primary leaves, they produce. I therefore selected a sample of seed whose capacity for producing anomalies I already knew. This was 15%; the sample was derived from a single seed-parent. I separated the seeds into three categories, small, intermediate and large. All in all there were 217 seeds of which 17 did not germinate. The characters of those which did are as follows:

	Number of divisions in the primary leaf.			
	1	2	3	2–3
Small seeds	31	9	16	12.5 %
Intermediate seeds	50	2	1	1.5 %
Large seeds	88	2	1	1.5 %
	169	13	18	15.5 %

It will be seen that almost all the abnormal seedlings are derived from the smallest seeds. The seedlings from the large seeds had, with a single exception which was an abnormal one, unfolded their primary leaves in May, and fourteen days after the seed had been sown; the same is true of the intermediate seeds with the exception of four, two of which were abnormal. 22 normal plantlets developed from the smallest seeds in the same time; the 9 other normal and the 25 abnormal ones did not unfold their first leaf until the third week.[1]

These facts show further that the number of seedlings with abnormal primary leaves does not depend simply on the degree of fixation of the variety. It depends mainly on the proportion of small seeds. This, however, in its turn, depends on the size of the harvest.

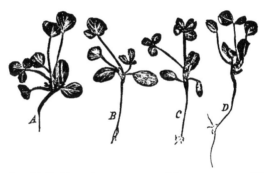

Fig. 48. *Trifolium incarnatum.* Monstrous seedlings from the smaller seeds. A, B, D, with two to four primary leaves; C, with a double leaf with broad flat peduncle.

In the 41 samples which composed the harvest of 1898 there were 8 with 8-27% abnormals; these samples consisted of from 0.3 to 1.5 cc. of seed. The remaining samples consisted of from 2 to 5 cc. of seed and the number of abnormals produced ranged between 0 and 5%. From these facts we see that the weaker individ-

[1] In stocks also the seeds which produce plants with double flowers and those which give rise to "singles" have different rates of germination, as is well known. An investigation of the seeds of inconstant varieties, or, as they are usually described, varieties which have not "*yet*" been fixed, would be certain to reward the inquirer with many interesting discoveries.

uals, which gave a poorer harvest, gave rise to the largest number of abnormals amongst their offspring.

I have repeated the same experiment with the harvest of 1899, with the seeds of four separate seed-parents, but as I did not know their capacity for producing abnormals in advance, the difference was not so striking. The large seeds gave rise to 2-4%, the small ones to 3-13% abnormals. Altogether seedlings from 2758 large and from 617 small seeds were examined.

Two questions present themselves in connection with the interpretation of these experiments: (1) Can the position in which the small seeds are chiefly produced on the plant, be determined?[1] (2) Are the germs of the small seeds perhaps the better nourished ones; is there, for instance, just as much nutriment brought to them as to the large seeds, but must they, for want of room or for other reasons, utilize it in some other way?

I recommend these problems for further study, and may perhaps in the mean time record a few facts bearing on them which I have observed. In the crimson clover, monstrosities occur much more frequently among the seedlings from small than among those from large seeds. The latter are almost all perfectly normal. The small seeds often produce plants with supernumerary cotyledons, or with two or more primary leaves (instead of one) or with divided peduncles, symphyses in the leaves and other malformations (Fig. 48). Unfortunately it is often difficult to keep these individuals alive and to bring them to flower.

Let us now cast a final glance over the whole course of the experiment.

---

[1] In stocks, according to CHATÉ, *Culture des Giroflées*, the seeds which produce double-flowered plants arise chiefly from the lower half of the pods of the strongest racemes of the plant.

A crimson clover plant with some quadrifoliate leaves was obtained by the selection of tricotylous and tetracotylous seedlings in conformity with previously studied laws of correlation. The anomaly proved to be heritable and has maintained itself until now, during six generations (1895-1900). It was improved by selection but only within very narrow limits. Plants with more than five leaflets per leaf have not as yet arisen, nor have plants bearing ten or more 4- to 5-foliate leaves, and it is nearly always the "small" seeds which give rise to seedlings with compound primary leaves.

But the chief result is that the desired race, rich in 4-foliate leaves, "*T. incarnatum quinquefolium*" analogous with *Trifolium pratense quinquefolium,* did not arise.[1]

§ 23. RANUNCULUS BULBOSUS SEMIPLENUS.

Double flowers are common phenomena amongst the buttercups.[2] They occur not only in the cultivated Ranunculi (*R. asiaticus*) but also in several wild species. The doubling may be either complete and brought about by petalomania as in the *Ranunculus acris* shown in Fig. 40, Vol. I, p. 194; or it may be more or less incomplete when caused by the transformation of a varying number of stamens into petals (*R. acris, R. auricomus, R. Philonotis, R. repens* etc.).

In *Ranunculus bulbosus,* the bulbous buttercup, the

[1] The same thing no doubt occurs also in other cases. The attempt to breed from occasional anomalies a constant race endowed with the particular variation, in some cases succeeds, but in others does not. For instance I have for many years endeavored to raise from the occasional polycephaly in *Papaver commutatum* a race with as beautiful crowns as those which characterize the familiar *Papaver somniferum polycephalum* (see Vol. I, p. 138, Fig. 27), but in vain.

[2] See PENZIG, *Pflanzenteratologie,* Vol. I, pp. 181-189.

stamens are often (either all, or only some of them) transformed into petals with the result that dense double flowers are produced.[1] These have been described by

Fig. 49. *Ranunculus bulbosus semiplenus.* A, the bulb; A' and A", its leaves from the axils of which the flowering stems S arise; E, terminal flower of the main stem; S, secondary flowers partly broken off; T, tertiary flowers. (See p. 256).

various authors.[2] In the neighborhood of Amsterdam this variety does not occur, so far as I know. On the

[1] *Loc. cit.*, p. 185. Fasciated stems with broadened terminal flowers are also met with occasionally in the *Ranunculus bulbosus* in Holland.

[2] Compare the *Ranunculus bulbosus Aleae* of Naples, described by TERRACCIANO, *Nov. Atti d. R. Instit. Napoli*, 1895, Vol. VIII, No. 7.

other hand on plants growing as they often do in sandy localities, the flowers often possess a slightly increased number of petals.

In these abnormal flowers there are usually six or, rarely, seven petals, very seldom more than 10-12. They are as a rule ordinary petals, but there sometimes occur some that are much smaller and narrower and are obviously metamorphosed stamens. This metamorphosis is often only partial, and the familiar intermediate stages are exhibited. The abortive stamens are usually to be found among the most peripheral ones; but they are not necessarily the outermost ones, directly adjoining the corolla.[1]

Fig. 50. *Ranunculus bulbosus semiplenus.* A flower with 31 petals (partly petalodic stamens); the only one amongst 4425 flowers. It occurred on a quaternary branch in my culture of October, 1892. See the series of figures on page 252.

The potentiality of this doubling is therefore present in a semi-latent condition in the wild plants of this species growing in this neighborhood. I regard this race, therefore, as a half race in contradistinction to the normal double race which is only known to me from the published records. Obviously the two possess the same character; which is, however, active in the one case but latent or semi-latent in the other.

It seemed to me important to discover whether it was possible to obtain the double from the half race by selection. According to the views advanced in this work this should be possible, but not every attempt need necessarily be successful. But if it does succeed the change

[1] See GOEBEL, *Jahrb. f. wiss. Bot.,* Vol. XVII, pp. 217-219.

must be brought about suddenly, and, under ordinary conditions of culture, be effected in the course of a few years. In this way the double variety may have arisen from time to time in the wild state; and in the same manner the present half race may perhaps, in the course of time, undergo this change.

This transformation, however, cannot be simply the result of careful selection. A mutation is needed; and we know as little about the causes of mutations as about the method of inducing them artificially. Mutations are known to occur with moderate frequency both in breeding experiments and in nature, but, up to the present, their occurrence has been a matter of chance (§§ 10 and 11, pp. 95-103).

In my experiment such a mutation did not occur, although it extended over five generations.[1] The half race was distinctly improved by repeated and very stringent selection. It became at the end very rich in extreme or almost extreme variants, but it was just in these that it proved to be so remarkably constant. In its five generations it reached a point which did not seem to me likely to be exceeded by further selection. It produced occasional flowers with more than 15 petals, and a single one with 31, but the mean number of the petals in its selected individuals did not exceed 9-10.

The double variety did not arise from it, in spite of every effort.

---

[1] The fluctuating variability of the semi-latent character in *Ranunculus bulbosus semiplenus* seems to cover a much wider range of forms than in *Trifolium*. There the extremes are 3 and 7 leaflets; in the buttercup they are 5 and 31 and perhaps more petals. From this it does not, however, follow that the variation is greater in the one case than in the other, but only that the variation is expressed by a larger number of divisions in the latter case, i. e., that there are more scale characters in the curve.

I conclude, therefore, that in this case the half race cannot be transformed into the double race by simple selection; but only by an internal change—a mutation—the external causes of which are still unknown to us.

Proceeding now to the detailed description of my experiment, I begin with the half race in the conditions in which I originally found it in nature.

I found the half race growing in 1886 and 1887 in a sunny and sandy spot not far from Hilversum, where I have often seen it since. The bulbous buttercup grew there in abundance; most of the flowers were normal, but a considerable number had more than five petals. I shall refer to these latter for convenience of expression as pleiopetalous.

For several years I have recorded the flowers in that locality. I give the records of 1886 and 1887, each of which relate to 300-400 flowers. The data are given as percentages.

Number of petals	5	6	7	8	9	10	11	12	13	14
Flowers in 1886	91.5	5.5	1.2	0.6	0.6	0	0	0.3	0	0.3
Flowers in 1887	90	7	2	0	0.5	0	0.5	0	0	0

The two series[1] agree as closely as could be expected and the records for the other years also fall in line. The maximum of the curve (see Fig. 51 H) is over the normal number of petals; and from it the curve falls rapidly. It is a so-called half GALTON-curve. Flowers with less than five petals do not occur in this locality.

The great steepness of this curve is due to the fact that on many of the plants no pleiopetalous flowers were found on the days when the observations were made. But this does not mean that the half race is mixed with

[1] *Ueber halbe Galton-Curven als Zeichen discontinuirlicher Variation.* Ber. d. d. bot. Gesellschaft, Vol. XII, 1894, p. 197, where some of the series of figures given below can also be found.

a pure race with five petals only. For the plants in question were either weaklings, or exhibited pleiopetalous flowers on other days. I was often able to observe that on many plants six-petalled flowers occur on one day but not on another. The 6-7-petalled flowers are found from the beginning of the flowering period, but the higher figures do not occur till later, as is also known to be the case in other instances of double flowers.

In 1887 I moved some plants in which the abnormality was well developed, to my garden, where they flowered again in the following summer and set seed. These plants constitute the first generation of my experiment. Since then I have sown seed every year, but only part of the plants, sometimes one-half, sometimes two-thirds, produced flowering stems in the first year, and I have always confined my attention to these, throwing away those which did not bloom during the summer. I have sometimes kept some of the best examples of the half race through the winter for secondary experiments, but I shall return to these later on.

During the period 1889-1892 the second to the fifth generation of the half race were grown in this manner, the extent of the cultures being gradually increased. I always harvested my seed from the most abnormal individuals, which I selected by simply cutting off the flowers with five petals from all the plants. The numbers of these on the individual plants were recorded in some years but not in others. Pollination was left to the bees, but no definite effects of cross-fertilization have been traceable in the results of the experiments.

The first two years of the experiment (1889 and 1890) need only a brief reference. Plants without pleiopetalous flowers or with only very few, were removed

Ranunculus Bulbosus Semiplenus.

as soon as possible, or were deprived of their flowers; of the rest, only the seeds of flowers with six and more petals were saved. But this process is not one of selection, as will be shown by means of some special experiments which were instituted later.

The result of selection could be seen in 1891 in the best examples of the half race, but in 1892 (the fifth generation of the culture) in nearly all the plants. The number of petals increased in every respect. The apex

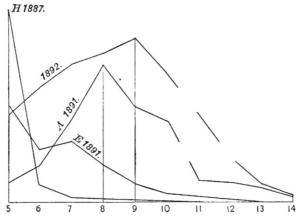

Fig. 51. *Ranunculus bulbosus semiplenus.* Experiments in selection during the period 1887-1892. H 1887, curve of the wild form; E 1891, curve of the abnormal plants in 1891; A 1891, curve of the selected seed-parents in 1891. 1892, curve of the whole crop in August 1892. The numbers at the base refer to the number of petals per flower.

of the curve shifted to 9 and 10 petals and even further; that is to say, the mean of the half race (9-10 petals) was separated by selection from that of the pure species (5 petals), a point which is rather striking because this was not effected in my experiment with *Trifolium incarnatum*. The course of the whole experiment is exhibited graphically in Fig. 51 which is composed of four curves. The first (H 1887) exhibits the countings given above, which were made in the original locality. Then there are two curves for 1891. In this year I had a culture of about four square meters from which I removed, about the beginning of August all plants which had not produced any pleiopetalous flowers as well as those which had not yet bloomed. For two weeks I counted every flower which appeared on the remaining specimens. They amounted to 128 and the various degrees of the anomaly were distributed over them as follows:

Number of petals	5	6	7	8	9	10	11	12	13
Number of flowers	45	24	28	17	8	4	0	1	1

The curve E 1891 in Fig. 51 is based on these figures. It is a half curve like the previous one, but without the steep apex. The disappearance of this is due partly to cultivation and to the repeated selection, but partly also to the fact that the individuals with the smallest number of pleiopetalous flowers had been removed before the counting took place.

After these data had been determined I carried out a still further selection. Several plants had not produced a single flower with more than seven petals. These were removed in the middle of August and observations on the rest were continued. There were 18 plants, all of which were selected for seed-parents as being the best representatives of the race. I counted all the flowers which bloomed from August 15 to 31, and obtained the following numbers:[1]

Number of petals	5	6	7	8	9	10	11	12	13	14
Number of flowers	9	17	39	64	45	37	15	9	6	2

[1] In the preliminary account of this experiment, mentioned above, this series is given separately in two curves, one of which refers to the most abnormal plant, the other to the rest.

Total 243. The curve (Fig. 51, A 1891) has become two-sided. It has no maximum at 5 but a very definitely pronounced one at 8. It is composed of observations made on 18 plants which differ little from one another. Individuals with these characters occurred neither in the original locality nor at the beginning of my experiment.

The sowing, in 1892, of the seeds of these selected individuals gave rise to above 300 plants which were coming into flower from July 21 to August 31. The curve for 1892 in Fig. 51 refers to these. Those which flowered later were examined separately and will be described afterwards. On all the flowers which opened between the dates named the petals were counted, and the numbers entered in my notebook separately for each plant. I give the totals, which relate to 4425 flowers. The numbers of petals were distributed over these as follows:

Petals	5	6	7	8	9	10	11	12	13	14	15	16–31
Flowers	409	532	638	690	764	599	414	212	80	29	18	20

The curve which is now an index of the degree of development of the *whole* race, agrees fairly closely with that of the selected seed-parents of the previous year (1891), as can be seen from a comparison of the curves A 1891 and 1892 in Fig. 51. The apex of the curve, however, has advanced a whole petal. There has been no regression as is the case in the selection of active characters, but a progression such as is usually characteristic of the selection of semi-latent characters.

The change in the right half of the curve is also important although not given in Fig. 51. It consists in the occurrence of more extreme variants. In the previous generation there were no flowers with more than 14 petals. Now there are 38, distributed as follows:

Petals	15	16	17	18	19	20	21	22	23	31
Number of flowers	18	8	5	2	1	1	1	0	1	1

It should be noted, however, that they were found amongst a group of 4425, and therefore only amount to about 1% (0.86%). But as not a single one was found in 1891 amongst 243 flowers, a genuine, although only a slight, advance has taken place.

The great majority of the 295 plants which were flowering in August and formed the 1892 culture, had

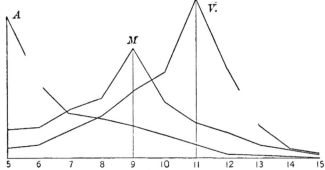

Fg. 52. *Ranunculus bulbosus semiplenus.* Composition of the fifth generation in 1892. A, the curve of some "atavists"; M, the curve of intermediate individuals; V, the group of extreme variants. The figures refer to the number of petals per flower.

individual curves whose maximum was at 9. But amongst their number were variants and extreme variants also. On the one hand there were "atavists" with a maximum on the ordinate of five petals, i. e., with a one-sided curve, as in those from the original locality; on the other hand there were variants on the plus side which bore on the average eleven petals per flower. In one case even a mean of 13 petals was reached. These curves were two-sided, and not, as in the five-leaved race of the red clover, inversely one-sided. But we are dealing here merely with a cumulative effect within a half race and not with an isolated, fully developed race. I have chosen a number of variants from the two categories, have added up the number of their petals, and obtained the following data:

Petals:				5	6	7	8	9	10	11	12	13	14	15	16	17	18	23
Number of flowers			A:	66	34	21	18	15	11	7	2	0	0	1	0	0	0	0
"	"	"	M:	13	14	22	28	51	26	16	12	6	4	2	1	0	1	0
"	"	"	V:	9	11	26	39	62	79	148	84	30	8	4	3	2	1	1

These data are exhibited graphically in Fig. 52. They relate to three small groups of individuals, chosen in such a way that the curves of the individual plants did not exhibit any considerable deviations from the mean of the group. A is the curve of the twelve atavists extracted from the whole series of observations; the maxima of all their curves were at 5 petals. M is a curve representing ten plants grown from the seeds of a single seed-parent. V is the curve composed of all the plants the apex of whose individual curves lay above 10. There were 22 of them altogether; the apices of their curves were at 11, with three exceptions which were at 12 and 13, but these curves did not exercise any marked effect on the shape of the average curve of the whole group.

If Fig. 51 and Fig. 52 are compared a most remarkable similarity will be observed. The latter figure gives the composition of my race at the end of a process of selection extending over five generations, the former relates to the separate stages in this process. This mutual resemblance lies in the fact that the original half curve (Fig. 51, H 1887) continues to appear throughout the process, although it is a little flattened; it occurs in 1891 (Fig. 51, E 1891) and also in 1892 (Fig. 52 A). Atavistic fluctuation therefore is still exhibited by my race in spite of the repeated selection.[1] The curve M has a more normal shape than the corresponding curve "1892" in Fig. 51; which is obviously due to the fact that the former represents a homogeneous group whilst the latter is a composite curve embracing all the groups of this culture. Curve V is related to curve M as it would be if we were dealing with ordinary fluctuating variability; it is simply shifted to one side.[2]

It seems obvious that the race could still be improved by sowing the seeds of those seed-parents the apices of whose curves are at 11, i. e., that these curves could be shifted still further to the right. I have made some such sowings since 1892, but only on a small scale and not without interruption. They were not intended as a continuation of the experiment. The number of petals per flower increased slightly, but the type itself was not essentially altered. I especially never saw a trace of anything like the origin of a double flower.

In order to find out whether there was any likelihood that the type of my race would in the near future manifest an improvement I made the following calculation. The 295 plants of which the culture of 1892 consisted, arose from the seeds of 21 seed-parents. I selected the ten best of these parental groups and plotted the curves for all the offspring of each seed-parent. The curves proved to differ very little from one another. Their apices all fell over nine petals, with one exception, which was over ten. I should say that in making the calculation I have left the groups which contained less than 300

---

[1] Whereas as a result of the selection of active characters the whole curve is shifted; see Vol. I, p. 73, Fig. 18, and the third part of the first volume.

[2] See Vol. I. Fig. 116, on page 536.

flowers out of consideration. But even these did not manifest any notable differences. I then compared these ten curves with the part-curves determined from the parents themselves (i. e., with the number of flowers counted on the seed-parents) and found no correlation. As a matter of fact the seed-parent with the smallest number of pleiopetalous flowers had the offspring with the largest number. The following four curves of the offspring of four seed-parents are extracted from my records. Under M are given the numbers of petals in the seed-parents of 1891.

M	NUMBER OF PETALS PER FLOWER													
—	5	6	7	8	9	10	11	12	13	14	15	16	17	18-23 Totals
C 5—10	37	47	81	81	85	102	47	31	6	3	4	1	0	0 525
C 6—10	25	67	80	75	117	77	75	45	30	10	6	1	2	3 613
C 6—11	54	53	62	78	87	60	59	37	10	4	4	1	1	1 511
C 7—11	52	57	76	77	95	64	26	13	0	0	0	0	0	0 460

Another fact which points in the same direction is that the plant which seemed to be far the best in the summer of 1891, inasmuch as the apex of its curve was over 11-12 petals, had offspring whose character corresponded exactly with that of the whole culture of 1892. The improvement on the seed-parent therefore did not justify the expectation of a real advance.

For these reasons I then discontinued the experiment. It seemed to me that *the impossibility of raising the double race from my half race by simple selection was placed beyond all doubt.* This result could only be expected from a further mutation.

The extensive material afforded by these cultures has been utilized to find out how far the number of petals per flower in the half race is determined, apart from selection, by internal causes, and how far by external.

I found it to be dependent only to a very slight degree on the former but in a high degree on the latter.

The first question that presented itself was: To what principles of distribution does the number of petals on the individual plants correspond. Is this number determined by the situation of the flower or by external factors or by both? With regard to the position of the flower BRAUN in his *Verjüngung* distinguishes between a strengthening and a weakening system of branching. In the former the branches increase in strength, though often but slightly, with each new degree of division; in the latter each secondary branch is weaker than the branch on which it is borne. *Ranunculus bulbosus* belongs to the former category (see Fig. 49 on page 244). The main stem (E) is surpassed by the vigorous lateral shoots (S), which arise directly from the tuber, and these in their turn are excelled by their own (tertiary) branches (T in the figure). The same thing continues with further growth until ultimately the process is reversed and weaker branchlets are produced. The more vigorous a branch is, the larger and stronger, as a rule, will be its flower.

If we now compare the number of petals on the flowers of this half race, with their position on the branches of the various orders, we are often struck by an apparently definite correlation. But this is only due to the fact that such cases produce a more vivid impression than the opposite ones. When a detailed record is made the latter are found to be just as numerous as the former. In September 1892 I determined the number of petals, and the position on the plant, of 1197 flowers on 82 plants; and plotted a curve for each position. Here however I only give the means of the curves.

	FLOWERS	NUMBER OF PETALS	PER FLOWER
A. On the main stem:			
1. Terminal flower	75	697	9.3
2. Secondary flowers	221	2005	9.1
3. Tertiary "	134	1237	9.3
B. From the tuber:			
4. Secondary flowers	259	2419	9.3
5. Tertiary "	397	3716	9.4
6. Quaternary "	111	1014	9.1
	1197	11088	9.3

It is evident that the number of petals in the various groups is practically the same.[1]

Even the seeds of pleiopetalous flowers are by no means better. In gathering the harvest of 1891 I collected the seed on each plant in a separate bag with reference to the number of petals of the flowers. The flowers were labelled for this purpose at the time of flowering. In the culture of 1892, therefore, the plants were arranged in groups, first according to their seed-parents, and secondly according to the petal-number of the flowers from which the seed had been gathered. I then grouped all of the figures by the latter character and obtained the following result:

NUMBER OF PETALS IN THE FLOWERS WHICH FURNISHED THE SEED	MEAN NUMBER OF PETALS OF THE OFFSPRING[1]	NUMBER OF FLOWERS COUNTED
C 5-7	8.3	932
C 8	8.7	1072
C 9	8.5	1217
C 10-11	8.6	1420
C 12-14	8.7	919
Average	8.6	Total 5560

[1] In the weakening system of branching on the other hand the contrary seems to be the rule; so for instance in my cultures of *Saponaria officinalis* with 5-10 petals, in *Chrysanthemum segetum* (§ 18) etc.

[2] The mean is slightly lower here than in the previous table because that only refers to countings made in September. (See later.)

Here again there is no discernible correlation. I have obtained the same result in other years. From this we see that in this case at least selection must not be founded on the different flowers of a plant but on the individual plants. However, the possible influence of the various grades of branching independently of the number of petals remains to be investigated.

But whereas no internal causes were found which determined the pleiopetaly in the individual flowers, the external causes could be discovered the more readily. This character follows the general rule; for the higher the nutrition and the more favorable the environment the more petals are produced per flower. The following experiments and observations will prove this.

I shall first refer to an observation for which unfortunately I can give no numerical corroboration, but which may throw some light on the independence of the character of the flowers, of the order of branching. In the summer of 1892 when I examined all the flowers of my culture, and recorded the number of their petals twice a week, I was struck by the fact that the high numbers fell on particular days whilst on other days only low or intermediate numbers were observed. This would seem to indicate that during the development of the flowers in May and June pleiopetaly is influenced by weather conditions, in such a way that flowers which are in the susceptible period of their development during fine weather will produce more petals, quite independently of the order of the branch which bears them.

This conclusion is supported by another set of observations. In September 1892 the flowers, on the whole, produced more petals than they did in August of the same year. Or, to be more accurate, the number was greater on those plants which opened their first flower in September, than on those which had already begun to flower in July and August. The number of individuals of the former group was 77; they produced 1134 flowers during the period ending with the beginning of November, when I stopped recording. In the other group there were 295 plants which flowered, and they produced 4425 flowers. The distribution was as follows:[1]

Petals:	5	6	7	8	9	10	11	12	13	14	15	16–31
Oldest plants.	409	532	638	690	764	599	414	212	80	29	18	20
September plants:	40	52	126	165	204	215	177	104	35	8	4	0

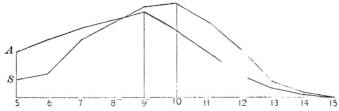

Fig. 53. *Ranunculus bulbosus semiplenus.* A, curve of the plants flowering in August; S, curve of those flowering in September. The figures at the base refer to the number of petals per flower.

These figures are exhibited graphically in Fig. 53; they have been reduced for convenience of comparison so that the numbers in the two groups are about the same. The apex of the curve of the early flowering plants is over the 9; it is the same curve which has already been given in Fig. 51 on page 249 for the year 1892. The other curve has its apex over the 10, and also remains above the other curve in the right half of its course.

The cause of this difference can only lie in the retarded germination. Either the seeds which germinate later are intrinsically more productive of pleiopetalous

---

[1] See above, p. 252 and Fig. 51 (1892).

flowers[1] (like the small, late-germinating seeds of the crimson clover), or germination in the height of the summer in better and particularly in warmer weather favors development in such a way that the flowers are richer in petals; for the plants which flowered in July and August, germinated for the largest part during the cold and unfavorable weather experienced in May shortly after they had been sown.

I first made an experiment to determine the influence of nutrition on pleiopetaly in 1890. I had wintered the selected plants of 1889, and in March transplanted half of them on a bed of pure sand, and the other half on a bed of ordinary garden soil. Only two-thirds (i. e., 12) of the plants of the former lot flowered, whilst all of the latter did. On the sandy bed I counted the petals of all the flowers and about twice their number on the control bed by simply picking off all the open flowers on alternate days. I examined in all 75 and 147 flowers respectively. The following is the result reckoned in percentages for convenience of comparison:

Number of petals:	5	6	7	8	9	10
On the bed of sand·	73	23	4	0	0	0
On garden soil:	53	26	14	5	1	1

The plants on the better soil produce distinctly fewer five-petalled and more 7-10 petalled flowers. It is perhaps permissible to conclude from this that the steep drop of the curve from the wild locality, where the soil was sandy is, to a large extent at any rate, due to low nutrition. For presumably the same plants would exhibit a higher degree of pleiopetaly if grown on better soil and

[1] With regard to this, it would be of great interest to find out in this and other plants the degree of development of the anomaly in such individuals which do not germinate until two or three years after the sowing of the seed

so give rise to a less steep curve, just as in the experiment under consideration.

I made a corresponding experiment in the summer of 1891, on the effect of manured and unmanured garden soil, with the race which was by that time considerably improved (Fig. 51 and page 250). The manuring was done with guano; the two beds lay next to one another and were of the same size. On each was sown half of the harvest of several plants which had been very productive of pleiopetalous flowers in 1890. In the course of the summer 159 flowers on the unmanured bed opened and were recorded and 376 on the manured. The relation between these two numbers is the best measure of the effect of the manure. The results, reckoned in percentages, are as follows:

Petals:	5	6	7	8	9	10	11	12	13	14
Without manure:	12	15	25	21	12	10	3	1	1	0
With guano:	14	15	17	21	14	9	4	3	2	1

Without manure the apex of the curve was over the 7 and there were very few flowers with more than eleven petals: with manure the apex was over the 8, and there were distinctly more pleiopetalous flowers.

In both the above experiments the control material consisted of other individuals than those used for the experiment itself. It is possible, however, to subject the same plant alternately to favorable and unfavorable influences, and when this is done the same result is obtained as in the previous cases. With this object I transplanted a series of the best plants of 1892 to a very dry bed in the spring of 1893. I left them there, and did not water them although the weather was continually dry. They suffered visibly under this treatment and some of them even produced fewer flowers than in the previous summer. I have a record, which has been already alluded to, of the number of petals of all the flowers of each of the plants of 1892; these were recorded in the same way in 1893. But I only give here the mean numbers of petals per flower.

PLANTS	NUMBER OF FLOWERS		MEAN NUMBER OF PETALS PER FLOWER		DIFFERENCES
	1892	1893	1892	1893	
No. 1	25	14	11	9	2
No. 2	43	19	9	5	4
No. 3	9	14	10	6	4
No. 4	44	5	8	5	3
No. 5	12	18	10	8	2
No. 6	16	21	9	8	1

The anomaly was thus diminished on every single plant as the result of transplanting to dry earth.

The results of all these experiments prove that the production of more than five petals in a flower is independent of the position of this flower on the plant, but on the other hand is dependent in a high degree on the external conditions under which the particular flower passes its early stages, i. e., the most susceptible period of its existence. The number of petals varies directly with the vigor of the plant, the moisture and richness of the soil, the warmness of the weather and even the amount of sunshine during this susceptible period.

*C*ultivation in the garden is therefore bound to convert the steep half curve of the wild locality (Fig. 51 for 1887) into a flatter one which will gradually extend to higher numbers of petals and will ultimately develop a new apex.

This process, however, takes place more conveniently and more certainly, if the cultivation is combined with selection (see the same figure). The latter process picks out the plants which manifest the anomaly most abundantly and most strongly; these must, however, according to the facts given, as a rule, be the best nourished ones, i. e., the most favored by their environment. For on the same bed, even if it has been uniformly prepared with the greatest care, the conditions under which neighboring plants grow are often very different. One seed may germinate in a place in which moisture is better retained; another may germinate in almost dry soil. Some germinate on warm and fine days and are in consequence ahead of their less favored brothers for their whole lives; and so on.[1] And so it is that the several plants from seeds of the same seed-parent sown on the same day and on the same bed, are necessarily exposed to diverse conditions of life. Amongst them selection picks out the best and therefore, at least as a general rule, the most highly nourished ones. Selection, so to speak, only precipitates the operation of these external factors; as we have pointed out before in connection with *Papaver somniferum polycephalum*.[2]

Selection and cultivation have, therefore, worked in the same direction in my experiment for four generations. They have about doubled the mean number of petals per flower, having brought it, in fact, to 9-10; they have produced, amongst several hundred plants and several thousand flowers, no more than three flowers with more than twenty petals (C 21, C 23 and C 31), i. e., not essentially more than would be expected according to QUETELET's law from the actual mean and the amplitude of variation. These flowers occurred perfectly fortuitously on plants which were not particularly favored otherwise, the means of the curves being only 10 for each

[1] See Vol. I, p. 138.   [2] See Vol. I, p. 140.

of the three plants. We are thus justified in concluding that by the selection of these plants as seed-parents the mean of the race might further be slightly improved during the course of some years, but that these extreme variants afforded no more hope than did the others, of the attainment of the double race.

Cultivation and selection cooperate in the direction of the desired end; they lead the half race measurably further on this line, but it is not through them that the object can be attained. The half race remains a half race, in spite of every effort and care, the semi-latent character expresses itself oftener and oftener, but it does not succeed in becoming the equal of the normal active characters, i. e., in constituting the mean character of a new race, independent of the continuance of selection and favorable cultural conditions.

To arrive at this result a process of an entirely different nature is evidently required. According to the current theory of selection the goal would be reached if the experiment could be continued for tens or hundreds of years. But the course of the experiment we have described does not support this view; it shows, on the contrary, that all that can ever be gained by nutrition and selection has already been secured in these five generations. The actual result is the production of an elite race which has a mean number of 9 petals in the flowers, under the favorable conditions of culture which obtained; and gives rise, according to environmental conditions, on the one hand, to better variants (with a mean of about 11-13, or perhaps a few more, petals) while, on the other, it throws off atavists with a half GALTON curve (see Fig. 52 on page 252).

It is my opinion, however, that if the culture of the

half race were still continued, the double race would some day appear quite suddenly, and that it would then, after a short but sufficient isolation, persist as a constant, though highly variable, race.[1]

§ 24. VARIEGATED LEAVES.

Variegated plants have long been great favorites in the garden, and their great instability has contributed largely to the development of the horticultural conception of a variety, for the variations in their color pattern are practically unlimited. Hardly any two leaves are alike, and many species have a whole series of dappled and flecked varieties. They also possess the striking property of continually and conspicuously reverting to the species to which they belong. Such reversions occur either amongst seedlings or as bud-variations, and since on shrubs and trees these latter often remain for many years and not rarely in more than one instance on the same plant, they can be seen by every one. In this way these bud-variations have come to be regarded as a sufficient proof of the idea that varieties are derivative and unstable structures, which always tend to revert to their parent species.

Especially in the first half of the eighteenth century were plants with speckled and striped leaves very much sought after.[2] About that time the well-known English gardener THOMAS FAIRCHILD possessed more than one hundred varieties of them in his garden, and afterwards SCHLECHTENDAHL published a list from which it can be seen that variegation is distributed over the whole

---

[1] I. e., as an eversporting variety with a wide amplitude of variation which however would not alter in the course of the generations.

[2] MEYEN, *Pflanzen-Pathologie*, 1841, p 282.

vegetable kingdom and occurs in all the larger groups and especially in most families of flowering plants.[1]

At that time some of the most widely cultivated forms were the ribbon grass, *Phragmites arundinacea variegata,* and the variegated holly, *Ilex Aquifolium.* Both are still much grown in gardens, the ribbon grass being relatively uniform, the holly highly variable. Of the latter there is a variety with white-edged leaves, besides the ordinary one with flecked leaves. *Phragmites* is different in many respects from genuine variegated plants and is much less variable in its character. The *Ilex,* however, is highly variable and often bears green shoots which may soon supersede the others on account of the greater facility with which they can obtain nourishment. A fine variegated bush of this species, or of any other, may become entirely green, whenever the green branches are not cut away every year. Thus it is probable that many specimens of the holly, which are now quite green, were originally variegated and were bought and planted as such. On closer examination we often find on them an occasional variegated twig which proves the correctness of this supposition. This is also the case with the horse chestnut, of which many older trees still living were planted at a time when the variegated variety was in special favor. Since then their foliage has become green and their original character is no longer seen. But an occasional checkered branch, or even the numerous small twigs with white leaves along the main stem, betrays the original variegated condition of the specimen. In the same way many cases of single variegated twigs on green bushes and trees are not to be regarded as the indi-

[1] SCHLECHTENDAHL, *Linnaea,* 1830, V, p. 494. Very little seems to be known about variegated mosses and thallophytes.

cation of something new but as a reminiscence of times long past when these varieties were in general favor.

Variegation is classified under several headings. In the first place there are the yellow and the white varieties. In the former the chlorophyll is only insufficiently produced, but in the latter even the xanthophyll or carotin is lacking;[1] and a more or less abortive development of the chloroplasts is usually correlated with the absence of these pigments.[2]

Further we distinguish marginate, flecked and striped sorts. The former seem to constitute a variety for themselves and are much rarer than the latter; they appear to be good races, that is, to be as constant as any ordinary garden variety, but I shall have little to say about them in this part. The most characteristic and best known example of them is the white bordered holly to which we have already referred.[3]

Whether a plant is flecked or striped depends as a rule on the mode of venation of the leaves. Many variegated monocotyledons have striped leaves (*Agave, Convallaria majalis, Phormium tenax, Tradescantia repens,* etc.) whereas the dicotyledons are usually flecked or streaked.

The incomplete development of the chlorophyll ob-

[1] See T. TAMMES, U*eber Carotin*, Flora, 1900.

[2] For further information on this point see the elaborate anatomical studies of A. ZIMMERMANN, U*eber die Chromatophoren in panachirten Blättern*, in Beiträge zur Morphologie und Physiologie der Pflanzenzelle, Heft II, 1891, pp. 81-111, and Ber. d. d. bot. Ges. VIII, 1890, p. 95. Also H. TIMPE, *Beiträge zur Kenntniss der Panachirung*, Inaug.-Diss., Göttingen, 1900.

[3] Marginate forms are commonly supposed by gardeners to be more stable than flecked ones. This fact was noted by MORREN in 1865. (*Hérédité de la panachure*, Bull. Acad. roy. Belg., T. XIX, 2d series, p. 225). VERLOT however maintains the opposite opinion (*Des Variétés*, 1865, p. 74). For information relating to variegated varieties of *Ilex* see FOCKE, *Abh. d. Naturw. Ver. zu Bremen*, Vol. V, pp. 401-404.

viously results in an insufficient assimilation of carbonic acid gas. Thus the variegated parts grow less vigorously and are less resistant than the corresponding green ones. The *Cyperus alternifolius* of our greenhouses, the *Aspidistra elatior* and a number of other favorite varieties show this clearly. *Arundo donax* often attains a height of three or more meters whereas its striped variety is scarcely half that height. Leaves of the variegated *Aspidistra* very often have one of their longitudinal halves green, but the other colorless. In such cases the leaf is distorted owing to the insufficient growth of the colorless half. The same thing happens in many other cases.

The yellow leaves and parts of leaves, however, are not entirely without the green coloring matter, nor wholly without the power of assimilation. Most of them give a green extract when put into alcohol, and if examined under the microscope patches of green tissue can be found here and there, especially near the veins. The power to sustain life, however, is often lacking and the leaves die shortly after their growth is completed. Therefore, a high degree of the anomaly is not in favor, because the plants which possess it often become disfigured by the edges of their leaves turning brown. Many plants in which the variegation has gone too far die in their very early stages, while others have not sufficient strength to flower and bear seed. This latter circumstance is of special interest because it follows that plants with a high degree of variegation as a rule can have no part in the propagation of the variety.[1] In the opinion of some

[1] It is perhaps scarcely necessary to state that these remarks do not apply to brown and purple leaves or those with red spots. For information on this point see STAHL's excellent article *Ueber bunte Laubblätter,* Ann. Jard. Bot. Buitenzorg., Vol. XIII, Pt. 2, 1896, p. 137.

authors another fact is connected with this, viz., that varieties which have both variegated leaves and double flowers are much rarer than would have been expected from the prevalence of these two anomalies in horticulture.[1]

In variegated plants, as is well known, not only the leaves are flecked. Their stems and calices are also often variegated, and the same is true of the fruits (pears, grapes, the siliquae of cabbage, *Barbarea vulgaris, Cheiranthus Cheiri, Alyssum maritimum, Acer, Ilex, Aegopodium, Ligusticum,* etc.).[2] I have also sometimes found galls on variegated oaks to be variegated, especially in the case of the beautiful orbicular galls of *Cynips Kollari.*

I shall now proceed to the important question of the inheritance of this abnormality or the degree of fixing as it is usually called. As already stated I shall exclude from consideration the white-flecked[3] and the marginate forms of variegation, and shall confine myself to the ordinary cases of yellow variegated leaves. I shall give the numerical proofs of my conclusions later, and shall now proceed to deal with the question whether variegated sorts are half races or intermediate races (see Chapter II of this part).

In my opinion the great majority of the variegated garden varieties are intermediate races, as for instance *Barbarea vulgaris*; whereas wild plants which occasionally present this character represent half races. Their

---

[1] B. VERLOT, *Sur la production et la fixation des variétés dans les plantes d'ornement*, 1865, p. 75. Also MORREN, *Hérédité de la panachure, loc. cit.,* p. 226.

[2] MORREN, *loc. cit.,* p. 233.

[3] I have not myself made any observations on this phenomenon (*Albicatio, Albinismus*) and the published records of it are very scanty. The fine white-variegated *Humulus japonicus variegatus* would be well worth experimenting with.

multiformity and instability corroborate this view. It is only the commonness of variegated sorts and the great interest which attaches to them which brings them to be regarded as analogous to the best constant varieties. Moreover this view is supported by the general opinion that a complete development of the yellow color would characterise the supposed constant variety, but that it would at the same time of necessity lead to the destruction of the plants. In this conception variegation is regarded as an incomplete anomaly whose complete condition would involve its own destruction; but this view is incorrect.[1] Complete yellow varieties are not only possible and capable of existence but actually well known in horticulture, although the number of such forms is small. Instances can be found in seedsmen's catalogues; e. g., *Sambucus nigra aurea* and *Fraxinus excelsior aurea*, also the *aurea* varieties of *Chrysanthemum carinatum, Mirabilis Jalapa, Scabiosa atropurpurea, Humulus japonicus* (*lutescens*) etc. These plants, so far as I know, are all either yellowish-green or golden-yellow.[2] They also appear to be very constant and never or very seldom to revert to the green type. I have made a number of experimental sowings on a large scale of the seeds of the ordinary golden-yellow variety of *Chrysanthemum Parthenium*[3] (*Matricaria eximia nana compacta foliis aureis* Hort.) and did not find amongst the many hundred examples a single atavist; neither green nor variegated seedlings occurred. But amongst other commercial seeds I have not found so great a degree of purity, the admixture

[1] See § 3 of this part (pp. 18-26).

[2] I have not grown all the above forms myself; and it should be noticed that the name *aureus* does not always relate to uniformly colored sorts, e. g., *Agave striata aurea*.

[3] VILMORIN, *Blumengärtnerei*, Vol. II, p. 509.

of green plants, however, not being larger than might as a rule be expected from commercial seeds. For instance, *Stellaria graminea aurea* gave only 28% and *Myosotis alpestris compacta foliis aureis* only 3% of green seedlings. But even in these cultures there were no variegated plants.

The fact that the *aurea* varieties give a green extract in alcohol and contain sufficient chlorophyll for their nutrition does not need special mention.

Fig. 54. *Thymus Serpyllum*. The ordinary Thyme; a plant with a variegated branch B.

The *aurea* varieties and the yellow variegated sorts owe their character to the masking of the green pigment by the yellow which is developed in the former case all over the leaf, and in the latter only in certain tracts. The majority of variegated plants are analogous to those numerous half races which manifest their anomaly (which may be doubling, pitcher formation, the production of quadrifoliate leaves, etc.) only in isolated organs and parts of organs. Some sorts I regard as analogous to the double varieties, whilst the *aurea* varieties are probably just as constant as the *Varietates discoideae* and as the best elementary species.

The very general occurrence of variegated plants points to the conclusion that the latent capacity for variegation is widely distributed throughout the vegetable kingdom. Moreover the fact that branches and whole plants with this character are met with every year in new species both in the garden and the field points in the same direction. In this connection I may mention the fact that forms with white variegated or white- or yellow-edged leaves occur only rarely. I observed an instance of the latter in a wild specimen of *Oenothera Lamarckiana* (1887, see. *Vol.* I, p. 480) and of the former I found specimens in *Spiraea Ulmaria, Calluna vulgaris, Trifolium pratense, Lychnis diurna* in 1886 and 1887 in the neighborhood of Hilversum. In the above mentioned years I found yellow variegated plants of *Plantago major, Phalaris arundinacea, Rhinanthus major, Erica Tetralix, Urtica urens, Hypericum perforatum, Trifolium pratense, Hieracium Pilosella, Rubus fruticosus, Polygonum Convolvulus* and *Geum urbanum.* In 1869 I found a beautiful variegated specimen of *Arnica montana* in the Thuringian Forest and later one of *Plantago lanceolata* in Saxon Switzerland, and one of *Thymus Serpyllum* near Wyk aan Zee in Holland (Fig. 54), and I have since frequently found occasional variegated specimens of other wild species. In the same way they appeared in my own cultures where there can be no question but that they have been preceded by many generations of purely green ancestors; so for instance in *Chrysanthe-*

mum segetum, Antirrhinum majus, Polygonum Fagopyrum, Linaria vulgaris, Silene noctiflora etc.

The large scale on which I have conducted my experiments with *Oenothera Lamarckiana* has enabled me to watch the origin of variegated forms in that species more closely. Here they appear almost every year from green ancestors, and in the most widely different experimental families and elementary species.[1] Instances of it I found in the main *Lamarckiana* families, first in the original wild locality, then in 1889, 1890, 1892, 1895, 1898 and 1899 in my cultures, arising from series of seed-parents which were in every case green plants; also in *O. rubrinervis* in 1891, 1893 and 1894; in *O. laevifolia* in 1891, 1894 and 1899; in *O. sublinearis* in 1896; in *O. lata* in 1890 and 1899; in *O. nanella* in 1890, 1896 and 1899; in *O. scintillans* in 1898 and so on; also from the crosses *O. lata* × *O. cruciata* and *O. Lamarckiana* × *O. Lamarckiana cruciata* and others. In 1899 only eight variegated plants arose in my whole cultures which consisted of over five thousand plants of *Oenothera*, that is, between 0.1 and 0.2%. But in the field the anomaly was evidently much rarer.

One of the most striking phenomena presented by variegated plants is the so-called twig or bud-variation. From a bud a branch arises which is unlike the whole of the rest of the plant in the character of its variegation, and in this case both variegated plants bearing green twigs occur and conversely plants which have hitherto been green may bear stray variegated branches. In both cases a latent potentiality is manifested.

The appearance of green branches on variegated plants is generally regarded as a case of atavism, that is

[1] See also Vol. I, p. 480.

reversion to the parental form. It is especially common on woody species and in shrubs. *Evonymus japonica, Quercus pedunculata, Weigelia amabilis, Cornus sanguinea* and many others afford well-known examples. Others are found amongst perennials and perhaps best of all in *Arabis alpina*. I may cite as further instances partly from the literature on the subject and partly from my own observations: *Castanea vesca, Kerria japonica, Aesculus Hippocastanum, Yucca pendula aurea, Ulmus campestris, Zea Mays, Rubus fruticosus* and so on.

The green branches can obtain nutrition better than the variegated ones. Therefore they grow more vigorously and become stronger during the course of years, and very often overgrow the others. As a rule all their leaves and branches are pure green, and they look as if they had entirely lost the capacity for variegation. But this is not the case, for sometimes we see single variegated twigs on these green branches. *Arabis alpina* is especially instructive in this connection, for it often gives rise to variations from its buds, and since it is easy to separate these and cultivate them further. Analogous cases of this double reversion, as it may be called, were observed by me in 1893 in *Castanea vesca variegata* and *Kerria japonica variegata* which bore a little variegated twig on a green branch; and the same has been observed in other cases.

The deficient nutrition frequently makes the variegated leaves smaller than the green ones. If the pigment is mainly absent in the margin of the leaves this becomes too small for the middle area and the whole leaf becomes crumpled. A unilateral checking of the growth leads to a corresponding bending. It is due to these circumstances that the habitus of variegated plants is often so different from that of the typical form, but as soon as reversion occurs through bud-variation all these secondary characters are dispersed at once, the green leaves becoming flattened out, assuming the normal form, and often attaining twice the size of the variegated ones. In this way the reverted branch easily strikes the eye. I observed this most beautifully in *Castanea vesca* and *Ulmus campestris,* but *Kerria japonica* and many other species show it as well.

The question which buds are most likely to give rise to atavistic branches has been much discussed, and the general opinion seems to be that the rhizome and the adventitious buds on the roots are most prone to reversion. Thus *Glechoma hederaceum variegatum* often produces green runners[1] whereas the variegated *Tussilago Farfara* breeds true from its runners. For the last ten years I have had a variegated plant of *Rubus fruticosus* which has produced both green and variegated plants from its radical buds in proportions which vary according to conditions, and to the year. It seems to me probable that the weaker buds are most likely to give rise to atavists; but since this results in the production of green branches which grow much more vigorously than the neighboring variegated ones, it is not easy to decide this point.[2]

Variegated branches on green plants are almost as common. It is the general idea amongst gardeners that the numerous variegated varieties of woody plants have, with few exceptions, arisen in this way. One of these

---

[1] VERLOT, *loc. cit.,* p. 78.

[2] In papers on this subject we often come across an expression of the opinion that it is the strongest branches which become green; but this view, no doubt, is largely due to a misapprehension of the relation between the cause and its effect, as explained in the text.

exceptions is *Weigelia amabilis variegata* which was raised by VAN HOUTTE[1] from the seed of the green variety; another is the variegated grape raised by KNIGHT.[2] In many cases a record of the original discovery has been preserved. Thus WOLFF[3] states that he found a variegated branch on a bush of *Spiraea opulifolia*; the leaves were whitish green with a sulphur yellow margin, marked here and there with dark green flecks. The new form was easily multiplied by cuttings and appeared on the market as *Spiraea opulifolia heterophylla fol. aur. marg.*

In nature bud-variations of this kind are also occasionally found, and it seems that this is almost the only bud-variation which is met with amongst wild plants, for usually this phenomenon is observed as a case of reversion on cultivated varieties or on hybrids. I myself have found very beautiful and large variegated twigs in *Quercus pedunculata, Betula alba* and *Fagus sylvatica* in the forests near Hilversum; in each case there was one large variegated branch on an otherwise green tree amongst hundreds or even thousands of perfectly green individuals.

On the variegated branches the variegation often appears unilaterally. The anomaly is developed laterally or unilaterally, or to use a more accurate expression, sectorially. For in the vertical projection of the branch there is usually only one sector which is variegated; sometimes one-half, sometimes one-third, and often even a smaller section of the circumference of the stem being affected.[4] The sectorial variation behaves in the same

[1] VERLOT, *loc. cit.*, p. 74.
[2] DE CANDOLLE, *Physiologie*, II, p. 734.
[3] *Gartenflora*, Vol. XXXIX, 1890, p. 9.
[4] A study of sectorial variation in relation to the divisions in the

way in this case as in that of the striped flowers. The buds in the axils of the leaves on the variegated sectors usually produce variegated branches, but those of the green sectors green ones. Breeders take this fact into consideration in the choice of buds for use in the multiplication of variegated forms, as we have already seen.[1] It seems that the bud-variations, that is to say both the progressive (producing variegation) and the retrogressive or atavistic ones, are generally the result of a preceding sectorial variation. But in most cases all traces of the latter are soon lost. In *Quercus pedunculata* I observed, as I have already mentioned, a variegated twig on a green bush. In this case the variation extended on to the bark and the color of the branches of the preceding years could thus still be recognized. The main branch proved to be unilaterally variegated and the variegated twig arose from this side. The branches at the other side were green (Hilversum 1886). When the leaves are arranged in two rows as in *Castanea vesca, Ulmus campestris* etc. the leaves on one side of the branch may be variegated and on the other, green. In such cases I found the lateral twigs on the older parts on the green side to be entirely green and those on the variegated side entirely variegated; but I might repeat here that there is still a tendency in the green branches, even if only a slight one, to produce variegated leaves. The contrast between green and variegated is therefore not nearly so great as would appear at first sight.

We have now to consider the question of the influence of external conditions on the degree of variega-

apical cells would be of great interest, especially in the Conifers and vascular Cryptogams (e. g., *Juniperus, Adianthum, Selaginella* etc.).

[1] See SALTER's method, Vol. I, Part I, p. 147.

tion.[1] On this point the literature is rich in contradictory information. This contradiction is probably mainly due to the fact already mentioned that the green parts are so very much stronger than the variegated ones. This strikes the eye so forcibly that the idea easily arises that the strongest parts of the variegated plants are most liable to become green and the weakest branches of green plants most likely to become variegated. In my opinion, however, this conclusion is incorrect. The relative vigor is determined by the anomaly, but from this it by no means follows that the anomaly, in its turn, is determined by it. So far as my experience goes the reverse is the case, and variegation forms no exception to the general rule for semi-latent characters, that favorable conditions increase the intensity of the anomaly.

The best instance that I know is furnished by the variegated horse-radish (*Cochlearia Armoracea variegata*), which with unfavorable treatment is almost green, but under glass or in a cold frame may even become entirely white. Plants growing in the open in a sunny position are often beautifully variegated, whereas in shady positions they are a much darker green. The same is true according to SCHLECHTENDAHL of *Plectogyne variegata* on the leaves of which a greater or lesser number of white stripes can be induced at will by merely transplanting it.[2] *Fragaria indica variegata* is a favorite plant for hanging-baskets. If one wishes it to be nicely variegated it must be planted in good dry soil, not too loamy or calcareous.[3] The same is true of the striped sorts of the ordinary strawberries, in which, as VER-

---

[1] E. LAURENT, *Sur l'origine des variétés panachées*. Bull. Soc. R. Bot. Belgique, Vol. XXXIX, 1900, pp. 6-9.

[2] *Bot. Zeitung*, 1855, p. 558.

[3] VILMORIN-ANDRIEUX, *Fleurs de pleine terre*, p. 408.

LOT[1] says, "*La panachure peut s'obtenir pour ainsi dire à volonté,*" by merely growing them in a dry position. A dry position is however at the same time as a rule a sunny one, whereas a damp one, as a rule, is shaded. Experiments which I have conducted with these and several other variegated sorts of various species, in order to determine the influence of daily and profuse watering in full sunlight on variegation, have been without any positive result. On the other hand I succeeded with *Tradescantia repens* in controlling the proportion of yellow and green stripes. In this experiment I cultivated the plants in pots and simply removed the pots to better or less well-lighted parts of the greenhouse without altering the soil or the amount of water I gave them. The more intense the light the more variegated were the new leaves that were formed.[2]

On variegated shrubs we often see that in the better lighted parts variegation is more intense and in the shaded ones less pronounced. Even variegated conifers such as the *Juniperus,* may show this, and it is well known to be the case with *Sambucus nigra.* The variegation in myrtles with striped leaves is also dependent on nutrition;[3] and various authors and gardeners hold that the soil and position exert a more or less important influence on the degree of variegation.[4] *Pelargonium zonale, Convallaria majalis, Mentha aquatica, Phalaris arundinacea, Phlox decussata* and others are given as instances.[5] Such

---

[1] VERLOT, *loc. cit.,* p. 76.

[2] For facts relating to the influence of galls on variegation in *Eupatorium cannabinum* see Vol. I, p. 407.

[3] MEYEN, *Pflanzen-Pathologie,* p. 287.

[4] As for instance SALTER, quoted in DARWIN, *Variations,* II, pp. 263-264.

[5] DARWIN, *loc. cit.,* I, p. 390; II, p. 263.

plants are often entirely green during years of improper treatment, but with due care can be restored to the variegated condition.[1]

The degree of variegation is not only dependent upon the conditions of growth but also on the time of the year. If we look closely at variegated plants in green-houses we shall see that the branches which arise in summer are beautifully flecked, whereas those which arise in winter, when they get less light, are much greener and sometimes even quite green. This at least is true in our climate; but one must remember that the leaves formed in summer remain on the shrubs through the winter, and undergo no further change in their variegation. Therefore it is not the general appearance which is dependent on the time of the year. *Quercus pedunculata argenteopicta* is green in spring, but becomes white or variegated later on.[2] Young plants are often still green in spring even though later they may become variegated, as for instance, *Symphytum, Barbarea vulgaris*, etc.[3] I have observed in a culture of *Geum urbanum,* which I have kept up during several years, that the variegated specimens gradually develop green leaves in the autumn and lose the variegated ones. In winter they were almost completely green, but as soon as life awoke in the spring they began to develop flecked leaves again, and throughout the whole summer they were fully variegated. They behaved in this way throughout every winter of their life. On the other hand a variety of the ornamental curly cale with yellow-veined leaves is beautifully varie-

---

[1] VERLOT, *loc. cit.,* p. 75.

[2] L. BEISSNER, *Knospenvariation,* Mitth. d. deutsch. Dendrolog. Gesellsch., No. 4, 1895.

[3] VERLOT, *loc. cit.,* p. 76.

gated in late autumn and winter but becomes completely green in summer.[1]

In all these observations there was no question of bud-variation. Of the causes of this phenomenon little is known. On the other hand it is generally recognized that if resting buds on variegated plants are allowed to grow vigorously they often develop into completely white or yellow so-called chlorotic branches;—I mean those buds which on green sorts would develop into strong succulent shoots under similar inducements, but the chlorosis soon puts an end to this development. Adventitious buds which arise from the stem a little above the soil frequently give rise to chlorotic branches, either after the stems have been cut down or when the leaves have been eaten by snails, or from some other cause. *Aesculus Hippocastanum* is the best known example, so also are *Evonymus japonicus, Pelargonium zonale, Azalea japonica, Aucuba japonica, Ilex Aquifolium;* also *Spiraea callosa, Kerria japonica, Vinca major*,[2] *Hydrangea hortensis*,[3] *Fagus sylvatica*,[4] *Ulmus campestris*,[5] *Cornus sanguinea*,[5] *Sambucus nigra*,[5] *Myrtus communis tarantina*,[6] *Zea Mays* etc.

The inheritance of variegation through seeds is one of the most interesting phenomena presented by this

[1] H. Molisch, *Ueber die Panachüre des Kohls.* Ber. d. d. bot. Gesellsch., Vol. XIX, 1901, p. 32.

[2] Verlot, loc. cit., p. 75. Here also will be found information relating to *Glechoma hederacea.*

[3] Morren, *Hérédité*, loc. cit., p. 230. Here also *Pelargonium inquinans.*

[4] According to Schleiden, after being damaged by snails, cited by Morren, loc. cit., p. 227.

[5] *Ulmus, Cornus* and *Sambucus* according to my own observations. Moreover I have seen such branches on almost all the above named varieties.

[6] G. Arcangeli, *Bull. Soc. Bot. Ital.*, 1895, pp. 16-18.

whole set of facts. The variegated sorts are nearly all more or less constant; sometimes to a very small and sometimes to a very large extent. This character has been subjected to what we may call an automatic selection, for every gardener naturally plants out only variegated specimens neglecting the green ones; also it is customary to cut away the green twigs which arise by bud-variation. Here we have a sort of unconscious selection which has however been exercised in the same direction for many years, and in many cases through some centuries.

And what has been effected by this continued selection? Absolutely nothing. At least, so far as we know, nothing more than maintaining the variegated varieties and keeping them in a fairly pure condition. But nothing like fixation has resulted; that is to say, the varieties continue to produce atavists when grown from seed, and moreover, the pure and constant varieties which correspond to them have not been obtained. For in this case these varieties, as we stated above, would have to be the pure yellow ones, such as are known to gardeners under the name of *aurea* forms.

All in all there are in our gardens, perhaps twenty or thirty, or even a few more of these *aurea* forms; and this number is as nothing compared with the almost unlimited series of variegated forms. Moreover exactly those variegated forms which have been cultivated most carefully and for the longest time have not given rise to *aurea* varieties.[1]

From this discussion I draw the conclusion that contintined selection with variegated plants will not of itself lead to the production of constant forms. For this,

[1] See the list on page 270.

something else is necessary, and this something chance alone can provide. What we want is the transition from one race to another, a transition, which according to my opinion, cannot be effected gradually, but takes place suddenly from as yet unknown causes; we have, in fact, to wait for a mutation.

The longer a variegated plant has been in cultivation the more strongly does the fact that it has not progressed support this conclusion. The best instance is afforded by the familiar Rocket[1] which is one of the oldest, favorite and most widely distributed variegated plants in cultivation, and which is often seen to escape from gardens (*Barbarea vulgaris variegata*). The plant is cultivated almost solely for its variegated leaves, although it is a kind of cress. It is biennial and sufficiently constant; it is usually grown from seed, although it can also be propagated by division. Only a small percentage of the seedlings are found to be variegated. Amongst a thousand grown from seeds which I had harvested myself from isolated variegated plants, I found only one per cent variegated and ninety-nine per cent plants whose cotyledons and first leaves were pure green. No white or yellow seedlings occurred. Of the greens a large proportion developed later into variegated plants, as was to be expected.[2] But the variety can by no means be regarded as fully fixed.

Very many variegated varieties, especially of annual and biennial plants, come true to seed. MORREN, CARRIÈRE[3] and other authors have drawn up lists, and much information relating to the subject can be gathered from

[1] VILMORIN-ANDRIEUX, *Fleurs de pleine terre*, p. 387.
[2] According to MORREN, *Hérédité*, loc. cit., p. 229, from 70-90% of the seedlings become variegated in later life.
[3] E. A. CARRIÈRE, *Production et fixation des variétés*, 1865, p. 14.

seedsmen's catalogues. Such statements relate, of course, only to practical and not to absolute constancy. It suffices that the harvests justify a reasonable hope that a certain number of variegated individuals will occur amongst the seedlings. Information as to the magnitude of this proportion is rarely given. GODRON found *Acer striatum variegatum* to repeat the anomaly in only one-third of its seedlings.[1] VIVIAND-MOREL found only occasional variegated specimens amongst five hundred seedlings of *Hedera Helix variegata* and only one amongst fifty of variegated *Yucca,* the majority being green.[2] PÉPIN states that the seeds of *Sophora japonica foliis variegatis* always give rise to more variegated than green plants;[3] but in the case of these and similar data we know nothing, as a rule, as to whether the seeds have been derived from individuals which had been isolated. POLLOCK sowed the seeds of a variegated plant of *Ballota nigra* which he had found wild and obtained thirty per cent variegated seedlings. In the next generation the seeds of these, however, gave 60% of variegated individuals.[4] The plant is now on the market and from the commercial seed I raised 25% variegated and 75% green plants. The seeds of a variegated specimen of *Chrysanthemum inodorum* found near Amsterdam produced 65 plants in my garden, of which 5% were variegated whilst 17 produced spotted leaves during the course of the summer, and the rest were green (1893). From the seeds of a variegated *Lunaria biennis* I raised green plants only, (1893) and I obtained the same result in

---

[1] *Mém. Acad. Stanislas,* 1873.
[2] *Lyon horticole,* 1893, p. 144.
[3] VERLOT, *loc. cit.,* p. 75.
[4] DARWIN, *Variations of Animals and Plants,* I, p. 409.

1896 from some self-fertilized variegated *Oenothera Lamarckiana*, although these two sorts are ordinarily constant from seed. Variegated *Oenothera rubrinervis* gave rise to 20% variegated seedlings (1892), but on a repetition of the experiment with another plant (1893) all the offspring were green.

In sectorial variegation we might expect the seeds of the variegated sectors to give rise to more variegated plants than those of the green ones. The only information relating to this question as far as I know is due to HEINSIUS.[1] He found a stem of *Dianthus barbatus*, one of the longitudinal halves of which was variegated, whilst the other was colored in the ordinary way. During the flowering period the plant was protected from insects by gauze and artificially fertilized, each flower being pollinated with pollen from another in the same longitudinal half. On the one half the capsules were white, on the other green; both produced ripe seed. The seeds of the white fruits produced seedlings without chlorophyll but the seedlings from green capsules were the normal green. In 1888 I myself collected the green and the variegated fruits of a sectorial main stem of *Oenothera Lamarckiana* separately. The seeds of the former gave rise almost exclusively to green plants, those of the latter to a large proportion of variegated ones. In the summer of 1895 I saved the fruits from a green and from a variegated branch of the same plant of this species, but both sets of seeds gave about the same very small proportion of variegated specimens, viz., 2%.

In the summer of 1898 I conducted a more exhaustive research with sectorial variegation in *Oenothera La-*

[1] H. W. HEINSIUS in the Proceedings of the *Genootschap ter bevordering der Natuur- Genees- en Heelkunde te Amsterdam*. Meeting of May 7, 1898.

marckiana. In the normal families of my cultures some specimens that happened to be variegated had, after artificial self-fertilization, given rise to no more than two per cent of variegated offspring and in the next generation the same proportion was repeated, the conditions being the same. From these I selected in 1898 the four finest young plants, planted them out a meter apart, and thereby obtained strong, richly-branched individuals, of which some were slightly, and others strongly, variegated. On all of them the flowers from which I intended to save seed were artificially and purely fertilized with their own pollen. On each of the four plants I first fertilized flowers on the pure green and afterwards some on the variegated lateral branches. Amongst 675 seedlings of the former and 1300 of the latter group the seeds produced the following percentages of yellow and variegated seedlings:

PLANTS	PERCENTAGE IN VARIEGATED SEEDLINGS	
	GREEN BRANCHES	VARIEGATED BRANCHES
No. 1	0–0 %	1 %
No. 2	—	3
No. 3	0–0	4–12–18
No. 4	0–0	6–9–45-100

Each number refers to a separate branch The six greens gave rise, as we see, to green seedlings only, but the variegated ones to a larger or smaller number of seedlings with this character. The figures 1, 3, 4, and 45 in the last column relate to the slightly variegated branches; the rest to those with this character more strongly developed. The latter therefore gave a higher percentage of variegated offspring. The variegated seedlings had either yellow or flecked cotyledons, or green cotyledons and flecked leaves, and of these three groups there were 68% of the first, 12% of the second and 20%

of the third group. The more yellow seedlings there were in a group the more variegated specimens did it, as a rule, also contain. I collected the seeds of a yellow fruit separately; only eleven of these germinated but these had all pure yellow cotyledons. On the other hand striped fruits had percentages of variegated seedlings which varied greatly, and this was also true of the striped parts of capsules when their seeds were harvested separately. Lastly the seeds of green capsules produced only green seedlings.

The color of the seedlings is therefore to a large extent determined by the color of that part of the mother plant which produced the seed (and also the pollen).

I made a further investigation of the seedlings from seeds of green and variegated branches of individual plants in various other species, after artificial isolation had been secured, either by means of parchment bags, or by planting the plants some distance apart, or by making them flower at different times.[1] I obtained the following percentages of variegated and yellow seedlings:

	SEEDS OF	
	GREEN BRANCHES	VARIEGATED BRANCHES
A. Commercial variegated races:		
Arabis alpina	2–10 %	90 %
Helianthus annuus	0 "	100 "
B. Occasional finds:		
Lamium album	0	3 "
Geum urbanum	0.3 "	4 "
Silene noctiflora	(5 ")	(34 ")

The high percentage of green plants in *Arabis alpina* corresponds presumably to the readiness with which this species produces bud-variations, variegated branches be-

[1] In *Silene noctiflora* only was fertilization left to the free agency of insects.

ing easily produced by green plants and green branches by variegated ones.

Now let us consider the yellow seedlings of variegated plants. They appear, it is true, to be mutants, but, as a matter of fact, they are the extreme variants which, however, do not attain to their goal but perish in the attempt, for they are too poor in chlorophyll and are thereby destined to die early. Nearly all of them die without so much as having unfolded their first leaves, or sometimes even their cotyledons. They constitute the extreme limit of a long series of variegated forms, but have, so to speak, followed a wrong direction. They are by no means rare; for instance they are well known in the holly, *Ilex Aquifolium,* and they often result in a very considerable loss amongst the seedlings raised from the seed of variegated plants.

It is not, however, variegated plants only which produce such seedlings; green plants do so only too often, and this even occurs in families cultivated for experimental purposes when the cultures are pure green and have been so for many years or did not produce more than an occasional variegated leaf or twig. If in such cases the seeds of the single seed-parents are sown separately the proportions in which variegated seedlings occur in the various groups are found to vary greatly.

Some species appear never to produce them, for instance the tricotylous races of *Cannabis sativa, Mercurialis annua,* and *Phacelia tanacetifolia* which I have cultivated, although I have sown the seeds of several hundreds of individual plants separately in the course of some years. In other species they are very rare; in some, however, the percentage of yellow seedlings is so considerable as to become a real nuisance. Thus, for instance, the highest numbers (not to mention the numerous smaller ones) that I found amongst the seedlings of individual seed-parents were as folows:

	YELLOW OR WHITE SEEDLINGS
Antirrhinum majus	5-6 %
Clarkia pulchella	9-13 "
Papaver Rhoeas	15-30 "
Polygonum Fagopyrum	8-12 "
Scrophularia nodosa	10-15 "
Trifolium incarnatum	4-6 "
Chrysanthemum segetum	13 '
Linaria vulgaris	25
Trifolium pratense	13
Oenothera Lamarckiana	20

In many other species I have, as yet, found not more than one or two per cent of yellow seedlings from the seeds of individual parents. Therefore I presume that this extreme variation is brought about, besides by the heritable potentiality, by causes similar to those in operation amongst the variants in the small seeds of *Trifolium incarnatum* (p. 239).

In some cases, as for instance *Polygonum Fagopyrum* and *Trifolium incarnatum* it struck me that the higher numbers were more frequent than some of the lower ones. This was especially the case in *Papaver rupifragum,* amongst the offspring of a single parent plant. This plant was selected as being a

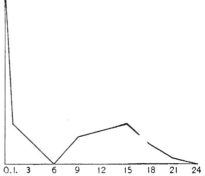

Fig. 55. *Papaver rupifragum.* Proportion of yellow seedlings among the seeds of 54 green plants. These plants themselves were the offspring of a single green parent plant. The two first ordinates are slightly reduced. The figures 3, 6, 9, etc., signify 2-4, 5-7, 8-10% yellow seedlings and so forth.

tricotylous specimen and had been raised from seed obtained in exchange; it flowered in 1893 in complete isolation and produced 6% yellow seedlings amongst its offspring. I planted out about 60 of the green plants which grew to healthy individuals in 1899. I left the pollination to insects but saved the seeds of each plant separately and then counted the number of yellow seedlings for each in a lot of 300.

Y. S.	0	1	2	3	4	5	6	7	8	9	10	11	12	13	14	15	16	17	22	24	27	30
Ex.	27	6	1	1	1	0	0	0	1	1	2	3	1	1	2	4	0	3	1	0	0	1
Or	27	6		3			0			4			5			6		3	1	0	0	1

"Y. S." signifies the yellow seedlings in each lot of seed and "Ex." the number of plants which exhibited this proportion amongst their seedlings. In the lower line from 2% onwards these are arranged in groups of 3 in order to emphasize the general result. The numbers of this last row are plotted in Fig. 55. This curve is similar to the curves of monstrosities which have been subjected to a selection extending over many years.[1] Even as these, it consists of a half curve and of a bilateral curve. It indicates therefore the selection of a latent character which in this particular case started with the choice of a tricotylous plant which happened to be a particularly suitable one.[2]

The observations and experiments which we have described or passed in review show that variegated plants constitute a group of forms which in spite of being selected for years or even for a century have manifested no further improvement in the quality and constancy of

[1] See the second section and *Sur les courbes galtoniennes des monstruosités*. Bull. sc. de la France et de la Belgique, published by A. GIARD, Vol. XXVII, April 1896, p. 396.

[2] See *Ueber eine Methode, Zwangsdrehungen aufzusuchen*, Ber. d. d. bot. Ges., Vol. XII, 1894, p. 25.

their peculiar character than many of the most recently arisen variegated sorts. They are highly variable and give rise in many cases almost every year to green descendants, on the one hand, and to pure yellow, on the other. The former are regarded as atavists, the latter, however, are only variants and not mutants, since so far as the observations extend they give no hope that they will ever form the basis of a pure yellow race. These, the true *aurea* varieties, have only arisen in relatively very rare cases; possibly from variegated types but without showing any evidence to support this supposition.

The capacity for producing variegated leaves or yellow seedlings is more widely distributed in the vegetable kingdom in a latent and semi-latent condition than perhaps any other character.

§ 25. ALTERNATING ANNUAL AND BIENNIAL HABIT.

One of the strongest pieces of evidence for the doctrine of mutation is the phenomenon in beets which is called bolting or shooting. It can be observed in almost every field of beets. Occasional plants are seen to develop a stem in the first year, to flower and to bear seed. They store no sugar or other food-stuffs or at any rate only to a very slight extent in their roots which become correspondingly woody. They are useless for practical purposes. On good fields about 1% of the plants ordinarily behave like this, and more rarely a smaller percentage of the whole crop. Under unfavorable circumstances, however, their number often increases considerably; reaching for instance from ten to twenty per cent and sometimes still more.

No farmer uses the seeds of such annual plants for

sowing. They obviously offer too great a prospect of a repetition of the evil. Moreover the seeds of these bolters cannot, either by chance or carelessness, get mixed up with those of the biennial beets because they ripen a year earlier. Thus in every generation an absolutely rigid selection of biennial examples as seed-parents takes place, and must have taken place as long as the culture of beets has proceeded on rational lines.

Nevertheless these bolters have not disappeared. Stringent selection has failed to eliminate them. Moreover as far as historical data enable us to decide, the proportion of bolters remains about the same. In reference to this case at any rate we are therefore fully justified in stating that selection cannot effect in the course of a long period of time what it fails to bring about within a few years.

This belief is widely and firmly held by beet-farmers. They are always in search of new means of combating this evil; but the mere selection of biennial beets is considered to be without prospect of success. RIMPAU has endeavored to attain this end by raising a triennial race, by selecting the so-called laggers, i. e., plants which have not flowered in the second year;[1] but most agriculturists content themselves with making the conditions of culture as unfavorable as possible to this evil.[2]

These laggers are in a sense analogous to the bolters, inasmuch as they have been eliminated by the normal process of selection since the time when beets were first

---

[1] W. RIMPAU, *Das Aufschiessen der Runkelrüben*, Landwirtsch. Jahrbücher, Vol. V, 1876, p. 31, and Vol. IX, 1880, p. 191. By the same author, *Das Samenschiessen der Rüben*, Deutsche Landw. Presse, Jahrg. XXI, No. 102, Dec. 22, 1894, p. 984.

[2] A list of the most important papers on the subject is given by RÜMKER, *Die Zuckerrübenzüchtung der Gegenwart*, Blätter für Zuckerrübenbau, 1894, pp. 22-23.

cultivated and have nevertheless not been extirpated.

The general opinion of botanists is that the representatives of the main line of the evolution of plants have been for the most part perennials. From these the annual and biennial forms must have arisen independently in the various families and groups; and it is further natural to suppose that the biennials arose first and that the annuals arose from them. If this is true the production of a biennial from an annual or of a perennial by one of these two would have to be regarded as a phenomenon of reversion.[1] Instances of such atavism seem to occur very generally in the vegetable kingdom, but progressive transitions, that is to say, those that take place in the opposite direction, are also by no means rare.[2]

From the abundant literature on this point I select two cases which seem to me the most important. *Phaseolus multiflorus* (*Ph. coccineus* L.) is, with us, an annual plant, producing occasionally, however, a bulbous root which can be wintered and by means of which the plant can be perpetuated. Von Wettstein,[3] to whom we owe our knowledge of this phenomenon, has obtained plants which lived four years, and in my own experimental garden I have succeeded in wintering several such *Phaseolus* tubers. Von Wettstein's view is that we are dealing here with the transformation of a peren-

---

[1] Many, however, hold the opposite view. See Darwin, *Variations*, II, p. 5; and Rimpau, *loc. cit.*

[2] See the works relating to this subject by Irmisch and Warming. Also Hildebrand in Engler's *Botan. Jahrb.*, II, 1882, pp. 51-135; with regard to different sorts of beets: F. Schindler in *Bot. Centralblatt*, 1891, Nos. 14 and 15, and the literature cited there.

[3] R. von Wettstein, *Die Innovationsverhältnisse von Phaseolus coccineus* L. (= *Ph. multiflorus Willd.*), Oesterr. bot. Zeitschrift, 1897, No. 12, 1898, No. 1.

nial species into an annual one.[1] The careful experiments of BRIEM lead to the same conclusion; for he succeeded in wintering the sugar beet after it had borne seed and in inducing in this way the same plant to bear seed a second and sometimes even a third time.[2] All that was necessary to bring this about was that the beet in question should continue to increase in thickness[3] and accumulate, in its new rings of tissue, the necessary quantity of sugar and other food-stuffs.

It is known that summer wheat can be changed into winter wheat[4] by a selective process, and that the converse process may also take place;[5] also that a perennial sort of rye is occasionally raised in Russia besides the ordinary annual rye.[6] Numerous annual species also give rise to biennial and perennial forms such as *Arabis dentata* and *Delphinium Consolida*;[7] and as a general rule interferences of various kinds with the normal vital processes of the plant are considered to be the causes of these changes.[8]

[1] *Loc. cit.*, p. 11.

[2] F. STROHMER, H. BRIEM and A. STIFT, *Ueber mehrjährige Zuckerrüben und deren Nachzucht*, Oesterr.-Ungar. Zeitschr. für Zuckerindustrie, Pt. 4, 1900, with Plate XV.

[3] For facts relating to this growth in thickness see *Die abnormale Entstehung secundärer Gewebe* in PRINGSH. Jahrb. für wissensch. Bot., Vol. XXII, 1890, p. 35; and Plate III, Fig. 14.

[4] Numerous illustrations of the question dealt with in the text are furnished from agricultural experience by C. FRUWIRTH, *Die Züchtung der landwirtschaftlichen Culturpflanzen*, 1901, p. 146.

[5] DARWIN, *Animals and Plants*, I, p. 333.

[6] A. BATALIN, *Das Perenniren des Roggens*. A very important paper dealing with these questions is H. C. SCHELLENBERG'S *Graubündens Getreidevarietäten*, Ber. d. Schweiz. bot. Gesellschaft, Part X, 1900.

[7] THEOD. HOLM, *On the Vitality of Some Annual Plants*, Amer. Journal of Science, Vol. XLII, 1891, p. 304.

[8] W. BARTOS, *Zeitschrift f. Zuckerindustrie in Böhmen*, Vol. XII, 1898, p. 456.

Conversely many perennial plants which under normal conditions flower the second year after germination for the first time, can be induced by favorable cultivation to flower in the first summer, though this does not always succeed with every individual. In this way many perennial species are treated in horticultural practice as annuals, and I myself have cultivated a whole series of plants more or less regularly as such; for instance, *Achillea Millefolium, Hesperis matronalis, Lychnis vespertina glabra, Picris hieracioides, Trifolium pratense quinquefolium* and others.

Let us proceed now to our more immediate subject, the phenomenon of the occurrence of many species partly in annual and partly in biennial specimens. Such plants are regarded by descriptive systematists as biennials, as, for instance, the name *Oenothera biennis* indicates; for, under the less favorable conditions which usually obtain in the field the great majority of the specimens will be biennials.

In my opinion this view is quite correct, but the biennial species in question must possess the capacity of growing as annuals, in a semi-latent condition. Moreover this capacity does not seem to be universal, but to be confined to particular races. For instance, KOCH's *Synopsis Florae Germanicae et Helveticae* (3d ed. 1857) and GRENIER and GODRON in the *Flore de France* (1852) give *Dipsacus sylvestris* as an annual, whilst I myself have hitherto only been able to raise biennial races of it from seeds derived from numerous different sources, and in spite of the fact that I modified the culture in every conceivable direction in the hope of making them annual. It is highly probable that many species exist in certain regions as pure biennials, in others as annuals, and in still others in a mixture of these two forms.[1]

Inasmuch therefore as the biennial habit is to be regarded as the character of the species and the annual habit as the anomaly, the latter is likely to follow the general rule according to which the development of the anomaly is favored by improved conditions of life. And the experiments which I propose to describe in this section prove the correctness of this view.

However, there is an apparent contradiction, for, as is well known, RIMPAU has shown in the case of the beet that every retardation or interruption of the growth, whether it occurs during germination or just after the seed comes up or at a later stage of the development of the plant, favors the production of the seed in the first year of the plant's life.[2]

But in this case it only appears that we are dealing with conditions favorable to the production of the anomaly whereas in reality we are concerned with the stimulus necessary for the manifestation of this bolting. As it is not very easy to make this difference clear I shall select an instance of a pure biennial race[3] which lacks the power of giving rise to annual specimens. I refer to my cultures of *Dipsacus sylvestris*. This race can be sown at any period of the year, and the plants will always remain rosettes until the end of the next winter and develop a stem in the spring of their second year. According to whether the sowing was made in the spring or in the summer or not till autumn are the rosettes vigorous or

[1] Instances of this are given by J COSTANTIN, *Les végétaux et les milieux cosmiques*, Paris, 1898, pp. 28 f.

[2] *Landw. Jahrbücher, passim,* 1880, p. 194.

[3] *On Biastrepsis and Its Relation to Cultivation*, Annals of Botany, Vol. XIII, No. LI, Sept. 1899, p. 395.

weak, but this treatment has no effect on the period at which the stem will be developed. If the seeds are sown in March in the greenhouse and the seedlings are picked out early into pots and planted out in May or June, we get vigorous rosettes with abundant leaves, but not a single stem in the first year. If the seeds are sown in September in the greenhouse, soon after harvesting, the rosettes remain weak until winter, but nevertheless develop a stem in the following spring. By sowing the seed in late autumn in the open ground, however, the plants will develop only a single pair of leaves above the cotyledons and they can be induced to pass through the winter without producing their stems in the spring. In this case they pass through the whole of the summer as rosettes, become extraordinarily vigorous and do not develop a flowering stem until after the second winter.

These experiments show that a definite stimulus is necessary for the production of a stem. Under the conditions of my own experiments it seems to be the winter which exerts the stimulus and that it can do so at any age of the plant except the very young stages when only the first two leaves are unfolded. But without this stimulus no stem is formed.

The experience of beet cultivators goes to show that the chief cause of the bolting is the night frosts of the spring. Manifestly they exert an effect on the young plants similar to that produced by the winter. It is a fact generally known that the percentage of bolters is high in direct proportion as the seed was sown early; crops which have been sown late are sometimes perfectly free from this defect. RIMPAU showed that if a small section of a field which has been sown early is covered over with a sheet every night that threatens to be frosty, the occurrence of bolters is considerably diminished; in one experiment for instance from 7 to 4%.[1] Other results point in the same direction.

HEUZÉ, in his valuable little book on the oil plants,[2] says with regard to the rape (*Brassica Napus oleifera*), that in the north of France it should not be sown before the middle of July or after the middle of August, for in the latter case the plants will not be strong enough to survive the winter, and in the former too large a proportion will set seed in the first year. The same thing is true of a whole series of other biennial plants both cultivated and wild; those which germinate late become biennial; of those which germinate early a greater portion become annual, the earlier the sowing or the germination took place.

In these cases we are not concerned with the induction of bolting by night frosts, or by any other stimulus, but with a case of inherited variability. It is true that the beet possesses this variability also, but the general conditions in this species are much complicated thereby. That we have to deal with a phenomenon of inheritance is proved by the fact that the annual form can easily be fixed by selection, without, however, attaining a state of absolute purity. RIMPAU sowed the seed of bolters,[3] and by always selecting seeds ripened in the first year, he obtained in the fourth generation a race whose seeds when sown on the 31st of March produced annual plants only and which in the fifth generation, when sown on the 5th of April, was as constant an annual as the normal

[1] W. RIMPAU, *Das Aufschiessen der Runkelrüben,* Landwirtsch. Jahrbücher, 1880, p. 192.

[2] L. HEUZÉ, *Les plantes oléagineuses,* Bibliothèque du cultivateur, Paris, 2d ed., p. 16.

[3] *Loc. cit.,* p. 197.

beet, sown at the same time, was biennial. The same is true with other species. Seeds of the wild *Daucus Carota* saved from annual plants gave me a large proportion of annuals; but seeds from plants which had come through the winter gave a predominant proportion of biennials. On the other hand selection does not seem to lead to the production of annual races which would be free from occasional atavism. It is my custom now to cultivate my *Oenothera Lamarckiana* and its derivative species mostly as annuals. Many of these cultures have been continued for six or more generations by means of the seeds of annual specimens only. Nevertheless every year there occur occasional and sometimes several biennial plants amongst them.

Aster Tripolium[1] is usually given as an annual in the floras, but with us it is represented by specimens which pass through the winter as well as by plants which flower in the first summer. In experimental sowings in the garden I obtained roughly equal numbers of the two types; but if I sowed the seed in March or April in the greenhouse the plants developed stems in the first year almost without exception. They were, as a rule, covered with glass every night until June, and thus protected from night frosts, and they were well treated also in other respects, especially by transplanting them soon after germination into rich well-manured garden soil. For according to my experience one of the best means of inducing biennial plants to behave as annuals is to give them plenty of manure, provided of course that the

---

[1] KOCH, *Synopsis Florae Germanicae et Helveticae*, p. 361. GRENIER ET GODRON, *Flore de France*, Vol. II, p. 102; KARSCH, *l'ademecum botanicum* etc.

capacity to do this is present in them in a semi-latent state.[1]

With *Oenothera Lamarckiana* I have made some more extensive experiments on accelerating the development of the stem by improving the conditions of growth. In the wild state this plant consists chiefly of biennials, but partly also of annual and of triennial individuals. Under experimental conditions, however, the duration of its life seems to depend more on external influences than on the choice of seeds. I have especially tested the distance between the plants, the sunniness of the position, and the richness of the soil.

In 1888 I selected some seeds of my biennial stock plants of the *Lamarckiana* family of 1886-1887,[2] in order to investigate the effect of the degree of separation of the plants in the bed. For this purpose I selected four adjacent beds of similar contents with regard to soil and manure, sowed the seeds in the middle of April fairly thickly in rows, and weeded them out during germination in such a way that on two beds the plants stood at moderate distances, on one further apart, and on a third more closely. In the summer up to the middle of September I recorded the number of individuals with stems and the number of the rosettes. The sum of the two obviously affords a measure of the distances between the plants. The extent of each bed was 13 square meters. The figures are:

---

[1] *Sur la culture des monstruosités*, Comptes rendus de l'Ac. d. Sc., Paris, January, 1899; *Sur la culture des fasciations des espèces annuelles et bisannuelles*, Revue générale de botanique, Vol. XI, 1899, p. 136; and *Ueber die Abhängigkeit der Fasciation vom Alter bei zweijährigen Pflanzen*, Botanisches Centralblatt, Vol. LXXVII, 1899.

[2] See the pedigree in Vol. I, p. 224.

BED	PLANTS	PLANTS PER SQUARE METER	PERCENTAGE OF PLANTS WITH STEMS
No. 1	1350	100	23 %
" 2 and 3	630 + 650	50	43 "
" 4	380	30	58 "

That is to say, the closer the plants are together, and the less room each one has, as a result of this, for the unfolding of its leaves, the smaller is the number of annual plants.

In the following year I repeated this experiment, but this time with the seeds of annual plants. The result was, however, the same. There were 1188 plants on one bed of 13 square meters, that is, about 90 per square meter; of these 20% were annuals. On the other bed of the same size there were 348 plants (or 27 per square meter) and 54% developed stems.

I repeated the same experiment once more, in 1890, with the seeds of an annual plant of 1889. On the one bed there were 40 plants per square meter, of which 17%- were annual. On the control bed there were only ten plants in the same area, and of these 72% produced stems in the first summer; the extent of the bed in both cases was 5 square meters.

In 1891 I investigated the influence of the distance between the plants in an experiment with *Oenothera laevifolia*, raised from the seeds of an annual race which had been selected for three generations.[1] The two beds were of the same size, had the same aspect and the same soil, and both received a similar and liberal dressing of guano. They were sown in the middle of May on the same day, but at the end of July they contained 195 and 638 plants respectively (per each 6.5 square meters). As a result of this, the bed in which the plants were far

[1] See the pedigree in Vol. I, p. 224.

apart had 162 plants which developed stems, whilst on that in which they were close together there were 145. The difference between the two reckoned as a percentage of the whole culture is of course more striking, viz., 83% as against 20%. More important, however, is the fact that per each square meter in absolute numbers more annual individuals are produced when the plants are grown far apart than when they are grown more densely and therefore in larger numbers. When viewed in this way the result points to the great importance of sowing seeds thinly in experimental cultures.

Experiments with shading are met with the difficulty that the young plants cannot stand it very well, even when, as in my experiments, the shadow is that of trees. The experiment was conducted at the same time as that of 1890, referred to above, on a similar scale and by growing the plants far apart; it produced about 46% annuals as against 72% in the control experiment already mentioned.

By far the best means, however, of increasing the proportion of annual plants or even of securing their exclusive production, is to sow the seed and keep the young plants under glass. In doing so the seed can be sown in March or April in un-manured sterilized soil, and the seedlings may be pricked out singly into pots containing richly manured soil after the appearance of the third or fourth leaf. In this condition they remain under glass until the end of May, at least during the nights and on cloudy days, and can then be turned out of the pots without breaking the ball of earth round the roots and transplanted to the place where they are to grow. Treated in this way almost all the plants behave as annuals, and of late I have grown all my cultures by this method or by some slight modification of it.

In order to determine the effect of the soil on the development of the stem I have compared the difference between plants grown on manured and unmanured beds, and also the difference between plants grown on barren sand and on fertile soil. The first of these two experiments I have made with the *Oenothera laevifolia*. I used seeds which I had saved in 1890 from the third annual generation of my culture (see Vol. I, p. 273). The seeds were sown in the middle of May on three beds of 3¼ square meters each. They were adjacent to one another, had the same soil, a similar exposure, and so forth. The seedlings were thinned out early, to 100 per bed, in such a way that the distances between them were as uniform as possible. The sole difference lay in the kind of manure which they received, which in No. 1 was nothing, in No. 2 a quarter of a kilo of guano, and in No. 3 a quarter of a kilo of hornmeal. In the second bed, therefore, the manure was rich in phosphates and in the third in nitrogen. On the 30th of July 1 recorded the plants with the following result:

	PLANTS	ANNUALS
No. 1. Without manure	100	77%
No. 2. With guano	98	90%
No. 3. With hornmeal	108	94%

In spite, therefore, of the fact that the race had been selected for three years the proportion of annual plants on the bed without manure was only 77 per cent, whilst this proportion was considerably increased by the addition of manure, and more by the addition of nitrogen than by that of phosphates. Further experiments with different quantities of the same manure showed that the amounts employed in this experiment (about 80 grams per square meter) should not be exceeded, that is to say, that the result cannot be improved by still heavier manuring.

For the experiment with sand I dug in my experimental garden a bed of 13 square meters in extent and one-half a meter deep, and filled it with ordinary fine sand. On this bed and on a neighboring one of the same size I sowed seed of *Oenothera Lamarckiana* in the summer of 1899. The control bed was not manured but contained a very fertile soil; the seed was sown in the middle of April.

The sand of the bed bordered immediately on the rich soil of the path which surrounded it.[1] Therefore the plants at the margin could thrust their lateral roots into this, and thus obtain richer food than the more central rows. This circumstance showed very important results during the course of June, for while many flowering stems were produced towards the outside of the bed, hardly any occurred in the middle. It was not until the middle of July that the development of stems set in here also. Curiously enough this occurred in almost every instance at exactly the same time. In the middle of August among the 82 plants of the outer rows about 60% had developed stems, whilst in the middle there were 133 rosettes amongst 203 plants, that is to say about 24% of annual specimens. We see that the distances between the plants in this experiment were very considerable, for on 13 square meters there were only 285 plants. Even at the end of the summer they hardly touched one another. In the control experiment in which

---

[1] In subsequent years I have separated the sand from the earth by boards.

the distance between the plants was practically the same there were about the same number of plants that developed stems as there were on the margin of the sand bed, in fact a little less, 53% amongst 348 plants.

Our main result therefore is that the proportion of plants which developed stems in the center of the sand bed is 34% as against the 53% and 60% amongst the plants on the margin of this bed and in the control bed respectively. Equally striking was the sudden change in the behavior of the central plants in July. This pointed to some special cause. I suspected that it was connected with the growth of the roots and that these about this time had penetrated the layer of sand and reached the fertile earth beneath it. When I dug up the roots at the conclusion of the experiment I found that these were, as a matter of fact, longer than half a meter and had branched freely below the level of the sand.

In order to find out whether this was the real cause of the development of the stems I made an experiment in 1891 with a bed in which the layer of sand was much deeper (one meter). A part of the original sand bed which was only one-half meter deep, and a neighboring bed filled with ordinary good garden soil served as control. This time the bed was surrounded by boards and, consequently, there was no difference in the behavior of the central and marginal plants. For this experiment I used the seeds of a culture of *Oenothera rubrinervis* which had been cultivated as an annual for two generations (seeds of 1890 of the pedigree of *V*ol. I, p. 273). The sowing took place in May 1891. At the end of July I recorded the plants on the three beds, each of which was 3 square meters in extent.

	PLANTS	ANNUALS
Sand-bed, 1 meter deep	161	21 %
" $\frac{1}{2}$ " "	226	50 "
Garden-soil	131	98 "

On the control bed the distances between the plants were somewhat greater, but as practically they did not touch one another on the sand bed this fact does not signify.

The seeds employed in this experiment gave a larger proportion of annual specimens than did those of the previous one. The main result is that the proportion of plants which produce stems in their first year can be reduced to about one-half by cultivation in a bed with half a meter of sand, and to less than a quarter by cultivating in a meter of sand.

The results of the foregoing experiments prove that biennial species which possess, in a semi-latent state, the capacity to produce annual specimens, can be induced to manifest this anomaly to a much greater extent by supplying them with more food. Crowding of plants, shading, lack of manure, or cultivation on sand, favor the production of biennials; but the more space, light and nourishment in the soil there is at the disposal of the individual plants the greater will be the number of those which will produce stems, flower and ripen their seed in their first summer. The stimulus of the winter or spring frosts, which in other cases induces the young plants to develop stems, is without effect here; for under the described conditions even seeds sown in the middle of May in the open ground may give rise almost exclusively to annual plants.

Continued selection, however, fails either to fix the biennial races and to free them of annual specimens, or to free the annual races of biennial individuals.

VIII. NUTRITION AND SELECTION OF SEMI-LATENT CHARACTERS.

§ 26. INCREASED NUTRITION FAVORS THE DEVELOPMENT OF THE ANOMALY.

Fluctuating variability is a phenomenon of nutrition, whereas mutability is the result of hitherto unknown causes (Vol. I, p. 575). This statement, which is perhaps the sharpest expression of the contrast between fluctuating or continuous variability on the one hand and occasional sudden transitions from one species into another on the other, has been discussed more than once in this work. It is equally true for the variability of semi-latent characters as for that of normal ones. This side of the statement has also been already alluded to, and I have cited many instances in order to prove its truth. Everywhere nutrition and variability are so intimately connected that the physiology of the latter phenomenon can hardly be dealt with without discussing its relation to the former.

Artificial selection is the choice of the better nourished individuals, except of course, when selection is carried out in the negative direction (Vol. I, p. 142). In the first volume I cited as a proof of this generalization an experiment with a semi-latent character. The number of accessory carpels of *Papaver somniferum polycephalum* was shown to be dependent on selection and, to a no less extent, on nutrition. In the third part of the first volume, the curve representing the length of the fruits in *Oenothera* and the curves of the rays of certain Umbelliferae and Compositae also proved that these two factors operate in the same direction. Active and semi-latent characters are thus shown to behave in the same way with regard to these two factors.

Since, however, the extraordinary variability of semi-latent characters (of which an account has been given in § 2 of this part, p. 9), is one of the strongest supports of the doctrine of selection, it seems to me that it is worth while to attempt to make the relation between this phenomenon and nutrition specially clear. In this last chapter I will therefore deal with a series of facts gathered partly from the literature of the subject and partly from my own observations, which all point more or less definitely to the conclusion that semi-latent characters are largely dependent on the external conditions of life.

External influences exert their effect on the development of organs during their youth, that is to say during the so-called susceptible period. After the character of the organ has been definitely established in this period, the further development cannot affect it. The number of leaflets in a clover leaf, of the petals of *Ranunculus bulbosus,* of the accessory carpels of *Papaver* are finally determined in this period; but the conditions of life at the critical moment are not the only factor. The accumulated effects of previous influences have also combined to determine the individual strength of the organ or of the individual; and the part which this latter factor plays in the determination of the degree of development of the deviating character is sometimes greater and sometimes less than that of the immediate external conditions.

The most susceptible stage seems to be that of the young embryo in the ripening seed, for external influences show their greatest effect on seed-plants. But they also play a part in vegetative methods of propagation and operate in the same direction, though as a rule with less intensity.

The gap between anomalies and normal active characters is bridged by a complete series of normal, latent characters. These too are dependent on external conditions in just the same way as the other two are. As a general rule we may state that, within the specific range of its character, the form of an organ is determined by external physical influences.[1]

As an instance let me cite the result of some experiments with the germination of potato seeds.[2] The first leaves of the seedlings are simple (Fig. 56), and the following ones gradually approach the peculiar pinnate form of those of the grown plant. If the germination takes place in the garden, in full sunlight, the plant develops quickly and the various steps in the development of the leaf-form

Fig. 56. Seedling of potato grown under unfavorable conditions of light. From 1 to 6 the form of the leaf increases in complication, but from 8 to 12 it decreases again. Cultivated in a room in 1876.

[1] *Keimungsgeschichte der Kartoffelsamen* in Landwirthsch. Jahrbücher, VII. Jahrg., 1878, p. 35.
[2] *Loc. cit.*, p. 35.

follow rapidly on one another. But if the conditions are unfavorable, as in a room, differentiation proceeds more slowly. The internodes tend to become abnormally long, to produce too little wood, the leaves develop small pinnules only, and in very unfavorable conditions I have sometimes observed an interruption in the series of leaf-forms on the stem. Above the lyrate leaves simple ones were again formed, the series turning backwards.[1]

These phenomena are much better illustrated in those cases in which the first leaves are more compound than the later ones; for instance in the species of *Acacia* which produces phyllodes in reference to which GOEBEL's important investigations have thrown so much light on the relation between embryonic forms and external conditions.[2] I have already referred to this above; but I might now mention a figure of a seedling of *Acacia verticillata* which, after it had already reached the stage of producing phyllodes, was induced to repeat the bi-pinnate form of the embryonic leaves by unfavorable conditions. In the same way the production of linear or arrow-shaped leaves of *Sagittaria sagittifolia* and that of the perforated leaves of *Monstera deliciosa* and others was shown to be dependent on external conditions. Insufficient nutrition tends to bring about a recurrence of the embryonic form, and it seems to be a secondary question whether this is the simpler or the more complicated. The *Campanula rotundifolia* studied by GOEBEL, the flowerstalk of which changed from the linear to the heart-shaped form of leaves,[3] is perhaps the best

---

[1] See also E. ROZE, *La transmission des formes ancestrales dans les végétaux*, Journ. d. Bot., Année X, Nos. 1 and 2, 1896.

[2] K. GOEBEL, *Organographie der Pflanzen*, I, p. 150, Fig. 105.

[3] GOEBEL, *Flora*, 1896, Vol. LXII, Pt. I.

known example. In the case of the Conifers BEISSNER has also shown that insufficient nutrition, for instance by cultivation in pots, can lead to a protracted retention of the embryonic form.[1] In *Eucalyptus Globulus* and *Acacia cornigera* stems which have been cut down produce branches which repeat the embryonic form of leaves, which are sessile in the one species and thornless and destitute of the so-called ant-bread appendages in the other.[2]

Exactly the same general conditions obtain in the development of anomalies, that is to say of those characters which are only exceptionally or never developed in the normal life of the species. Here again their precise nature seems to be a matter of indifference, that is to say, whether they are harmful or harmless; in either case they are under the influence of external conditions. Instances of deleterious characters are furnished by variegated leaves and by flowers and flowerheads which have become sterile by doubling (see §§ 19 and 24). The same is also true of real monstrosities, such as fasciation and twisting, as we shall see in the next chapter; and of new characters, reversionary phenomena, progressive and retrogressive variations of which I shall give a series of instances in the following section (§ 27). It is true both of half races and of middle races; in both it is the older or specific character which is intensified by unfavorable conditions, whilst the anomaly or the younger character is intensified by favorable ones. Obviously there is only a small step from these two races characterized by the semi-latency of the former or the

[1] L. BEISSNER, *Handbuch der Nadelholzkunde*. See also *Bot. Zeitung*, 1890, p. 539.
[2] F. HILDEBRAND, *Botan. Zeitung*, 1892, p. 5.

latter character to the true elementary species in which the character of the parent species has become completely latent, for in this case the variation of the new character of course conforms to the general laws of variability.

We have studied this relation of variability to nutrition from various points of view in our researches with the half races of *Trifolium incarnatum quadrifolium* and *Ranunculus bulbosus semiplenus,* but especially with the true middle races, *Trifolium pratense quinquefolium*, and *Chrysanthemum segetum plenum* as well as with analogous groups.

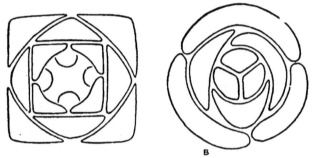

Fig. 57. *Lysimachia vulgaris*. Transverse sections of two buds which were to develop into upright stems in winter. A, quaternary, and B, trimerous symmetry.

It should be remembered that in all these cases we are dealing with variability in the restricted sense and not with mutability. How one race is transformed into another, we do not know. The phenomenon is as yet far too rare and has not yet been adequately investigated. The variability of the eversporting races is of the transgressive kind; but it does not lead, as a rule, to mutability.

Finally I wish to illustrate by means of a scheme the relation between the variability of semi-latent characters and the external conditions of life, and I select as material

Nutrition Favors the Anomaly.

for this the disposition of the leaves of *Lysimachia vulgaris* (Fig. 59). This species has opposite leaves as a rule, but often occurs with quaternary and ternary whorls. With regard to this character the species behaves as a half race, in this neighborhood at any rate.

If we examine the rhizome in the spring we find the vegetative buds growing vertically upwards under the top soil or moss and ready for sprouting. At this period it can easily be seen that all of the buds have their scales

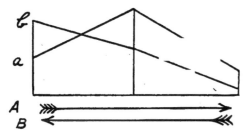

Fig. 58. Scheme to illustrate the relation between external conditions and anomaly. Shifting of the apex of the curve. A, the effect of high nutrition; B, the effect of unfavorable conditions. The (*a*) and (*b*) curves illustrate the disposition of the leaves of the stems of *Lysimachia vulgaris*; *a*, after high, *b*, after low nutrition. (See the figures on page 314.)

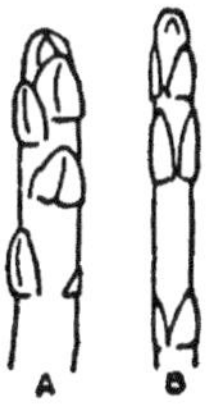

Fig. 59. *Lysimachia vulgaris*. Rhizome buds which would develop into stems. A, thick, with quaternary bud; B, thin, with ternary whorls in the interior of the bud. The visible scales show a decussate arrangement in both cases. (For diagram see Fig. 57.)

in a decussate arrangement at the lower end (Fig. 59), but within the bud the disposition of the leaves is different. In this region the structure is that which will be manifested by the growing stem in the summer.

The thickness of the future stem is correlated with that of the upright subterranean shoot, and on it depends the disposition of the leaves. The weakest shoots have the leaves arranged in opposite pairs, the stronger ones in trimerous and the strongest of all in tetramerous whorls. This character of the stem, which must be definitely determined by microscopical examination, can, however, be predicted with sufficient accuracy by the thickness of the shoot. Of course, the same relation obtains in the summer when the contrast strikes the eye far more forcibly, and then we see that the lateral branches of ternary and quaternary stems have almost without exception decussate leaves, obviously because they are the weakest of all.

If the plant is taken into cultivation it is easy to modify the proportions of the various sorts of stems by suitable treatment, although vegetative methods of propagation alone are employed; but it must be remembered that the buds for the following year are already completely laid down in October, and that the disposition of their leaves is therefore finally decided at that time. The actual disposition in any given year is therefore determined by external conditions which prevailed in the previous year.

The curves in Fig. 58 are plotted from observations which gave the following results. In March 1890, I planted out a group of rhizomes in a favorable position in my garden, and in June I noted the two-, three- and four-whorled stems, which therefore had been laid down under the influence of the unfavorable conditions which obtained in 1899. In the following summer I repeated the observations and thus formed an estimate of the effect of growth under more favorable circumstances (better soil and more light). The results were:

RECORDS FOR	WHORLS 2-	3-	4-MEROUS	TOTALS
Summer, 1890	35	21	2	58
" 1891	17	40	10	67

The apex is seen to be shifted from stems with opposite leaves to stems with trimerous whorls (Fig. 58 *b* and *a*).

The arrow A in Fig. 58 indicates the result of improved conditions, the arrow B that of more unfavorable ones.

The figure therefore not only illustrates a particular case but is a graphic representation of our thesis that high nutrition favors the production of the anomaly.

This scheme can be applied to a long series of cases, both of anomalies that occur in the field and of highly variable horticultural varieties. The best known example is furnished by the tricolored pansy *Viola tricolor maxima*. Here, as every gardener knows, the spring and early summer flowers have larger, much broader and more intensely colored petals than those of late summer.[1] The greater drought and the gradual exhaustion of manure around the plants are the causes. The same generalization applies to numerous cases to some of which we will refer in the following section.

§ 27. THE INFLUENCE OF EXTERNAL CONDITIONS AND OF MANURING.

J. COSTANTIN has dealt with the relation between the plant and its environment in a book devoted to this question.[2] In it he treats of the influence of the environment both on the normal characters of the plant and also on varieties and anomalies. An immense range of facts and observations is thus made accessible to the student.

---

[1] V. B. WITTROCK, *Viola-Studier*, Acta Horti Bergiani, Vol. II, 1897, Nos. 1 and 2. See also VERLOT, *loc. cit.*, pp. 46-47.

[2] J. COSTANTIN, *Les végétaux et les milieux cosmiques.* Bibl. scientif. internationale, 1898. The earlier writings of this author have been dealt with in our first volume (p. 99).

We shall here confine ourselves to true anomalies, that is, to semi-latent characters,[1] and will start with some instances of plants which are propagated vegetatively. First let us look at the continued growth at the top of the inflorescence of the white clover (Fig. 60), an anomaly which is fairly rare in this region, but which has been intensified considerably by favorable cultivation in my garden. Fig. 60 affords instances of this anomaly selected from a vast number of available cases.[2] Elongation of an inflorescence into an ear-shaped peduncle, proliferation or formation of two inflorescences, one above the other, on the same stem, and the appearance of small clusters in the place of the individual flowers accompanied by an increase of the bracts are some of the more important instances.

In the summer of 1890 I found near Hilversum a specimen which bore a single flower on the elongated axis of one of its inflorescences. I transplanted it to my garden, sowed the seeds in the following year and obtained a few "perumbellate" inflorescences. Again I collected the best seeds and sowed them in 1891. When the plants flowered about 2% of the several thousands of inflorescences had proliferated, most of them belonged to the type shown in Fig. 60 B, others to the rarer types A, C and D. I then selected the best plant, isolated it completely, and made certain that all the branches really belonged to it. After this I divided it, planted out the parts, and let them grow as strong as possible. In this

[1] The methods of cultivation suitable for producing pure white flowers on colored varieties of *Syringa* in winter, and the well-known blue coloration in the *Hortensias* are widely different. (See VERLOT, *loc. cit.*, pp. 60-61).

[2] The anomalies in question have long since been dealt with in the literature of the subject, and have been collated by PENZIG, in *Teratologie*, I, p. 387.

way I filled two beds of about two square meters each in the summer of 1892. They gave a very characteristic half curve when the inflorescences were plotted according to the number of flowers on that part of the axis which had proliferated. These numbers varied from 0 to 10 and were distributed as follows (August 1892):

Inflorescences	NUMBER OF FLOWERS ON THE PROLONGATION OF THE AXIS											Totals
	0	1	2	3	4	5	6	7	8	9	10	11	
On the 1st bed	325	83	66	51	36	36	18	7	6	1	1	0	630
" " 2nd bed	403	97	62	35	46	20	20	14	11	3	2	1	721

Fig. 60. *Trifolium repens perumbellatum.* Four different inflorescences from the same culture, 1891. A, with thick continuation of the main axis; B, the continuation thin and sparsely covered with flowers; C and D, with three-fold continuation.

We see that one-half of the inflorescences on this plant had proliferated this time, partly as the result of selection, partly on account of improved cultivation. Most of the perumbellate inflorescences appeared in July. Before July there were 21%, in July 47%, in August 38%, and this last record was made on over 500 inflorescences.

I continued the experiment by planting out some of the creeping stems of this plant into two other beds, of which one consisted of good garden soil, the other of dry sandy soil. In this year under less favorable conditions of growth, the proportion of anomalies was less . Amongst about 300 inflorescences in each bed 6% had elongated on the sand bed and 12% on the control bed. On the former the plants were small, their leaves being smaller and paler and less numerous than the normal.

This experiment shows that the proportion of anomalous inflorescences is to a very large extent determined by external conditions even when the method of propagation is purely vegetative.

In the literature of this subject we sometimes come across the view that anomalies are favored by improved external conditions because more nutriment is necessary for their development, and the fasciations, and the multiplication of the number of leaves, leaflets, etc., are given as instances. As a matter of fact even when the anomaly consists in a reduction of parts the same relation obtains. This is shown by the second instance that I shall give. This is *Potentilla anserina,* of which I found some plants with occasional tetramerous flowers near Hilversum in 1889 among the ordinary pentamerous ones. I planted them out and divided them, and in the summer of 1891 I selected the best specimens and made sure that all the runners were still organically connected. In 1892 I planted out one-half of this on a manured bed and the other half on an unmanured bed adjacent to it.

From the middle of July to the beginning of August,

I counted the sepals of all the flowers. There were about 2500 on the manured bed, and about 1500 on the unmanured. Amongst them were many with five and four sepals, and about 20 with 3, but none with fewer than 3 or more than 5. Here again, a pronounced half curve was the result. I have reckoned together the proportion of 3-4-merous flowers for the individual counts, and at each stage in the counting collected all the flowers which had opened since the preceding one. The counts were made when possible every fourth day, or, when the number of flowers was too small, at greater intervals. The result was as follows:

PROPORTION OF 3-4-MEROUS FLOWERS IN %.

	June			July							Augus.
Day:	19	23	27	1	5	9	13	17	21	25	29	2
Manured:	7	13	24	28	34	39	50	65	49	49	43	27 %
Unmanured:	—	—	7	—	20	33	39	—	42	49	46	44 %

We see that the proportion of anomalies increased on both beds gradually throughout the summer, reached its maximum in the second half of July, and then sank again. On the manured bed, however, this proportion amounted to 65% of the whole and on the unmanured bed to 49% of the flowers counted (160 and 224 flowers in the two cases respectively). In this case, therefore, both the periodicity and the relation to the external conditions are in all essentials the same as in the case of the white clover. In this latter case a plant which I had raised from seed served as material; but in that of *Potentilla* a specimen which I had collected in the field.

I shall now deal as briefly as possible with a series of further instances, emphasizing as before that the external influences have the results in question only when the particular characters are already present in the semi-latent condition. On completely latent characters no effect can be produced; we are dealing solely with half and intermediate races. "In other specimens, however, this is obviously not the case," says GOEBEL, "they retain their normal form even when grown in rich soil; the high nutrition operates on the malformation, not as a cause, but as a releasing factor."[1]

It is a familiar fact that many garden plants deteriorate if they are allowed to remain for a long time in the same place. They exhaust the soil and must, therefore, be moved from time to time. This is true, for instance, of Pansies, Anemones,[2] Dahlias, Petunias,[3] the crested forms of many ferns[4] and so on. MORREN planted out a specimen of *Saxifraga decipiens* which had hitherto borne normal flowers on stony ground, into good garden soil. In this it grew very vigorously, formed larger flowers than before, and manifested at first a slight transformation of its stamens into petals which, however, increased gradually during the course of the summer until ultimately the flowers became entirely double.[5] In *Hedychium coronarium* the structure of the flowers is also shown to be dependent on nutrition.[6] Wild apples and medlars lose their thorns in a few years if they are transplanted to gardens,[7] and *Carlina acaulis* becomes the so-called *Var. caulescens,* in rich soil, a fact which has already been recorded by WOLFF in his *Theo-*

[1] K. GOEBEL, *Organographie*, I, p. 159. Various instances are also given by BURKILL, *Journ. Linn. Soc. Bot.,* Vol. XXXI, 1895, pp. 218ff.

[2] VILMORIN-ANDRIEUX, Les fleurs de pleine terre, p. 87.

[3] HILDEBRAND, *Ber. d. d. bot. Ges.,* Vol. XIV, 1896, p. 327.

[4] LÖWE, cited by GOEBEL, *loc. cit.*

[5] *Bull. Acad. R. Belg.,* Vol. XVII, Pt. I, p. 424.

[6] FR. MÜLLER, *Flora,* 1889, Pt. III, pp. 348-352, Pl. 16.

[7] DE CANDOLLE, *Physiologie végétale,* II, p. 721.

ria generationis. The branching of the ears of *Triticum turgidum compositum* (Vol. I, p. 125) and the carpellomania of *Papaver somniferum* (Vol. I, p. 138) are to a very large extent dependent on external conditions. Double poppies become almost single under unfavorable conditions; I have observed this in *Papaver somniferum nanum album* in my own cultures. Again the double *Saponaria officinalis plena* seems to become single after transplantation, but subsequently to regain its double character.[1] The ordinary *Saponaria officinalis* often forms hexamerous flowers in this neighborhood. These were, however, more numerous in my cultures under good than under bad treatment. The studies of PEYRITSCH on the influence of the cutting down of woodbands on peloric varieties has already been referred to (§ 21, p. 225). On a lime tree in the vicinity of Baarn pitchers are produced every year; but I always found them on the open sunny side and never in parts of the tree where the branches were shaded by neighboring trees.

The color is also well known to be very dependent on external conditions. *Achillea Millefolium rosea* will only form its fine red inflorescences in a sunny position; if this is shaded the color is pale or absent, as I have myself often observed. Inflorescences grown in the dark remain quite white, even when they would otherwise have been red. The same is true of *Begonia semperflorens atropurpurea Vernon,* whose brown red foliage cannot, so to speak, stand the smallest amount of shadow. In this case I have succeeded in making some plants almost quite green by shading them during their youth. The

---

[1] MUNTING, *Waare Oeffeningen der Planten,* 1671, p. 588. Also in my garden.

color of *Amarantus tricolor,* the variety of whose color is its only claim to popularity, is dependent on external conditions.[1]

Zea Mays forms more bi-sexual panicles and ears when the seed has germinated at a high temperature. *Ranunculus bulbosus semiplenus* (§ 23, p. 258) forms more petals if it germinates in the summer than if it germinates in spring. Summer wheat can, as is well known, be transformed into winter wheat by sowing it in autumn, although, as it appears, only in a small proportion of individuals.[2]

Amongst the cultivated Begonias we sometimes find bi-sexual flowers which are the result of the appearance of stamens in female flowers; in them the inferior fruits become more or less completely superior and other anomalies make their appearance.[3] For the last 12 years I have grown such a specimen of *Begonia Sedeni* (*B. boliviensis* × *B. Pearcei*) which I have gradually multiplied by dividing its tubers. In the summer of 1890 I marked the tubers which produced the smallest numbers of such transformed flowers and planted them out in 1891 into a bed which was more richly manured and better situated than the rest. As a result of this they produced a considerably larger number of anomalous flowers than the control specimens. *Lupinus luteus* sometimes produces twisted inflorescences.[4] Seeds of such flowers collected in the field and sown in the garden did not repeat the anomaly; but in the next generation it reappeared as the

[1] VILMORIN-ANDRIEUX, *Les fleurs de pleine terre,* p. 64.

[2] MONNIER, cited by DARWIN, *Variations,* I, p. 333.

[3] P. MAGNUS, *Sitzber. d. bot. Ver. d. Prov. Brandenburg,* XXVI, 1884, p. 72, Table II, and PENZIG, *Teratologie,* I, p. 500.

[4] *Monographie der Zwangsdrehungen,* Jahrb. f. wiss. Bot., 1891, Vol. XXIII, p. 107, Pl. IX.

result of heavier manuring and better treatment (1890-1892).

Of a kindred nature, probably, is the well-known fact that anomalies are more abundant in certain years than in others. MUNTING records this for *Lilium cruentum plenum*[1] and KICKX for pitchers, of which there was in the neighborhood of Gent, in Belgium, something like an epidemic in 1848 in the tobacco fields, and in 1851 in *Rosa Gallica* and *Rosa centifolia*.[2] In the neighborhood of Freiburg there was an extraordinary abundance of floral malformations in the summer of 1866.[3] In France the hot and dry summer of 1893 brought out a large number of these, and GAGNEPAIN records a long series of anomalies which he observed at that time.[4] The year 1845 was a great year for peloric *Calceolarias*, 1862 for central umbels in *Auricula* (in England). For ten years I have observed the formation of pitchers in *Magnolia obovata* and that of hermaphrodite flowers in *Salix aurita*. In both these cases the frequency varied greatly with the year, although the specimens which were examined closely every year, were growing in our garden. I shall not extend this list which the reader may easily complete either by personal observations or from the abundant literature on the subject.

§ 28. THE PERIODICITY OF SEMI-LATENT CHARACTERS.

The immediate external conditions which obtain during the susceptible period of development do not constitute the sole factor which determines the greater or less

[1] MUNTING, *loc. cit.*, p. 501.
[2] J. KICKX, *Bull. Acad. Roy. Belgique*, Vol. XVIII, Pt. I, p. 591.
[3] HILDEBRAND, *Bot. Zeitung*, 1866, p. 239.
[4] *Bull. Soc. Bot. France*, Vol. XL, 1893, pp. 309-312.

visible development of semi-latent characters. Of almost equal importance is the individual strength of the young plant, which, however, is the result of the operation of external factors during preceding periods of time, which may be weeks or months or even years. The stronger a bud is, the more is it liable to produce anomalies.

This phenomenon is most clearly seen in the periodicity of the manifestation of anomalies by the same plant, and in the parallel between this manifestation and the gradual increase and subsequent decline in strength, either of the whole individual or of the succeeding orders of its branches.[1] This periodicity has been exhaustively studied in the five-leaved clover, and we have become familiar with instances of it in several other species. It remains now therefore to examine the nature of this process from a more general point of view.

For this let us select a concrete instance.

In gardens a double form of *Chelidonium majus* is often found in which, as a rule, the doubling is only slight, and seldom consists in the production of more than 16-20 petals per flower (Fig. 61). On the plants in my cultures this doubling regularly increases from the spring until the summer, both on the main stems of plants in their first year and on the lateral stems of plants that have been wintered. For instance in May all the flowers were single, i. e., with four petals (Fig. 61 A). With June the number began to increase, and many flowers with 6 and 7 and occasional ones with as many as 10 petals occurred: whilst in the second half of June the majority had 12 to 14 and some 15 and 16 petals. Every year

[1] *Over het periodisch optreden der anomalien op monstreuze planten.* Bot. Jaarb., Gent, Vol. XI, 1899, p. 46, and *Ueber die Periodicität der partiellen Variationen*, Ber. d. d. bot. Ges., Vol. XVII, 1899, p. 45.

the doubling was seen to increase in intensity in the same way.

BRAUN, in his great work on the rejuvenescence of plants, has discussed the part played by periodicity in normal development with great thoroughness and clearness.[1] In the whole plant and also in the separate orders of branches the vigor of life goes up and down. The in-

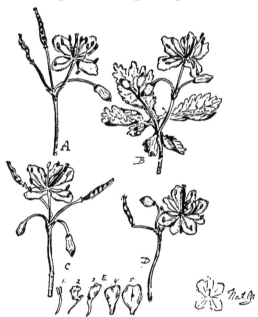

Fig. 61. *Chelidonium majus plenum.* A, a normal flower with four petals; B and C, flowers with five petals; D, a flower with eight well-developed petals and two petaloid stamens. 1-5, the transition between stamens and petals.

dividual strength of the plant is perpetually fluctuating, affecting the size of the leaves, the length and thickness of the internodes, the number of leaflets in compound leaves, the number of rays in the umbels, the ramifications in the inflorescences, the number of florets in the

[1] A. BRAUN, *Verjüngung*, pp. 23-55, 75-76, 90 etc. See further: HEINRICHER, *Biolog. Centralblatt*, Vol. XVI, No. 1, pp. 13-14. POKORNY, *Sitzber. d. Acad. d. Wiss.*, Vienna, 1875, Vol. 72, pp. 527-547.

flowerheads of composites, and so forth. Every shoot has its cycle. It begins with simple or atavistic forms, then gradually manifests the characters which are peculiar to the species in their full development, and gradually reverts. From the primary shoot this cycle is transmitted to the secondary shoots and from these to the branches of still higher orders. In this process the latter may become stronger or weaker in comparison with the main shoot, or even of equal strength. As a rule the branches are weaker than the main shoot; those which are not fall into two categories according to BRAUN, strengthened shoots and repeating shoots (*Erstarkungssprosse* and *Wiederholungssprosse*).[1] Spikes and racemes are the best known instances of weakened shoots; their apical flowers, if such are present, are in advance of all the rest in their development as for instance in the peloric *Digitalis*. In the red clover the main shoot consists of a short upright rhizome, whereas the lateral branches gradually increase in strength and develop into flowering stems. In *Tetragonia expansa* the main shoot is always short and erect, terminating with a flower. The secondary shoots often grow to more than a meter in length and are much stronger. Instances of strengthening shoots are the lateral branches which arise from the axils of the radical leaves of many plants (*Ranunculus bulbosus*), or from the middle part of the stem (*Chrysanthemum segetum* and *Trifolium incarnatum*); also the forked inflorescences as in *Saponaria officinalis*. Repeating shoots are often runners, and then we have what BRAUN calls "repetitional generations," as in *Valeriana officinalis, Lysimachia vulgaris,* etc.

See § 23, p. 256.

Space does not permit the extension of this list;[1] moreover all I am concerned with here is to show that this scheme also holds good for the distribution of anomalies on the plant. Unfortunately it often can not be exactly applied because it is complicated with the effect of external influences. Every shoot and system of branches has its susceptible period, during which the external factors which happen to prevail favor the production of the anomaly, or the type of the species, according to their nature. But apart from these numerous fluctuations the rule holds good where the material is sufficiently plentiful to justify the neglect of exceptions.

In a culture of *Specularia Speculum* which I had in 1892 I found the flowers to be partly tetramerous and partly pentamerous. The pentamerous ones occurred at the top of the stem and of the strongest branches of the second order, whilst all the remaining flowers were tetramerous. A similar difference between apical and lateral flowers is exhibited as a normal character by *Adoxa Moschatellina* and by many other plants. The *Pinus sylvestris* of this neighborhood often has its needles combined into groups of three instead of groups of two. Each such group of needles, as is well known, belongs to a single dwarf shoot or spur. I found that the trifoliate spurs occurred almost exclusively on the stronger branches, and chiefly at their upper end; but where they did occur they were numerous and closely packed. As a rule it is the main stem itself which bears them, but sometimes the strongest branches do so as well. There occurred up to 15 trifoliate spurs on the same one year's growth of the stem, all close to the apex, mixed with

[1] On the question of periodicity in the branching of cereals see SCHRIBAUX in *Journal d'Agriculture pratique*, 1899, and RIMPAU in *Landwirthsch. Jahrbücher,* Vol. XXIX, p. 589.

numerous bifoliate ones. *Pinus Pinaster* exhibits the same phenomenon. In different years the frequency of its occurrence varies greatly.

Camellia japonica with striped flowers, is striped mainly in November and December; but if it flowers in April, it produces only self-colored ones.[1] A form of *Trifolium repens* produced pitchers in my garden almost every year in no inconsiderable numbers and in great variety of forms. But they never occurred except in the spring;[2] just as the lime bears its pitchers chiefly on the first leaves of the branches and *Saxifraga crassifolia* on the lower abbreviated and leafy part of the flower peduncle.[3] *Ulmus campestris*, on the other hand, forms its pitchers chiefly from the strongest leaves in the middle and upper part of the branches.

It is in this middle region of the branches that anomalies are most commonly found, either exclusively or chiefly. Thus FRITZ MÜLLER describes a *Begonia* from Brazil, of the height of a man, which bore little appendages at the base of the leaf blade,[4] which were usually not more than 1-3, and sometimes from 5-50 mm. long. They were only found on the fourth to the tenth leaf, in one case from the second to the fifth leaf, of the upright stem; and occurred at the same height on every anomalous stem, both in the garden and in the field.

In May 1890 I observed a group of *Epilobium hirsutum*, the numerous and still young stems of which were for the most part forked. These divisions occurred always about the same height and did not recur during

---

[1] VERLOT, *loc. cit.*, p. 67.
[2] J. C. COSTERUS, *Bekertjes aan de eindblaadjes van Trifolium repens.* Botan. Jaarboek, Gent, 1892, p. 13, Pl. I.
[3] T. TAMMES, *Kon. Akad. d. Wetensch.*, Amsterdam, 1903.
[4] FR. MÜLLER, *Ber. d. d. Ges.*, Vol. V, p. 44.

the course of the summer. They were split fasciations, but the division had gone somewhat deeper, as was visible by the individual leaves being torn from below upwards with one half adhering to the one arm of the fork and the other half inserted on the other arm. Such leaves occurred on several shoots, but, as already stated, always at the same height on the plant.

The lower end of many racemose inflorescences is a favorite place for anomalies. Thus slightly double varieties of *Gladiolus* bear double flowers almost always in this position only. The racemes of *Prunus Padus* bear lateral racemes in this position almost exclusively; and, in other cases, it is also only in this position that tetramerous flowers are borne. Many double varieties are known to bear single flowers at the end of the flowering period, and sometimes also at the beginning. These flowers alone set seed, while the double ones are sterile.

It is well known amongst horticulturists that in multiplying perennials and bulbs by seed, the value of the plants cannot definitely be estimated in the first year in which they flower. It is not until the second or the third year of the flowering that their qualities are displayed to their full advantage. Many specimens of *Chrysanthemum indicum,* which when raised from seed, are only half double in the first year, will develop double flowers in the second year if grown from cuttings.[1] The varieties with tricolored leaves of *Pelargonium zonale tricolor* do not exhibit their full range of color until the second year after their seed is sown.[2] To breeders of tulips, hyacinths and other bulbous plants this rule is well known.

[1] REID and BORNEMANN's Catalogue, 1891, p. 20.
SUTTON's Catalogue, 1891, p. 77.

The rule for ordinary branching is that the anomaly diminishes with the higher orders of branching, omitting from consideration, of course, the strengthening and repetitional shoots. Every one knows the beautiful case of *Myosotis azorica Victoria* (*M. alpestris var.*), which has been on the market for many years, and was described by MAGNUS.[1] This heritable anomaly has a very much extended flower at the top of its main axis, often consisting of more than 10 and sometimes over 20 petals in one row. The number of sepals and stamens has correspondingly increased. The subsequent flowers of the inflorescence have become much less compound and the number of petals gradually diminishes during the course of the flowering period, until finally only pentamerous and hexamerous flowers are produced. *Chrysanthemum inodorum plenissimum* manifests a similar periodicity, and the number of petals in *Ficaria ranunculoides* and *Centaurea Cyanus* are in the same manner dependent on the order of branching.[2] *Veronica Buxbaumii*, according to BATESON and PERTZ, bears the largest number of anomalous flowers at the beginning of the flowering period, that is to say, just before it is at its height.[3] *Myosurus minimus* bears the more single flowers the weaker these are.[4] A number of similar cases have already been collected by MUNTING in the seventeenth century, and recently by BURKILL amongst others.[5]

[1] *Verhandl. d. Bot. Ver. d. Prov. Brandenburg*, XXIV, 1882, p. 119, Pl. IV.

[2] J. MAC LEOD, *Botanisch Jaaboek, Gent*, 1899, Vol. XI.

[3] W. BATESON and MISS PERTZ, *Notes on the Inheritance of Variation in the Corolla of Veronica Buxbaumii*, Proceed. Cambridge Phil. Soc., X, Part 2, p. 78 (1898).

[4] H. MÜLLER, *Nature*, Vol. XXVI, 1882, p. 81.

[5] A. MUNTING, *Waare oeffeninge*, 1671; J. H. BURKILL, *Linnaean Soc. Journ. Bot.*, Vol. XXXI, 1895, p. 216.

Tagetes africana and *T. signata* often first exhibit single flowerheads only, but subsequently double ones in increasing numbers on the same plants. The same thing can be seen in *Zinnia elegans,* and in *Althaea rosea* the flowers are sometimes seen to become more double as we proceed up the stem.

Weak lateral buds in variable races often exhibit a special tendency to atavistic variations. This is true of the branches of the highest orders, of the accessory shoots, which are formed in addition to the normal axillary bud, of resting buds, etc. But it should be remembered that it is just these buds which often develop into very strong branches and become the succulent shoots which tend to reproduce the anomaly rather than the specific type. I have already referred to this exception above.

Capsella Heegeri, the new species described by SOLMS,[1] produces reversions to *C. Bursa pastoris* on its weak lateral branches and, according to the same author, the same phenomenon is exhibited by those varieties of *Nasturtium palustre* which have been lumped together into the genus *Tetrapoma*. *Papaver somniferum polycephalum* produces flowers without any accessory carpel almost exclusively on the very weak branches which arise from the main stem towards the end of the year when the lateral branches have been cut away. The peloric flowers of *Digitalis purpurea* are monstrously developed at the apex of the stem and on the strongest lateral branches; but on the weaker ones they are simple and very regular (Fig. 42, p. 223).

Such buds can often be induced to develop by cutting

---

[1] H. GRAF ZU SOLMS-LAUBACH, *Cruciferenstudien*. Botan. Zeitung, 1900, Pt. X, p. 167.

down the main stem; and by this means we can often obtain the desired variations, as GOEBEL has shown, unless the anomaly is too much favored by a very vigorous growth.

§ 29. THE CHOICE OF SEEDS IN SELECTION.

From the periodical changes in the tendency of the plant to produce anomalies, we might expect a corresponding periodicity in the seeds.[1] The seeds of flowers which have bloomed under favorable circumstances should produce more anomalous individuals, those of the weaker branches, on the other hand, more atavistic ones. This expectation, however, is only to a slight extent fulfilled, and the latent capacity of the seeds seems to be regulated much more by factors other than those determining the periodicity of the parent plant. Our knowledge of this subject is, however, still in its infancy.

Nevertheless there is a whole series of facts which are worth collating because they are likely to serve as a stimulus to further investigation. In doing so, it should be emphasized that in the selection of seeds we are concerned with variability and not with mutability, and in most cases, in fact, with the extreme fluctuating variability of semi-latent characters.

It is only a strong, well-developed seed, furnished with a healthy germ and richly supplied with food material that can give rise to the very strongest plant in a healthy culture. Doubtless, of course, the differences between the individual seeds are to a large extent levelled down during the first weeks after sowing; but this is not always the case. Obviously the most favorable condi-

---

[1] C. FRUWIRTH, *Die Züchtung der landwirthschaftlichen Culturpflanzen*, 1901, p. 102.

tions can only give the best possible result when they are combined with the highest germinal capacity.

For this reason in agricultural practice and occasionally also in horticulture special attention is often paid to the individual seed. The points to which attention is paid are, on the one hand, the size and weight of the individual seeds and, on the other, the place of their origin on the parent plant. The practice of selection in cereals consists essentially in the choice of the largest and heaviest seeds, or more strictly, in the elimination of the smaller ones by winnowing machines or other devices.[1] When the object is to produce small families to serve as the stocks of new races, measurement and weighing of the individual seeds is recommended by the best authorities, and special trays for determining their weight have been devised.[2]

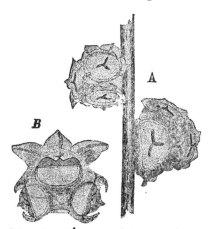

Fig. 62. Clumps of fruits of the sugar beet; half schematic. A, two ripe clumps on a stem; B, one of these cut longitudinally showing the three seeds in the three special fruits.[3]

An important advance in the method of selection has been made in recent years by VAN DE VELDE who

---

[1] See VON RÜMKER, *Getreidezüchtung*, 1889, and VON RÜMKER, *Der wirthschaftliche Mehrwerth guter Culturvarietäten und ausgelesenen Saatgutes*, Arbeiten der D. Landw. Gesellsch., 1898, Pt. 36, p. 127.

[2] VON RÜMKER, *Journ. für Landwirthschaft*, 39. Jahrg., Pt. 2, p. 129.

[3] The specimens from which this drawing was made I owe to the kindness of Messrs. KÜHN & Co., beet seedsmen in Naarden, Holland. They were taken from selected beets of the very high

has studied the relation between the size of the seeds and rapidity of germination.[1] As a rule the larger seeds germinate faster than the smaller ones; and thus their weight favors the production of strong plants in more than one way. Moreover, as has long been known, the seeds which germinate the first are the best and produce the finest plants. When harvesting of flower seeds has to be done in the field, the first seeds often fall out before the harvest; they germinate easily and usually do so at once; and we all know that from such seeds the finest specimens may be produced. In fact for certain definite species the handbooks recommend the practice of allowing them to sow themselves, since this self-sowing leads to much better results than are obtained even from seed carefully harvested at the proper time.

The question as to where the best seeds occur on the plant is one that has been much discussed. Let us consider the sugar beet as our first instance. In this species 2-5 or more fruits are associated to form a so-called clump (Fig. 62), in which the upper flower produces the largest seed Fig. 62 b); whilst the lateral seeds of the same clump are less fine and markedly smaller and differ amongst themselves very much in size.[2] This phenomenon has been recently subjected by BRIEM to an exhaustive investigation.[3] He sowed the clumps, and planted out their individual seedlings separately in a row. In

percentage value of from 18.9 to 20.1% of sugar, and with roots weighing 900-1100 grams.

[1] VAN DE VELDE, *Invloed van de grootte der zaden op de kieming.* Botanisch Jaarb. Gent, 1898, pp. 109-131.

[2] *Keimungsgeschichte der Zuckerrübe,* Landw. Jahrb., VIII, 1879, p. 14,

[3] H. BRIEM, *Studien über Samenrüben, einem Rübenknäuel entstammend.* Oesterr.-Ungar. Zeitschr. f. Zuckerindustrie und Landwirthsch., 1900, Pts. II, IV, and VI.

this way the plantlets derived from a single clump could be compared with one another throughout the whole course of their development. The largest seed becomes the largest seedling which produces the finest beet, and lastly the most fertile seed-parent. The relative weights of the five seedlings of a clump at the end of germination were as follows: 100—74—67—51—46. The heaviest seedling weighed 5.8 milligrams. The full-grown beets derived from a single clump weighed in one case 1156, 859, 574, 344, 310 grams, and they furnished respectively 241, 167, 202, 239, 104 grams of seed at the end of their second year.

E. SCHAAF considers a very important advantage of the so-called "cutting"-culture in beets to be the exclusive development of the largest seed of each clump which is effected by the close proximity of the plants during the production of the clumps.[1]

Amongst cereals the matter has been most thoroughly investigated and I refer the reader to the literature on the subject which I have already cited. The heaviest grains are situated at the middle or somewhat below the middle of the ear. FRUWIRTH showed this to be true for barley, rye, wheat, spelt, and also for maize.[2] There proved to be certain subordinate differences characteristic of the various varieties and species.[3] BRUYNING found that in oats the lower grains of the lateral ears are far

---

[1] E. SCHAAF, *Blätter für Zuckerrübenbau,* Jahrg. VII, No. 24, Dec. 1900.

[2] C. FRUWIRTH, *Ueber den Sitz des schwersten Kornes in den Fruchtständen beim Getreide,* in WOLLNY'S Forschungen auf dem Gebiete der Agric.-Physik, XV, 1892, p. 49.

[3] E. NOTHWANG, *Unters. über die Vertheilung d. Körnergewichtes an Roggenähren,* Diss., Leipsic, 1893; Bot. Centralblatt, 1895, II, p. 263.

better than the upper ones,[1] and the same is true of other species of cereals.

Lastly some reference should be made to those cases in which individual seeds possess the peculiarity of germinating late, and of remaining one or more years in the soil, as for instance the small seeds of various species of clover. In *Xanthium canadense* each fruit contains two seeds, one of which germinates after the first winter, the other not until after the second.[2]

When we are dealing with semi-latent or, in general, with highly variable characters, a selection of seeds either by their size and weight or by their place of origin on the plant is to be recommended in many cases, and the general rule seems to be that the place of origin of the best seeds will also be that of the desired variants. There are some cases in which this rule does not apply, such as we have seen in *Trifolium incarnatum,* where it is the smallest late germinating seeds which contain the best representatives of the four-leaved half race.

In *Chelidonium majus plenum* the single flowers bloom first, and the double ones later, as we have seen in the preceding section (§ 28). I have harvested the seeds of both and sown them separately; but found no difference in respect of the doubling amongst the offspring. BATESON and MISS PERTZ[3] also failed to find any difference in respect of doubling amongst the offspring of normal and abnormal flowers of the same plant with *Veronica Buxbaumii*. In *Oenothera Lamarckiana* I found about the same percentage of annual and

[1] F. F. BRUYNING, *Proefnemingen met havervarieteiten,* Wageningen, 1900.

[2] J. C. ARTHUR, *Proceedings Ann. Meeting Soc. Agric.,* Science, August, 1895.

[3] W. BATESON and MISS PERTZ, *loc. cit.,* p. 79.

The Choice of Seeds in Selection. 337

biennial individuals from the upper and lower fruits of the same spike. In *Viola tricolor maxima* the small summer flowers furnish the main quantity of seed. It never occurs to a seed collector to regard them as of less value than the first seeds. In many double plants, especially in *Begonia,* it is practically only the flowers which bloom last that produce pollen and set seed. This seed is always sown without any detrimental result to the degree of doubling of the variety.

On the other hand it is generally regarded as desirable to harvest the lower seeds of the inflorescence in the coxcomb, *Celosia cristata*; and in the case of the crested species of ferns (*Varietates cristatae*) the spores which are found on the dissected leaves, and still more on the tips of such leaves, are regarded as the best, although spores found on the other parts of the leaves will certainly repeat the anomaly.[1] Another instance which has been studied by many investigators is furnished by the stocks, whose double varieties have been known for a very long time, for a century at least, to consist in each generation, of double and single plants in about equal numbers.[2] The former are absolutely sterile, lacking pistils and pollen as a result of their petalomania (see Vol. I, p. 194), so that only the latter can play a part in the continuation of the race. There are certain differences between the seeds which produce the single and the double specimens. The latter are heavier and germinate more quickly,[3] and the young plants can be sorted out in the

[1] See the long list of references given by GOEBEL, *Organographie*, I, p. 158; VERLOT, *loc. cit.*, p. 97; CARRIÈRE, *loc. cit.*, p. 67; KENCELY BRIDGMANN, *Ann. Sc. nat.*, 4° Serie, Vol. XVI, p. 367; C. T. DRUERY. *Journ. Roy. Hort. Soc.*, Vol. XII, III, 1890, p. 517, etc.

[2] E. CHATÉ FILS, *Culture pratique des giroflées*, Paris, Biblioth. de l'horticulteur praticien.

[3] NOBBE, *Botan. Centralblatt*, Vol. XXXII, 1887, p. 253.

beds long before they exhibit buds.[1] The lower pods on the main stem and on the principal branches as well as the lower two-thirds of each pod furnish on the average more double than single plants; the upper sections of the pod and the pods of weaker branches yield more single ones. The proportion of double plants in the harvest can be increased to about 60% either by limiting the production of seed by means of culture in pots, or by pruning; and in the best nurseries the finer sorts are usually subjected to this treatment. If the seeds are kept through some years the proportion of double seedlings gradually increases, because the mortality is greater amongst seeds that were destined to produce singles.

It is stated that in the case of the Balsam and many other double flowered varieties the seeds are rounder and fuller and also smoother than those of the corresponding single sorts. The "double" seeds of *Petunia* are said to germinate later than the single ones, so at least I have been told by nursery men at Erfurt. In the Composites the central seeds of the disc in double varieties are said to be more likely to repeat the anomaly than the marginal ones. All these statements should, however, be regarded critically, and many of them are in need of experimental confirmation.[2]

Nevertheless the general rule is that the various seeds of a plant may give rise to offspring of widely different degrees of individual vigor, according to the place of their origin on the parent plant, their size and their

---

[1] This process which is carried out by children is called *ésimpler*, in France. The matter stands in need of closer investigation.

[2] PEYRITSCH has collected references to the earlier literature relating to this point. *Zur Aetiologie pelorischer Blüthenbildungen*, Abhandl. k. k. Akad. Wien, 1877, pp. 135-136.

weight; and that, in accordance with the rules which we have already enunciated, a greater or less development of the varietal character is correlated with these degrees of strength.

PART II.
THE ORIGIN OF EVERSPORTING VARIETIES.

I. TRICOTYLOUS RACES.

§ 1. THE OCCURRENCE OF TRICOTYLS AS HALF RACES AND INTERMEDIATE RACES.

In the chapter on latent and semi-latent characters in the first part of this volume, I have discussed the difference between half races and intermediate races. It is not in the possession of certain elementary characters that they differ one from another; in this respect they are identical. They possess exactly the same characters and in the same numbers. That feature, however, which constitutes the point of difference, is semi-latent in the half race, that is to say manifests itself only rarely and in occasional individuals, one in every thousand for instance. In the intermediate race, on the other hand, it is active and equivalent to the character to which in the half race it is, as it were, subordinate. Considered with regard to the features which distinguish them, both races, therefore, possess two elementary characters, which, however, cannot be expressed simultaneously in the same organ but are mutually exclusive.

In an ideal intermediate race, these two antagonistic characters would be of exactly equal value; that is to say, half of the individuals would exhibit the one, and the other half the other character. Whether such ideal races actually exist in nature is an open question, since as a rule one of the two characters is more or less easily manifested than the other. Moreover those cases would have to be excluded in which either the conditions of life or some selection of the race could have exerted an influence in one direction or the other; for, as we have seen, intermediate races are very susceptible to both these groups of factors.

Among the intermediate races known to me the tricotylous and syncotylous forms approach most closely this ideal picture.[1] For, in pure cultures, they furnish as a rule 50% dicotylous and 50% tricotylous or syncotylous seedlings. By altering the conditions of growth as well as by selection this proportion can be easily and greatly modified in both directions, almost to the exclusion of one or other of the two types. But such treatment leaves the essential nature of the intermediate race untouched. It neither reverts to the half race when subjected to selection, nor is it possible to derive a constant and pure tricotylous variety from it.[2]

As far as I know, there are hardly any references to tricotylous races in botanical literature, and the possibility of the existence of tricotylous intermediate races seems never to have been discussed. In this part, however, I shall describe some instances of such races in order to demonstrate their existence and to study their characters. In the period from 1892 to 1897 I succeeded in producing such races from half a dozen very different species.

I wanted also to include in this inquiry some pure tricotylous and pure dicotylous races, that is to say, races the seedlings of which were in the first case exclusively tricotylous and in the latter exclusively dicotyl-

[1] See my preliminary note *Ueber tricotyle Rassen* in Ber. d. d. bot. Ges., 1902, Vol. XX, p. 45.
[2] See the scheme on page 24.

ous. But hitherto I have not discovered a single instance of the former, and have only obtained one instance of the latter.[1]

Before I proceed to a detailed description of my races and cultures it seems desirable to give a general account of the manner in which tricotylous seedlings are found, and how the desired half and intermediate races may be most easily derived from them.

It is well known that amongst the seedlings of dicotylous species occasionally individuals are found with three seed leaves. It is only necessary to look over a seed-bed in the garden in spring in order to find instances of these. The more extensive the sowing and the more careful our search the greater will be the number of tricotylous seedlings found. Some species produce them in greater, others in smaller proportions; and they can often be found

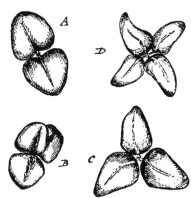

Fig. 63. *Antirrhinum majus.* A, C, D, seedlings with 2, 3, and 4 cotyledons. B, with a deeply cleft cotyledon.

even in the smaller pot-cultures of the greenhouse, but in many cases I have had to sow 10,000 or 20,000 seeds of a species before finding a single individual which showed any variation in this direction. But numerous species seem to produce one or several tricotyls in every hundred or thousand seedlings.

---

[1] My *Helianthus annuus syncotyleus* has not produced a single tricotylous plant in the ten years during which I have often counted hundreds or even thousands of seedlings every year. On the other hand they occasionally occur in *Helianthus annuus variegatus* and some other varieties of the sunflower.

If the variants are rare, they are as a rule normal tricotyls; but if they are more numerous the type is usually seen to be variable both in the *minus* and in the *plus* direction. For convenience of expression we may regard a tricotylous seedling as having arisen by the doubling of one of the two cotyledons of a dicotyledon by splitting, just as is so frequently observed in foliage leaves. Smaller degrees of the splitting would lead to variations in the *minus* direction; but if the splitting affects both cotyledons there arise variations in the *plus* direction, which, if the doubling is complete, result in the origin of tetracotyls (Fig. 63 D). A seedling with one normal and one split seed-leaf is called a hemi-tricotyl;[1] one with two split seed-leaves, or with three of which one is split, is called a hemi-tetracotyl. In the same way hemi-pentacotyls, and so on, may be found; but the deviations become rarer as they are more remote from the pure tricotylous type.

If we make a collection of all these forms it is easy to construct a continuous series which extends from the pure type, on one side, through stages characterized by more or less deep fission, to the dicotyls; and in the other direction in a similar manner to the tetracotyls and, if the material is extensive, even further, to the pentacotyls, and so on. Fig. 64 exhibits such a series derived from *Oenothera hirtella,* the unsplit seed-leaf of each plantlet being omitted. But obviously even here the forms figured are only a selection from a much more complete series. If we imagine those cotyledons which have been cut off in these figures to be split also, the series would represent the transition from the tricotyls to the tetracotyls.

[1] *Berichte d. d. bot. Ges.,* 1894, Vol. XII, p. 26.

A series of this kind is, however, a purely morphological one and neither physiological nor statistical. If we wish to obtain this, we must not merely pay attention to the forms, but also to the frequency of their occurrence. In doing so the first striking fact is that all the aberrant forms taken together, constitute only a very small percentage of the total number, even smaller sometimes than the figures already given. Therefore, if we construct a frequency curve the dicotyls produce a high peak and the curve extends from this only in one direction and is therefore a so-called half curve.[1]

Further, amongst the aberrant forms themselves, the various forms occur in widely different proportions. The hemi-tricotyls are far rarer than the tricotyls; the number of all the hemi-tricotylous types together often does not amount to as much as that of the pure tricotyls. Deep clefts are somewhat less rare than shallow ones; and we often see specimens which at first seem to be purely tricotylous but which, when the peduncles of the cotyledons gradually elongate, turn out to be deeply cleft. This is particularly evident in *Amarantus speciosus* and *Antirrhinum majus* (Fig. 63 B), in which the closer juxtaposition of two of the cotyledons betrays the fact that they arise from a common stalk. Hemi-tetracotyls are always much rarer than tricotyls in sowings from commercial seed as well as in selected races. Nevertheless some species seem to be richer in them than others.

If we plot such a frequency distribution we obtain a two-peaked curve which has a small secondary apex over the ordinate for the tricotyls, besides the main one for the dicotyls. Thus I found a crop of 800 seedlings of *Cannabis sativa* of 1894 to have the following com-

[1] See above, p. 26.

position, the proportion of tricotyls (that is to say all the aberrant forms of the series) being about 10%.

	DICOTYLS	HEMITRIC.	TRICOT.	HEMI-TETRACOTYLS AND TETRACOTYLS
	726	24	46	4
Or in %	90.8	3.0	5.7	0.5

If we wish to make ordinates for the various degrees of cleavage we are met by the difficulty presented by the choice of the limits of such arbitrary groups. This diffi-

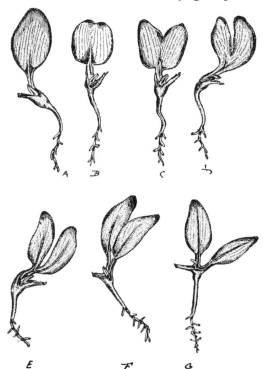

Fig. 64. *Oenothera hirtella.* Intermediate forms between the dicotylous seedling (A) and a tricotylous one (G). The normal cotyledon has been cut away from each plant. Cultivated in 1900.

culty is partly due to the fact that the stalks of the cotyledons continue to grow for some time after the cotyledons themselves have assumed their definitive form.

But as a rule we find the general relations to be such as are exhibited in Figure 65. This curve has a form which is also commonly seen in other anomalies.[1]

Whilst dicotyly is a character without variation tricotyly is one which exhibits a very high degree of fluctuation. The limit between dicotyls and the extreme *minus* variants of the tricotyls can always be easily and certainly observed; because the apex of the cotyledon is the first part of it to assume its definitive form, and a cleft in this apex, however small, can be clearly seen. There is therefore no fear that the rarity of such extremes might be due to imperfect observation. The limits between the remaining smaller groups are blurred and arbitrary. But this difficulty disappears in the evaluation of the degree of inheritance, because all the seedlings which exhibit cleavage in one or both cotyledons are united into a single group and treated as tricotyls in the larger sense.

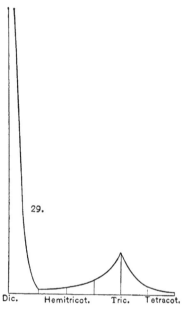

Fig. 65. Schematic representation of the fluctuating variability of tricotyly. The tricotyls and the dicotyls constitute the two apexes, the hemi-tricotyls and tetracotyls (together with the hemi-tetracotyls) constitute the remaining ordinates.

I shall denote the dicotylous seedlings of these races as atavists.[2] In the sowings

---

[1] See Chapter IV of this part, and *Sur les courbes galtoniennes des monstruosités*, in the Bull. Scientif. de la France et de la Belgique, published by A. GIARD, Vol. XXVII, 1896, p. 397.

[2] See above, p. 104.

of commercial seed the application of this term would not, of course, be justified; but in my cultures, which, almost without exception, were started by the selection of tricotylous individuals, this term is obviously fully justified. Moreover, in this way, the word dicotyl is left to its pure systematic signification. And, as in some other cases, atavism is here seen to be an oscillation between two empirically known extremes.[1]

So far as my experience goes, tricotylous seedlings are much commoner amongst cultivated species than amongst wild ones, and even amongst the latter they usually occur from seeds saved in botanical gardens and very seldom from those collected in the field. Amongst cultivated plants, again, they are commoner amongst species which are grown on a large scale than on a small one.

Thus I have obtained my intermediate races from amongst the former, partly from agricultural and partly from horticultural species. As an instance of the former I may mention the hemp and of the latter the snap-dragon and the wild poppy. But besides these I discovered a rich tricotylous race in my *Oenothera hirtella,* an entirely new species which was found quite by chance in my garden. It is not impossible that cultivation on a large scale favors the origin of new races.

A profusion of instances of some anomaly in a species either in the field or in cultivation, suggests the presence of an intermediate race;[2] a scarcity, however, that of a half race, as is especially well known to be true in the case of fasciations. Tricotyly conforms strictly to this rule. From seeds obtained from the trade or by exchange

[1] See page 108.
[2] See page 32.

Tricotyls as Half Races and Intermediate Races. 351

we rarely get more than low percentages; but it is obvious that we should not expect all the seed-parents in the field to furnish the same number of tricotylous seedlings. Commercial seed is almost certain to be a mixture and to yield a mean proportion, which may perhaps be much lower than the seed of those specimens would yield, which were the richest in tricotylous offspring. It follows from this that we can only form a proper estimate if we select a group of plants among our first sowing and harvest their seeds separately, if possible after taking the precaution of preventing too much cross-fertilization by insects. It is obvious that in such a culture a choice will have to be made, and as a rule the tricotyls, using this term in its widest sense, will be selected. From the separate harvest of these selected specimens, individual hereditary coefficients may then be obtained.

In spite of every precaution, the choice of the plants will depend mainly on chance, for, as we shall see later, tricotyls are by no means more likely to reproduce the anomaly than some of the dicotyls. On the contrary, plants much better in this respect sometimes occur among these; but we have at present no other means of discovering them except by growing their seed. Experience however has shown that if we allow a fairly large group to ripen their seed we may become independent of chance in so far that we can be fairly sure that at least some seedparents will give a relatively high proportion, provided of course that the original mixture contains such amongst its components. It will soon be seen that in spite of all apparent intermediate degrees, two main groups can be distinguished. In one of them we find only low individual values, from 0 to 3%, or very little over; in the second, on the other hand, besides these some higher ones which not rarely attain a value of 10% to 20 % and in rare cases even of 30% to 40 %.

Whenever the cultivation in the garden gives no ground for the assumption of special influences the difference between the values derived from the original commercial samples of seed and the self-harvested seed of the first year must be attributed almost exclusively to the fact that the commercial seed was a mixture whilst the self-harvested crops matured in isolation. But since mixtures of this kind have been the rule during the generations which preceded the purchase of the commercial, horticultural or agricultural, seeds, it is evident that those species which contain a number of rich seed-parents every year, will give a higher percentage of tricotyls in the mixture than others. Thus the percentage composition of commercial seed gives some idea of what may be expected from it by subsequent culture.

In the following sections (§§ 2-8) I shall describe my experiments on these two groups in detail. Here, however, I will give the main result. It is this:

By far the largest number of species contain only a half race in respect to tricotyly, but some few also contain, besides this, the intermediate race. If the latter is present in commercial seed or in seed obtained by exchange, it can be easily and speedily isolated; but if it is not present years of selection cannot bring it out. The half race and the intermediate race are, here as elsewhere, perfectly distinct things, which do not merge into one another, or if they do, they do so only by chance.

Sometimes, but on the whole very seldom, an indication of the likelihood of obtaining tricotylous intermediate races may be afforded in nurseries and in agriculture. This consists in species with a decussate arrangement of the leaves in the exhibition of trimerous whorls on the stems in the later life of the plant. I paid great attention to this point at the beginning of my experiments when I visited the great nurseries at Erfurt. Here the ternary individuals of *Antirrhinum majus* in the fields impressed me greatly. They were not considered by the gardeners as worth any attention, but they formed the foundation for my first tricotylous intermediate race.

The difference between tricotylous half and intermediate races lies in their percentage composition and not in the visible characters of the individuals. Neither the number nor the cleavage of the cotyledons on a single individual is decisive. As a rule tricotylous specimens of both races tend to produce a richer harvest of the tricotyls than the atavists of the same race; but experience shows that the difference is only a small one; and, further, that tricotyls, even those of a high productive capacity, are often surpassed in this respect by some of their atavistic brethren. The chief point is, however, that both the half race and the intermediate race are composed of both types of individuals; in the former the tricotyls are rare, whereas in the latter, under normal circumstances, both forms appear in about equal numbers. Moreover, both races contain all the stages of hemi-tricotyls, and, although these are rarer, of hemi-tetracotyls also.

It is not possible, therefore, to tell from a single plant to which race it belongs. Only its ancestry can determine this; and if this is unknown, we have to reach the decision by means of subsequent breeding. It is an extreme case of the transgressive variability which was discussed in the first volume.[1] The forms composing a

[1] Vol. I, Part II, § 25, pp. 430 ff.

half race occur in the intermediate race also, and as a rule all of them in every sowing, provided it is not too small. On the other hand, if the half race is cultivated on a sufficiently large scale, it will contain all the forms of the intermediate race. There is no morphological limit between the two, although the physiological one is perfectly definite, and in my experiments has never been

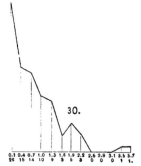
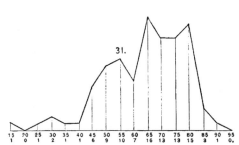

Fig. 66. *Oenothera rubrinervis*. Tricotylous half race. Curve of the hereditary capacities of the plants of 1894. The highest of these was 3.7%. The upper row of figures relates to the percentage composition in tricotyls, the lower to the numbers of seed-parents with this composition. The ordinates have the same signification as these latter figures.[1]

Fig. 67. *Oenothera hirtella*. Tricotylous intermediate race; curve of hereditary capacities of plants of 1896, including both atavists and tricotyls. The capacities oscillate between 15 and 95%. The meaning of the figures is the same as in Fig. 66.

overstepped. The tricotyls of the half race do not lead on to the intermediate race, nor do the atavists of this race lead on to the half race. The two races are just

[1] The percentage figures are calculated, in this experiment, from counts made on 300 seedlings from each plant; but the number of ordinates is reduced to one-third. Therefore 0.1 means, 0—0.2; 0.4, 0.3—0.5 and so on. The tricotylous races of *Dracocephalum moldavicum* and *Penstemon gentianoides* gave similar curves. (Harvest of 1894).

as sharply and unalterably separated as, in the first part of this volume, we saw was the case in numerous instances, and especially in the five-leaved race of the red clover. Nothing less than a mutation can effect the transition between the two, but I have not yet had the good fortune to observe such an occurrence.

In these experiments therefore the differences between individuals to which attention has to be paid are their hereditary values; and whether they themselves have two, or three or cleft cotyledons is a matter of secondary importance. In my cultures the selection of tricotyls as seed-parents has been the general rule, since this practice on the one hand increases the probability of excluding specimens with a low hereditary capacity, and on the other, of including those with the high; but the increase of this chance is only a small one, as the frequent cultures I have made from atavists clearly show (see below, § 6).

For a half race the curve describing these values is a half curve. The vast majority of individuals have either nothing, or little else, but dicotylous offspring; and the numbers of individuals with the larger numbers of tricotylous offspring decrease rapidly (Fig. 66). These curves may be improved in the same way as those of other half races, viz., by the selection of individuals with the highest value, as we have seen in *Ranunculus bulbosus semiplenus* (see § 23, of the first part of this volume, p. 249).

Curves describing these values in intermediate races usually have their maximum ordinates at 50%; they are, however, liable to be much altered by selection and external conditions. Fig. 67 is a curve of this kind for *Ocnothera hirtella,* whose apex is at about 65%. If from a group like this the plants with the smallest values are selected for the continuation of the culture, values which are equal to or even smaller than the best ones of the half race can be obtained. Here also the variability is of the transgressive kind; but it does not result in the transition from one race to the other.

Half races and intermediate races are, therefore, with regard both to the forms of their individuals and the magnitude of their hereditary capacities highly variable races. On the other hand they are perfectly constant inasmuch as neither race can be transformed into the other artificially. They behave like the majority of ever-sporting races of other anomalies.[1]

§ 2. TRICOTYLS, HEMI-TRICOTYLS, AND TETRACOTYLS.

The opinion, prevalent amongst gardeners, that hybrids are intermediate in form between their parents, and that intermediate types, therefore, should be regarded as hybrids, might easily lead to the assumption that the hemi-tricotyls are hybrids between tricotyls and dicotyls. Intermediate they are without any doubt, especially when the cotyledons are cleft over half their length. But the occurrence of a continuous series of intermediate forms between di- and tricotylous plants proves the incorrectness of this view, or at least indicates that it does not cover all the facts. The question has to be decided by experiment. For this reason I have made a series of crosses between dicotyls of a half race, and tricotyls of the corresponding intermediate race. Although I always paid especial attention to the occurrence of hemi-tricotylous seedlings amongst their results, they were just as

---

[1] See above, p. 22.

rare amongst the hybrids as in other cultures. The hybrids are almost without exception dicotylous, although of course occasional hemi-tricotyls and tricotyls occur amongst them, just as they do in the half races. There is therefore no ground for the supposition that seedlings with cleft leaves should be of a hybrid nature.

They are simply variants of the tricotylous type. The occurrence of seedlings in which both the cotyledons are cleft or doubled, favors this view. They cannot possibly be regarded as hybrids; they occur so regularly and abun-

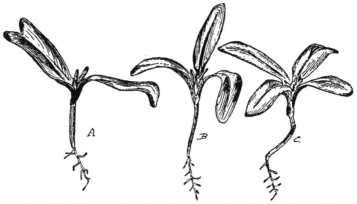

Fig. 68. *Silene odontipetala*. A, a hemi-tricotylous, B, a tricotylous, and C, a tetracotylous seedling, 1900.

dantly in tricotylous intermediate races that they must be simply regarded as *plus* variants of the same character. Moreover the fact that one or the other of the seed leaves in a tetracotyl may be cleft also, supports this view. Such a cleavage results in the origin of pentacotyls and hexacotyls which, however, will be, obviously, very rare. True pentacotyls, arisen by such a cleavage, I have seen, for instance, in *Scrophularia nodosa, Amarantus speciosus, Oenothera Lamarckiana* and *Papaver Rhoeas* (Fig. 69.)

Hemi-tricotyls and tetracotyls are, as we have already stated, as a rule rarer than tricotyls, even when we include all the *minus* and *plus* variants of these groups. This can be observed in bought samples of seed; but better if we sow the seed of plants grown from such a sample separately. Thus, for instance, I found in 1892 among the seeds of a plant of *Asperula azurea setosa* 3 hemi-tricotyls, 15 tricotyls and 3 tetracotyls among 1170 seedlings, and in the crop derived from another individual of the same culture 2 hemi-tricotyls and 5

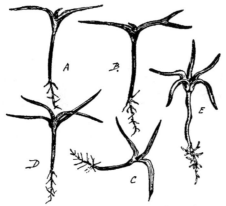

Fig. 69. *Papaver Rhoeas.* Semi-double cultivated form. Dicotylous, hemi-tricotylous, tricotylous, tetracotylous, and pentacotylous seedlings; from the seeds of 1899.

tricotyls amongst 550 seedlings. Further, I found in 1892, amongst 13,000 seedlings of *Amarantus speciosus* 202 hemi-tricotyls, 245 tricotyls, and 22 tetracotyls and hemi-tetracotyls. In the former group the cleavage was shallow in 47 cases, deep in 97 and intermediate in 58. The same general result was obtained in other plants.

To obtain a further knowledge of tricotyls and hemi-tricotyls we must cultivate them further, allow them to flower separately, and compare the composition of their progeny with that of the tricotylous individuals from the same culture. In doing so we find that they do not tend to reproduce their own type, but behave, as a rule, in the same way as tricotyls. Minor quantitative differences may occur, but of qualitative there are none. At any rate it seems impossible to isolate and fix these two subsidiary types and obtain pure hemi-tricotylous or tricotylous races.

I propose to deal first with the hemitricotyls. I have repeatedly isolated them and tested their hereditary capacity, especially in *Amarantus speciosus* and *Cannabis sativa*. In *Amarantus*, if we plant out some hemi-tricotyls and some tricotyls, the highest values are sometimes obtained for the former, and sometimes for the latter; but only with slight differences. For instance, in the harvest of 1892 the value of 20,000 seedlings was a mean of 2% for the former, and 3.5% for the latter. Moreover there were slightly more hemi-tricotyls than tricotyls amongst the former, the difference, however, was only 0.1% in 10,000 seedlings of each group, and in this figure the various degrees of cleavage were, moreover, lumped together. In 1892 to 1895 I then grew the hemi-tricotyls and the tricotyls separately for three further generations, always selecting the individuals with the highest hereditary capacities. In these four years the highest values obtained varied for the hemi-tricotyls from 4.2 to 8.5% with a mean of 5.5%, and for the tricotyls from 3.6% to 7.4% with a mean of 5.7%. From these figures it seems to make practically no difference whether hemi-tricotyls or tricotyls are selected as seed-parents. Of *Cannabis sativa* I had in the summer of 1894 a bed of each of the two types. The hemi-tricotyls gave values varying from 1 to 26%, and the tricotyls from 4 to 14%: with means therefore of 11 and 9%. In *Penstemon gentianoides* the cleft forms gave a mean of 2.8%, but the tricotyls a mean of 2.9%. On the whole, therefore, there are no essential differences in hereditary capacity between the hemi-tricotylous and tricotylous examples of the same races. Moreover we shall see later that this rule must be true, when we shall find that even the atavists in the pure races do not differ essentially in these values from the tricotyls (see § 6).

Fig. 70. *Acer Pseudo-Platanus.* A tetracotylous seedling, the axis of which splits above the cotyledon. In the cleft two leaves are seen, the lower part of whose stalks are grown together, back to back. (Spring, 1887).

We now come to the tetracotyls. Experiments here meet with the difficulty of distinguishing the true types from others. For, occasionally, double seedlings or twins occur. In these the axis is more or less deeply cleft; it looks as if two seedlings had fused together side by side. If the cleavage extends downwards into the hypocotylous region there are two separate groups of seed-leaves and these are frequently at different heights. There is obviously no danger of confusing such cases with tetracotyls, although the twin seedling does actually bear four seed leaves (Fig. 71). But if the division affects only the part of the axis above the seed-leaves, these stand in a whorl and the seedling cannot be distinguished from those in which the seed-leaves and not the axis have divided. Only after further growth its true nature can be decided. If, however, the doubling is continued without splitting of the stem, peculiar fasciated plants may be the result. In such cases the real state of affairs often remains hidden.

In *Amarantus speciosus* especially, I have often observed such twins (Fig. 71), and also in *Datura Stramonium, Acer Pseudo-Platanus* (Fig. 70) etc.[1] Fig. 72 represents a section of a stem of a tetracotylous plant of

Fig. 71. Seedlings of *Amarantus speciosus*. A, tetracotylous; B, twin, each of the split halves of the axis bearing two seed-leaves; C, twin, one of whose halves is tricotylous; D, hemi-pentacotylous seedling, the sole instance of this case in a culture of over 20,000 seedlings; E, Trisyncotylous plant. Harvest of 1893.

Amarantus speciosus, which split at a considerable height above the insertion of the cotyledons and bore in the fork two leaves the midrib of which had grown together on the dorsal side up to within a short distance of the apex. In the axils of these leaves small branches were seen with a little terminal inflorescence and a small leaf inserted below this. Above this point the two branches of the fork were of normal growth. Fig. 70 represents

[1] See L. J. LÉGER's exhaustive work on the anomalies in the seedlings of *Acer Pseudo-Platanus. Bull. Soc. Linn. Normandie,* 1889, p. 199, with plate.

a seedling of *Acer Pseudo-Platanus* with four seed-leaves which I found in the spring of 1887 in a forest and transplanted to my garden. Here it developed its stem. As soon as this had definitely split, and just before the cotyledons were about to fall away I dried this specimen in order to keep and photograph it.

It is obvious that such twins do not belong to the tricotylous race, that is to say, that their anomaly is due to some other elementary character. Therefore they should not be counted when recording the seedlings, nor be used as seed-parents. But as their nature can only be determined for certain in some cases, it is not always possible to take this precaution; and the fact that the hereditary values obtained from tetracotylous individuals are sometimes worse than those from the corresponding tricotyls may in part be due to this circumstance.

Fig. 72. *Amarantus speciosus*. Forking of a stem of a tetracotylous plant with two leaves in the fork which have grown together dorsally. The figure shows also their axillary twigs.

For the rest tetracotyls do not in any essential respect behave differently in inheritance from the tricotyls. From their seeds are produced, besides the atavists, mainly tricotyls, with hemi-tricotyls and tetracotyls in the usual diminishing proportions. My tricotylous half race of *Scrophu-*

laria nodosa produced, in the harvest of 1894, a mean of 2% and a maximum of 5.5% tricotyls per seed-parent. I harvested the seeds from two tetracotyls and obtained 0.5% and 3% tricotyls. Amongst the 2000 seedlings which these cultures contained, there were 30 tricotyls, 3 hemi-tricotyls and only 2 tetracotyls.

On the other hand a tetracotylous plant of *Asperula azurea* gave 7% and the corresponding tricotylous seed-parents only 2%, in 1892. Of this 7%, there were 5% tricotylous, 1% tetracotylous, and 1% hemi-tricotylous. I bred the tetracotyls of *Amarantus speciosus* for two generations, in 1893 and 1894. In the summer of 1893, 9 tetracotylous plants were left to flower; 3 proved to be fasciated, but the rest gave values varying from 1 to 7.5% with a mean of 5%. I counted for each seed-parent 500-1000 seedlings. The corresponding tricotylous culture gave values from 2.5% to 7.5%, that is, a mean of 4.5%; from each of the 15 seed-parents from 700 to 1000 seedlings being examined. We see there is practically no difference between the two cases. Together the tetracotylous parents produced only 6 tetracotylous offspring among 4000 seedlings, and the tricotyls 13 among 10,000; that is to say, they behaved in regard to this character as variants of the same race. I then selected the tetracotylous offspring of the tetracotylous seed-parents for a continuation of the race in 1894, but observed no further progress, the percentage in tetracotyls being only 0.2%.

The question suggests itself whether the proportion of tetracotylous seedlings, perhaps, simply obeys the laws of probability. The splitting of a cotyledon may be imagined to be distributed at random over a group of say 100 individuals, and we may ask, how many times a seedling will have two such divisions and so become a tetracotyl or hemi-tetracotyl. If, for instance, 50 divisions are distributed over 100 seedlings, with 200 cotyledons, how often may we expect a single plant to present two such divisions?

In the same way the expectation of pentacotyls may be calculated. Without going closely into this calculation, it is obvious that the proportion of tetracotyls will, on the whole, increase with that of the tricotyls independently of course of the nature of the species in question. As a matter of fact we do not observe such an independence. Some species are relatively rare in tetracotyls whilst others produce them more abundantly. Thus *Antirrhinum majus* never gave more than 1% to 2% of tetracotyls (Fig. 63 D, p. 345), although the proportion of tricotyls was as much as 79%. *Oenothera hirtella, Scrophularia nodosa,* and *Cannabis sativa* are also poor in tetracotyls. The latter produced only 1 to 3.5% of them, even when the whole value amounted to 63% (in 20 individual records). On the other hand, other species, or at any rate the races of them which I observed, produced tetracotyls abundantly.

I have grouped together well over 100 separate records from my cultures of 1894-1896, in which the hemi-tricotyls, tricotyls and tetracotyls were recorded separately for each sowing, which almost always consisted of about 300 seeds. From these I have especially calculated, besides the percentage composition in split-leaved seedlings, the proportion of these to the tetracotyls; and I give below the number of tetracotyls per 100 tricotyls in the wider sense of that term. This proportion varied in *Amarantus speciosus,* for 2-10% tricotyls, and in *Cannabis sativa* for 6-52%, between 1 and 7%. In *Mer-*

curialis annua for 8-86% tricotyls between 1 and 16%; in *Silene inflata* for 27-73%, in *Clarkia pulchella* for 6-16% and in *Helichrysum bracteatum* for 3-41% tricotyls, from 2 to 28%. In the individual records this ratio is obviously subject to considerable fluctuation on account of the small number of tetracotyls in the individual crops, and for a proper estimation of this ratio cultures on a much larger scale and especially designed for this end would be necessary. Here I shall content myself with giving an experimental series obtained with *Clarkia pulchella,* which shows roughly how the ratio of tetracotyls to tricotyls increases with the number of the latter.

Percentage ratio of tricotyls	6	7	14	16	27	55	62	63
Number of tetracotyls per every 100 tricotyls . . .	15	5	26	5	10	18	22	20

Similar figures were obtained with *Phacelia tanacetifolia, Papaver Rhoeas, Helichrysum bracteatum* and *Mercurialis annua.*

§ 3. THE INFLUENCE OF TRICOTYLY ON THE ARRANGEMENT OF LEAVES.

Elementary characters are not as a rule betrayed by a single external characteristic, but by several. In most cases one of these can easily be recognized as the primary one, and the rest are then termed secondary. In other cases a doubt may arise as to which should be regarded as primary and which as secondary. White-flowered varieties of red or blue species often exhibit the absence of color in the fruits as well as in the leaves or the stem. Moreover they can frequently be recognized as early as in the seedling stage by their pure green color. My new

Oenotheras differ from the parent species in several characters; nevertheless each arises suddenly with its characters complete. From this fact we conclude that all of them must be regarded as the expression of a single internal change. A single new elementary character can thus affect more or less profoundly a whole group of older internal characters.

We will regard tricotyly, for the present at any rate, as the primary expression of a definite internal factor which occurs in the latent state in large numbers of di cotyledons, though not necessarily in all. It also occurs, although as yet in a small group of cases, in the active

Fig. 73. *Scrophularia nodosa*. A tricotylous seedling with the first whorl of leaves which is ternary also; C, the cotyledons. From the harvest of 1899.

state alongside dicotyly. On this point of view the question suggests itself whether this internal factor will perhaps also betray itself during the later life of the plant. Tricotyly results in an abnormal arrangement of the seed-leaves, and thus it is only natural to expect that its internal cause may lead to anomalies in the disposition of the foliage leaves as well.

As a matter of fact, this is the case. In the first place, in species with decussate leaves, the arrangement of the leaves in whorls of three may continue upwards from the cotyledons (Fig. 73). This may be limited to the lower whorls, or continue in all of them. In the first case the transition between the two arrangements is often effected by intermediate stages, such as cleft leaves.[1] Other disturbances of the disposition of the leaves also can follow on tricotyly, amongst the most important of which are twisted and fasciated stems as well as the production of so-called terminal leaves. In the following paragraphs I shall describe some of the most important of these various phenomena which have occurred frequently, and in many cases almost regularly, in my experiments.[2]

It seems desirable to state beforehand that the anomalies in question exhibit an obvious genetic connection with the splitting or duplication of the cotyledons, although this relation needs closer investigation. Other malformations of structure no doubt are also met with amongst tricotylous individuals (for instance, variegated leaves, prolification of flowers and flowerheads), but not more abundantly than elsewhere. Moreover it is by no means a rule that in all species the same anomalies should occur amongst tricotyls. It appears, on the contrary, that certain species, (or at any rate, certain commercial races of them), have a marked preference for definite abnormalities, since both torsions and fasciations appear relatively abundantly amongst certain species, but rarely amongst others. In the same way subterminal leaves have hitherto been observed in quite a limited number of instances only.

On this relation between tricotyly and abnormalities in the disposition of the subsequent leaves I have based

---

[1] DELPINO, *Teoria della Fillotassi.*

[2] Further facts will be found in *Eine Methode, Zwangsdrehungen aufzusuchen*, Ber. d. d. bot. Ges., 1894, Vol. XII, p. 25.

a simple method of searching for such anomalies. Fasciations are so common in nature and in the garden that special means for obtaining them are not required; but twisting is much rarer and ordinarily it is only by a lucky chance that we meet with an instance of it.[1] If we wish to become independent of this chance we must have recourse to the culture of variations of cotyledons, because such will offer a greater likelihood of furnishing the desired anomaly than other examples of the same species. In the first, or at least in the second, generation we may count on finding them, if the extent of the experiment is sufficiently great, and once obtained, they can easily be further improved by ordinary selection. A single instance will suffice. MORREN found a very fine specimen of twisting in *Dracocephalum speciosum* in a meadow not far from Liège,[2] and when I read his description I became extremely anxious to investigate such a case of torsion in this species, or at any rate in this genus. For this purpose I selected *Dracocephalum moldavicum*, which, being an annual, seemed more suitable. In the spring of 1892 I selected a single hemi-tricotylous seedling found in a sowing of commercial seed (a little less than 20,000 seedlings), and from this bred a race which in the first year exhibited nothing remarkable but produced fasciations in the second year and traces of twisting in the third, and finally, in the fourth, some very fine instances of spiral torsion. One of these had the whole main stem transformed into a screw (Fig. 74). Fortunately in such experiments, the aim can be attained, as a rule, in a much smaller number of years.

[1] *Monographie der Zwangsdrehungen.* PRINGSHEIM's Jahrb. f. wiss. Bot., Vol. XXIII, p. 116.

[2] *Bull. Acad. Roy. Belg.*, Vol. XVIII, p. 37.

Tricotylous specimens of species with a decussate arrangement of their leaves very often produce the lower leaves of the stem in whorls of three. Sometimes this extends all the way up, or at least to the inflorescence,

Fig. 74. *Dracocephalum moldavicum.* Twisting of the main stem as the result of a breeding experiment extending over four years. (Compare below Fig. 82.)

sometimes, however, it reverts to the decussate arrangement sooner or later as we proceed upwards. Very often also the latter follows immediately on the seed-leaves (Fig. 76 B). All such cases can often be observed in the same culture from the seeds of a single seed-parent. This is especially the case in *Antirrhinum majus* and *Scrophularia nodosa,* in which species I have often preferred as seed-parents, the tricotyls, whose first whorls were trimerous. Nevertheless they have not, as a rule, proved the better qualified to continue the character of the race. Further instances are afforded by *Dipsacus sylvestris, Lychnis fulgens, Dracocephalum moldavicum, Dianthus barbatus* and so on. In the spring of 1887 I had some tricotylous seedlings of *Acer Pseudo-Platanus*; two of them are now high trees, whose trunks bear their branches in trimerous whorls.

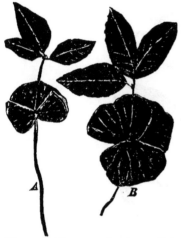

Fig. 75. *Fagus sylvatica.* Tricotylous seedlings. A, with a ternary whorl of the first leaves; B, with a leaf with two apices and a divided vein.

The lateral branches of ternary main stems tend, as a rule, to revert to the decussate arrangement. Subterranean runners, (for instance in *Valeriana officinalis*) and the secondary stems which are produced at the level of the ground (e. g., *Dianthus barbatus*), afford, however, numerous exceptions.

In tricotylous cultures, dicotylous individuals sometimes become ternary later. Thus I possess a plant with ternary whorls of *Aesculus Hippocastanum* (now 13 years old), which only had two cotyledons, and from the same crop a plant which was tricotylous but has since produced leaves on the decussate plan only. In both specimens the change in the disposition of the leaves took place before the cotyledons were dead. In *Dipsacus sylvestris torsus,* a race which is usually rich in ternary individuals, these are almost without exception decussate in early youth.

On the boundary between the 2- and 3-merous whorls 2½-merous ones not infrequently occur. I mean whorls with one normal and one more or less deeply cleft leaf. All degrees of cleavage (or symphysis) may occur. In the tricotylous races of *Antirrhinum majus* and *Scrophularia nodosa* they are particularly abundant, and merge

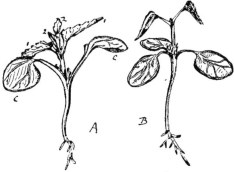

Fig. 76. *Mercurialis annua.* A, normal seedling; C, cotyledons, (1) the first, and (2) the second pair of leaves; B, a tricotylous seedling the first two leaves of which stand opposite one another (1900).

into the decussate arrangement in the first or second or in some later whorl. In the choice of seed-parents I have always paid attention to this point, although it has only a secondary effect upon the result. Further instances of cleft leaves in the transition from tricotyly to the later normal arrangement were afforded me by *Dianthus barbatus, Lychnis vespertina, Polygonum Fagopyrum, Collinsia heterophylla, Anagallis grandiflora* and in large quantities by *Fagus sylvatica* (Fig. 75) as well as by many other species. As the phenomenon is quite common when more extensive sowings are made and as, especially in perennial plants, the transition from one mode of arrangement to the other has been frequently observed on the same axis, as DELPINO has shown, I need not enter further into it now. Sometimes it happens also that the main stem of a tricotylous plant bears its leaves in quaternary whorls (*Scabiosa atropurpurea*).

Fasciations are a frequent consequence of tricotyly, though they sometimes do not appear until late in the life of the plant. *Mercurialis annua* and *Amarantus speciosus* furnished a series of instances nearly every year during several years of culture. In the former species they usually appear low on the stem; in the latter not below the inflorescence (See Fig. 83, p. 399). My tricotylous race of *Mercurialis annua* furnished almost all forms of fasciation. Fig. 77 represents a tricotylous plant whose stem was split from the first node after the cotyledons, and was therefore only fasciated in the epicotylous internode. Between this condition and a flat stem one centimeter broad, and much contorted, all intermediate stages are presented by this species. In the spring of 1887 I collected a large group of hemi-tricotylous, tricotylous and tetracotylous seedlings (See Fig. 70) of *Acer Pseudo-Platanus* not far from Hilversum and grew them for several years. Most of them I threw

Fig. 77. *Mercurialis annua.* A, tricotylous seedling with split stem; B, a seedling the first whorl of leaves of which was ternary; C, hemi-tricotylous plant (1900).

away as soon as they reverted to the decussate arrangement of their leaves, but seven of these trees are still alive, two with ternary stems (p. 370), two with a decussate arrangement of the leaves, and one with a much flattened main stem. This last one began with three cotyledons; it then became decussate, and in its second year (1889) became ternary again. In the autumn of the following year it began to flatten out, formed three five-leaved whorls and began to split when laying down the winter bud. I then broke off all the terminal buds except one, which in the spring of 1891, during subsequent growth, split into three flat twigs of which again two were removed. In the following summer the fasciation recurred, and also, after splitting in the winter, in the next year (1892), and again in 1893. Every time the forked branches were reduced to one. The divisions became much rarer, later on, and the older sections of the stem which were at first flat gradually became cylindrical, as usually happens in fasciation when it affects trees.[1]

On tricotylous specimens I also observed flattened stems in *Antirrhinum majus, Artemisia Absynthium, Scabiosa atropurpurea, Dianthus plumarius, Collinsia heterophylla, C. grandiflora, C. violacea* and *Tetragonia expansa* (Fig. 78); and amongst tetracotyls in *Scrophularia nodosa* and *Collinisia violacea* and other species.[2]

In many cultures I have observed that fasciations are more common amongst tricotylous plants than amongst dicotylous ones, but I shall only deal in detail with an experiment on *Asperula azurea*. In the spring of 1892 I selected the hemi-tricotylous, tricotylous and tetra-

[1] *Kruidk. Jaarb. Gent,* 1894, Plate XI, (*Abies excelsa*).
[2] *Ber. d. d. bot. Ges.,* Vol. XII, p. 38.

cotylous seedlings from commercial seed of this pretty little annual (Fig. 79). I cultivated them and saved their seeds and in the following year planted out the variants and the atavists separately. There were 37 of the former group and 15 atavists. From the former there arose 28 flattened stems and branches; from the

Fig. 78. *Tetragonia expansa.* A, branch split by forking: B, comb-like expanded terminal flower of a flattened main stem.

latter only four, that is to say, a proportion of 75% and 27% respectively. It should be mentioned that all the variants among the seedlings were planted out and that some of these gave rise to rather weak plants, whilst of the atavists I selected the strongest seedlings only. These, however, in spite of their greater individual strength and in spite of their origin from tricotylous parents, produced considerably fewer fasciations than the tricotyls.

We come now to the spiral torsions. These occurred in several of my tricotylous races nearly every year since 1893, and often in considerable numbers. As a rule they

Fig. 79. *Asperula azurea.* Fig. 80. *Melampyrum pratense.* Tetracotylous plant with spiral arrangement of leaves (1887).

consisted of fairly long, much twisted sections of the main stem, or of the stronger lateral branches which abutted above and below on the normal or decussate sections. They bore their leaves sometimes in a very steep spiral and sometimes in an unbroken line on one side of the stem.[1] I found these torsions not only in my own races but also on tricotylous individuals raised from commercial seed of *Anagallis grandiflora, Collinsia bicolor, C. heterophylla, C. violacea, Dianthus plumarius, Fedia scorpioides, Scabiosa atropurpurea nana, Silene noctiflora* and *Zinnia elegans*. Also in the second generation of *Asperula azurea setosa* and *Viscaria oculata*.[2]

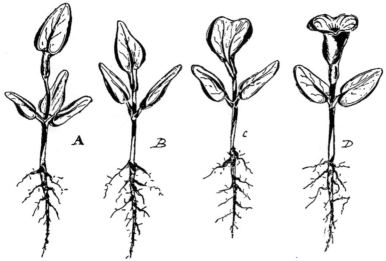

Fig. 81. *Antirrhinum majus*. Seedlings with terminal leaves A, tricotylous; the others are atavists from the same race. A, B, with a single terminal leaf; C, the two leaves of the first whorl fused laterally and placed terminally; D, a terminal pitcher formed of two leaves.

Lastly I wish to refer to a tetracotylous plant of *Melampyrum pratense* (Fig. 80), whose first leaves above the cotyledons, instead of being decussate, were arranged in a very irregular spiral. I found it in the summer of 1887 growing wild not far from 'S Graveland. It is important, because it shows that even a splitting of both

[1] Figured in *Ber. d. d. bot. Ges.,* Vol. XII, Pl. II, Figs. 9 and 10.
[2] *Ber. d. d. bot. Ges., loc. cit.,* pp. 32-35; see the figures on Plate II of this article.

the cotyledons and especially a splitting equally deep on the two sides may involve an alteration in the disposition of the leaves. This, of course, is by no means a necessary consequence, but the present case indicates that the internal cause is not necessarily limited in its operation to the cotyledons.

Terminal leaves have hitherto been regarded by teratologists as very rare occurrences, but my tricotylous races of *Antirrhinum majus* have afforded me the opportunity of observing them repeatedly and in hundreds of specimens (Fig. 81).

A well-known instance is the great terminal leaf of *Gesnera Geroltiana* described by MORREN and reproduced by MASTERS,[1] This specimen bears only one normal pair of leaves, and above these an erect leaf of double the normal size. In the figure we can see the node at which this is inserted; and the simplest explanation of this remarkable phenomenon is that, for some reason or other, the growth of the second leaf of the pair, as well as that of the terminal bud was impeded in early youth. BERNOULLI mentions a similar apparently terminal leaf of *Coffea arabica*, and was able in this case to confirm the correctness of this supposition by microscopical observation.[2] He also describes a shoot of *Fuchsia macrostemma* which bore a funnel-shaped leaf at the top.

The races of the Snap-dragon, which we have already mentioned bear every variety of these structures. These may consist of single or of double leaves, or of leaves grown together in the shape of a funnel. All intermediate stages between these and the normal plants oc-

[1] CH. MORREN, *Bull. Acad. Belg.*, Vol. XVII, Part II, p. 387; M. T. MASTERS, *Vegetable Teratology*, p. 88, Fig. 40.

[2] G. BERNOULLI, *Ueber scheinbar terminale Blätter*, Botan. Zeitung, 1869, p. 19.

cur. These anomalies are found especially on the seedlings and usually replace the first or second whorl of leaves and rarely the third or a higher one. In the spring of 1894 I had a large crop of the red flowered variety which produced about 1% of these abnormalities amongst many hundred seedlings. In subsequent years I had even larger numbers. In the spring of 1897, for instance, I had about 10% in very extensive sowings of a tricotylous race with striped flowers. The most important cases are, of course, those in which the rudiments of the opposite leaf and of the terminal bud can be clearly seen with the naked eye alongside the terminal leaf. I have frequently planted out such plantlets in the hope of growing them for my experiments, but usually without success. Either they did not develop a main stem at all, or only a delicate one; often there arose from the axil of one of the lower leaves or of one of the cotyledons a lateral branch which, however, remained weakly.

Sometimes the organ situated opposite the terminal leaf is somewhat better developed, but usually it cannot be seen without the help of a microscope. If the terminal structure has only one vein I regard it as a single leaf (Fig. 81 B). But if it has two points with a double or divided midrib (Fig. 81 C) it obviously represents the two leaves of a pair. Sometimes these are fused together laterally; the peduncle is, however, considerably broadened and its point of insertion clearly recognizable. Frequently, however, the peduncles are fused at their base, at both sides, and form a little tube which embraces the terminal bud. If the concrescence is of considerable extent terminal ascidiae are the result, which, in most cases, can be clearly recognized as consisting of two leaves (Fig. 81 D).

The fact that so rare a teratological phenomenon should occur so regularly in two well-known varieties of the same species,—the one uniform red, and the other striped yellow and red—means perhaps that the character in question has existed for a long time in the Snapdragon and will be found, after a close investigation, to exist in other cultivated varieties also and possibly even in the wild ancestral form. Of course the fact that I found them in a tricotylous race need not necessarily indicate a causal relation between this character and tricotyly, because, at the beginning of my cultures, I started by selecting the tricotyls and continued the race from their seeds alone. If such a relation did exist the fact that the anomaly occurs both on dicotylous and on tricotylous individuals would be very important, for it would show that it is not the visible tricotyly itself, but some corresponding internal character, which must be regarded as the cause. It is to be hoped that the abundance in which the anomaly can now be obtained will render possible a closer examination of this problem.

§ 4. TRICOTYLOUS HALF RACES.

Occasional tricotylous seedlings will be found among samples of seed in very many species. All that is necessary, therefore, to start a culture is to buy a sufficient quantity of seed and to sow it. The seed will either give no aberrant forms, or very few, or a considerable number. In the first case the possibility of obtaining tricotyls still remains open if a larger quantity of seed is sown. In the second case the variants can be used as the point of departure for the race; they offer the prospect of providing a half race. In the third case we may expect to obtain an intermediate race rich in tricotyls.

Almost every year I have made experiments of this kind, but I was especially engaged with them in the spring of 1895. At that time I sowed about 20 grams of each of 40 species of annual plants, or in the case of very small seeds a somewhat smaller quantity, so as to investigate several thousand seedlings of each kind. I shall now give a list of the species falling into the third category, species, that is, which gave so large a number of aberrant forms as to justify the expectation of an intermediate race—an expectation which has, as a rule, been fulfilled, as we shall see in the following section (§ 5).

TRICOTYLS FROM BOUGHT SEED.
(Spring, 1895.)

SPECIES	DICOTYLS	HEMI-TRICOTYLS	TRICOTYLS	TETRACOTYLS
Chrysanthemum inodorum plenissimum . . .	1000	3	32	0
Silene orientalis alba . .	3000	3	7	0
Papaver Rhoeas fl. pleno .	3000	1	15	1
Clarkia pulchella alba .	4000	5	5	0
Glaucium luteum . . .	16000	0	15	0
Nigella hispanica alba .	10000	0	15	0
Phacelia tanacetifolia . .	16000	8	18	0
Helichrysum bracteatum .	35000	9	16	3

With the exception of *Silene, Glaucium* and *Nigella*, I have raised intermediate races from all these sowings. As I have already mentioned, hemi-tricotyls and tetracotyls are seen to be rarer than the typical tricotyls. A sample of seed of *Lobelia Erinus*, grown in the spring of 1902, had a very high proportion of tricotyls, viz., 31 in 100 seedlings.

Very small proportions of tricotyls were yielded by *Silene hirsuta,* which only produced 3 in 80,000 seedlings. The following species produced from 1-2 specimens in every 10,000 seedlings: *Argemone grandiflora, Aster tenellus, Clarkia elegans, Godetia amoena, Hyoscyamus pictus, Silene Armeria,* and others. I observed no tricotyls at all in sowings of the same extent of *Argemone mexicana, Datura laevis, Hyoscyamus albus, Nigella damascena, Phacelia texana* etc. I tested 800 seeds of each of 8 species of *Cerinthe,* and only obtained a single tricotylous plant in *C. bicolor, C. gymnandra* and *C. major.*

The seeds mentioned were all obtained from the nursery of Messrs. HAAGE & SCHMIDT in Erfurt. It is not unlikely that if the seeds were bought from other nurseries, different results could be obtained, especially from firms who do not exchange seed with the nurseries at Erfurt.

Similarly the seeds of wild species occasionally produce tricotyls, but, as it seems, only in very small quantities, and they have hitherto given no promise of yielding an intermediate race. As instances I mention *Raphanus Raphanistrum* and *Epilobium hirsutum,* of which species I found 1-2 tricotyls in large crops. If the seeds of wild species come from botanical gardens the proportion of tricotyls is sometimes greater; for instance in *Silene noctiflora* (1892) it was about 20 in 10,000 seedlings. Amongst trees I have hitherto found tricotylous seedlings abundantly in *Acer Pseudo-Platanus* and *Fagus sylvatica* and also in *Robinia Pseud-Acacia* and *Ulmus campestris.*

In order to test the hereditary capacity of the cotyledon variants in such crops, I have frequently planted them out and allowed them to flower in isolation. Higher proportions are thus obtained: for instance in cultivated species in *Celosia cristata* 2%, in *Chrysanthemum Myconis* 1-2%, in *Oenothera longifolia* 1%, *O. mollissima* 1%, *O. undulata*, 1%, *Xylopleurum tetrapterum* 2%, *Podolepis gracilis* 2%, *Tetragonia expansa* 2%, *Veronica longifolia* 4%; and amongst wild species in *Chenopodium album* 1%, *Thrincia hirta* 1%, and so forth, the two latter having been grow for three generations. Further instances will be afforded by the beginnings of my cultures to be mentioned below.

If we compare the proportions just given, excluding those species which are so rich in tricotyls that they probably contain an intermediate race, we find from 0-2 tricotyls in about every 10,000 seeds, from material which has been bought or obtained by exchange or collected in the field, whilst the harvest obtained after the isolation of the tricotyls, contains from 1-2%. The original mixtures, therefore, must have contained the seed of many individuals without tricotylous offspring.

Besides the hereditary capacity of bought seed and of the tricotyls raised from it after artificial fertilization, we have to consider the question whether this capacity can be increased by a selection extending over several generations, or whether it maintains itself without changing. As I have already stated, the conclusion derived from my experiments is that the answer may fall into one of two categories. In some species selection may soon lead to a proportion of 50% tricotyls and more; in others, this does not take place even if the selection is continued for many years. Obviously this depends on the question whether an intermediate race is present in the given sample of seed, or not. If it is there, it can be isolated immediately; if it is not, no amount of selection will bring it about. The experiments in which isolation succeeded will be described in § 6; but the experiments which failed fall into two groups, according to whether it could be decided immediately, or not until much later, that an intermediate race could not be raised. In the former group fall those cultures in which there was no progress at all, or so small a one that I had to give them up after three or four generations. For if, after this time, a proportion of no more than from 1 to 2 or at most 4% is reached, or if in the case of higher proportions the ratio fluctuates greatly but does not exbibit a regular increase, how many years of work would it take before we can be certain that nothing can be attained? These briefer experiments will form the subject of this section.

In the second group fall two cultures which I have continued for a considerable time, namely *Amarantus speciosus* through nine generations and *Scrophularia nodosa* through ten. Neither now leaves any hope of ever becoming successful (see § 5); but, as I have already said in the first part of this volume (p. 227), it is just the experiments in selection that fail, which give us the deepest insight into the nature of elementary characters.

I shall now proceed to the description of my experiments; and I shall confine myself to those which were instituted with the express hope of breeding a tricotylous intermediate race, and were continued for 3 or 4 generations with this sole object in view, until it became evident that only a half race was present. The extent of the cultures varied greatly, according to the importance which I attached to them at the beginning. In the first place I shall deal with *Ocnothera rubrinervis*.

Ocnothera rubrinervis. In the pedigree of the *Laevifolia* family on p. 273 of the first volume, the origin of two specimens of *O. rubrinervis* is recorded for the year 1889. In the generations which were raised from these two mutants, no tricotyls were at first observed (1890-1891). They first appeared in the spring of 1892, and with them the culture of the tricotylous half race began. I selected the tricotyls from amongst thousands of seedlings and planted them out singly in pots with well manured soil. I obtained 22 strong plants which flowered freely in an isolated spot. The seeds were saved separately from each parent, each of which was labelled according to its individual vigor. The five strongest plants gave respectively 1.5, 1.9, 2.3, 2.6, and 2.8% tricotyls amongst from 700 to 900 seedlings. The remainder, the weaker plants, only 0-1% with an average of 0.7% amongst 8000 seedlings. The value 2.8% was obtained from a very vigorous plant; in the three following generations it has hardly been reached again, in spite of selection, and never was really exceeded.

The offspring of only the five best plants were planted out and in all cases only tricotylous individuals. There were about 70 of them and they were potted singly and well manured (1893). In the middle of May they were planted out into the bed at distances of about 30 centimeters apart, where they flowered in isolation in August, and could be mutually fertilized by insects. In the spring of 1894 the seed of each plant was sown separately, and when the cotyledons had completely unfolded the seedlings were recorded, 300 for each parent but 1000 or more in the 20 which appeared to be the best. Altogether 45,000 seedlings were recorded, and amongst these were 170 tricotyls, i. e., about 0.4%. Of these

12 were hemi-tricotyls and one was a tetracotylous specimen. There were also some few syncotylous ones. The ratio from the best seed-parents was no more than from 0.8% to 1.1%. The tricotylous offspring of these and of some with from 0.5 to 0.6% were planted out in the same way as in the previous year, provided they were strong little plants. This time the inflorescence of each plant was guarded against the visits of insects by means of a parchment bag and artificially fertilized in the hope of thus isolating individuals producing higher proportions.

On page 354 I have plotted, in Fig. 66, a curve of the harvest of 1894, based on the records made in the spring of 1895. There were two seed-parents which manifested an advance on the previous year, inasmuch as their ratio was 3.5% and 3.7%, but the difference was only a very inconsiderable one. In all, there were 87 seed-parents. As I had planted the offspring of the individual grandparents of 1893 together in groups in 1894, I could now make a selection not only between the parents but also between the grandparents. The grandchildren of those grandparents only, whose offspring had produced the highest mean proportion, were planted out. By means of such a selection of grandparents the pedigree becomes an individual one in spite of the size of the cultures, embracing, in each generation the offspring of one selected individual only. The method, therefore, unites this principle with the greater certainty that can be obtained by extensive cultures and a double selection. I have also applied it to a number of other cases.

From one grandparent with 1%, therefore, and from six of its offspring with from 0.9% to 2.1% 54 tricotylous specimens in all were planted out (1895). From the beginning of the flowering period all these were covered with fine gauze, guarded from the visits of insects and artificially fertilized every day or every other day. About 300 seedlings of each plant were recorded. The ratios were from 0 to 1.2% with an average of 0.8%. Two parents had 1.4% and one 2%, that is to say, that here again there was no progress. The offspring of the six parents fell into groups between which the differences were but slight (3 with 0.7%, and 3 with 0.8% on the average).

In the list which follows I have collected the highest values that were obtained in the course of these generations.

	SPRING OF			
	1893	1894	1895	1896
Highest values	2.8 %	1.1 %	3.7 %	2.0 %
Selected seed-parents	2.8 %	1.0 %	2.1 %	—

These figures show a fluctuation within fairly narrow limits, but no essential advance in the course of four generations. It seemed therefore to be useless to carry the experiment further. It is certainly probable that, in the course of time, further selection might have brought about some slight improvement; but obviously this would have been of little significance, and at any rate there was no prospect of ever obtaining a race with 50% tricotyls.

Chenopodium album. A tricotylous plant flowered in 1889, in isolation in my garden, and produced 1% tricotyls in the spring of 1890 amongst about 1000 seedlings. Four of these were cultivated further, but their seeds gave rise again to no more than 1%. The third generation was therefore not better than the second.

Dracocephalum moldavicum (Fig. 82). In the spring

of 1892 I obtained only a single hemi-tricotyl from about 20,000 bought seeds. The seeds of this plant gave rise in 1893 to five tricotyls and two hemi-tricotyls among 4000 seedlings, that is, about 0.2%. Their seeds were harvested separately, and as the expectation of tricotyls was a small one, great quantities of it were sown. I recorded from 800 to 2900 from each, i. e., altogether about 15,000 seedlings, and found the ratio of tricotyls (and hemi-tricotyls) to be from 0.2 to 0.4%. It did not, therefore, seem justified to make a selection among the individual seed-parents. In 1894, 12 tricotyls the stems of which had remained ternary, and twelve specimens with normal decussate stems were planted out. Several beautiful fasciations and occasional cases of spiral torsion occurred in this culture (p. 369). The seeds were harvested separately. In the spring of 1895 they again produced only from 0.1 to 0.4% of tricotyls. Seventeen specimens from the seed pans with from 0.2 to 0.4% were planted out about a meter apart, but the seeds produced, in 1896, scarcely any tricotyls and only from five seed-parents, the proportion being from 0.3 to 0.7%.

Fig. 82. *Dracocephalum moldavicum*. A whole plant.

In the course of four generations selection had, therefore, brought about practically no advance.

Lychnis fulgens. The tricotylous seedlings of this species are as a rule weakly; their culture, therefore, is difficult and their harvest poor. In 1892 I had a tricotylous plant whose seeds gave a proportion of 5%. From these I reached, in the spring of 1894, a ratio of 13% containing one tetracotylous plant; most of the tricotyls afterwards remained ternary. In 1895 they produced tricotyls in proportions varying from 3 to 11%, with a mean of 6%. In the next, i. e., the fifth generation (spring of 1896), I counted from 2 to 8% tricotyls per seed-parent, and from a particular individual 21 tricotyls amongst 110 seedlings, i. e., about 19%. But the number of seedlings in this case was too small to signify a real advance.

Penstemon gentianoides. In 1892 I had raised four tricotylous plants from bought seed. They produced respectively 0.3, 1.0, 2.6 and 3% tricotyls in 1893. I planted out the tricotylous seedlings of the best seed-parent, but only six managed to flower. Their seeds gave ratios varying from 4% to 12% (March 1892), with a mean of 7%. The tricotylous seedlings of those seed-parents only which had ratios above 10% were planted out. Of these 8 tricotyls, 6 hemi-tricotyls and 2 tetracotyls have flowered. The seeds of the former gave ratios of tricotyls, ranging from 0 to 3.3% with a mean of 2.8%; the hemi-tricotyls from 1.2% to 2.4% with a mean of 4.8%; and the four tetracotyls 10% and 11%, amongst which, however, only a single seedling had four cotyledons. The offspring of both of these tetracotyls and of the best of the remaining seed-parents were planted out in 1895. Only in the case of eight plants, however, was the harvest a sufficient one and gave a ratio which as a rule was between 0 and 12% and which attained its maximum in 15%. The latter occurred amongst the 300 seedlings from the seeds of a daughter plant of one of the two tetracotylous grandparents.

In the last three generations the maximum ratios were, therefore, 12, 11 and 15%, indicating no advance of any significance.

Polygonum Convolvulus. A tricotylous plant flowered in 1888, in isolation, in my garden. Its seeds gave rise to normal seedlings only (1889). From their seeds, about 4000 in all, 1450 seedlings were raised, and 12 of them were tricotylous, i. e., 1% (1890). I harvested the seeds of six tricotylous plants separately, and obtained ratios from 1% to 2.4% in sowings of about 1000 seeds each. Of these, 12 seed-parents produced only about 1% in the next generation in the spring of 1892; two of them, however, produced 1.5 and 2%. Seventeen plants were planted out. In their crop (April 1893) the proportion of tricotyls varied from 0.5 to 2% in lots of 200 and 400 seedlings, and twelve tricotyls succeeded in flowering. The next generation (April 1894) contained, in the best cases, 2.8% tricotyls; the next one, 0 to 2% from 8 seed-parents (1895); and the last, or ninth, again 2% only.

Silene conica. In 1892 I had a few tricotyls in flower from seed received by exchange from another botanical garden. Their seed gave 3 tricotyls amongst 1000 seedlings. I planted these out, together with some dicotylous seedlings, and in May 1894 I had from 0.2 to 1% tricotyls in every lot of 350 to 800 seedlings. Eight tricotyls were planted out and in the spring of 1895 their harvest gave a proportion of only 2% and less. From these I obtained in 1895 a fourth generation, the seeds of four seed-parents producing no tricotyls and those of one, 2 amongst 500 atavists.

Silene conoidea, like the foregoing species, was obtained by exchange in 1892 and gave a single tricotylous seedling. In the following spring I had about as many tricotylous seedlings as in the previous species; and in the succeeding generation (1894) 3% tricotylous plants amongst 407 seedlings from seeds of a single tricotylous seed-parent. Only these 12 tricotyls were then planted out; and their seeds were harvested separately in late summer. In every lot I recorded from 300 to 900 seedlings and only in two cases, where the harvest had been too small, so few as 200. The sowings contained, as a rule, from 0.5 to 1.5% and only in one case, amongst 316 seedlings, 3% tricotyls. All the vigorous tricotylous seedlings were planted out in such a way that the offspring of the individual seed-parents stood in groups. Their seed was harvested from each seed-parent and sown separately; and the proportion of tricotylous individuals was determined for each among 300 seedlings. This proportion fluctuated, for the 26 seed-parents, between 0 and 4.2%. The separate groups manifested no relation to the hereditary index of their several seed-parents. The offspring of parents with 0.5% and also of those with 2% and 3%, had, as a rule, 0.6%. On the other hand the offspring of seed-parents with a mean of 1% exhibited this mean ratio of 1% again, and the highest figure obtained in this experiment, 4.2% occurred amongst them.

The maximum values in the three succeeding generations were therefore 3, 3 and 4%, i. e., they were fairly constant.

In *Silene noctiflora*, also, I have studied tricotyly

through the course of four generations (1891-1894), and found it heritable to the extent of from 1 to 2%, but I have not made any detailed records.

Spinacia oleracea. The spinach also contains tricotylous seedlings. I employed the Dutch spinach with smooth round seeds without thorns, a fine and perfectly constant type. In 1892 I found a tricotylous specimen, and grew this amongst some normal plants, because the species is dioecious. When the harvest was first examined there proved to be five tricotyls and one hemi-tricotyl amongst 1000 seedlings, i. e., 0.6%. Then the best tricotyls were selected from several thousand seedlings; thirteen being met with. During the flowering period, several of them proved to be monoecious, a phenomenon which sometimes occurs in this species and has been observed also in *Cannabis sativa, Mercurialis annua,* and others. I harvested the seeds separately from 5 female or monoecious plants, and obtained ratios of 0, 0, 0, 1 and 2% tricotyls amongst the seedlings, of which only 4 managed to flower. They were one male and 3 female plants which latter set an abundance of seed. They gave ratios of 0, 0 and 2%, the latter occurring amongst 430 seedlings. In the summer of 1895 the tricotyls flowered, and I collected the harvests of the various female plants separately, and thus was able to determine the proportion of tricotyls for each parent in the spring of 1896. This fluctuated between 0 and 3% and in one case reached 4% (mean 1.5%).

Summary: The results of the experiments described show that, in the cases dealt with, a stringent selection, extending over three years, failed to effect any definite and certainly any considerable advance. The individual instances fall into two categories; as a rule the proportion of tricotyls was 1 to 2% and reached 3 and 4% so rarely that these numbers must perhaps be regarded as the extreme results of the errors of observation which are bound to occur in such countings. In two species the ratio was about 10 to 15%, but these were both perennial forms which, in my annual cultures set but little seed. They were *Lychnis fulgens,* from 1892 to 1895, with 5—13—11 and 8—19% tricotyls and *Penstemon gentianoides* in the same years with 3—12—11 and 15%. They should perhaps be excluded from further consideration. Summarizing my data therefore (with the omission of these cases) we obtain a very uniform picture of the inheritance of tricotyly in half races under continued selection.

INHERITANCE OF TRICOTYLY IN HALF RACES UNDER CONTINUAL SELECTION.

	FIRST GENERATION	MAXIMUM VALUES IN THE SPRINGS OF						
		1890	1891	1892	1893	1894	1895	1896
Oenothera rubrinervis . .	1892	—	—	—	2.8	1.1	3.7	2
Chenopodium album . . .	1889	1	1	—	—	—	—	—
Dracocephalum moldavicum	1892	—	—	—	0.2	0.4	0.4	0.7
Polygonum Convolvulus .	1888	1	2.4	2	2	2.8	2	—
Silene conica	1892	—	—	—	0.3	1	2	0.4
" *conoidea*	1892	—	—	—	3	3	4	—
Spinacia oleracea	1892	—	—	—	0.6	2	2	4

Thus we see that tricotylous half races exist which even under the most stringent selection can produce only small percentages of this anomaly. So far as we can conclude from indirect data, half races of this kind appear to be widely distributed in the vegetable kingdom. Samples of seed, whether they be bought or collected in the garden or in the field, which give rise to no more than occasional aberrant forms amongst about 10,000 seedlings, as a rule strongly suggest the existence of such half races.

§ 5. TRICOTYLOUS INTERMEDIATE RACES DO NOT ARISE BY SELECTION.

In the first part of the first volume I brought together a long series of facts from botanical and horticultural and, most important of all, from agricultural literature, which afforded sufficient proof that specific characters do not arise by selection.

Applied to tricotyly, the truth of this generalization is demonstrated by the results of the experiments described in the foregoing section (§ 4) and summarized in the table on the preceding page. This result is in complete agreement with my experience in regard to the origin of species and constant races in other cases. In contrast to these the so-called improved races of the agriculturists which have arisen by artificial and repeated selection are constant only to a small extent (VON RÜMKER). On the other hand the so-called heritable or constant races do not arise by selection, with however much acumen and perseverance it may be prosecuted.[1] The distinguishing terms are, I admit, not very practical and open to much criticism. The two kinds of races which they indicate are, however, absolutely distinct things, among wild plants as well as among cultivated species; but, unfortunately, if we only have a single individual before us we cannot, as a rule, determine to which of the two types it belongs. Only its progeny can decide, and even this, often not until after the lapse of

[1] See Vol. I, p. 106.

several generations. But, at the beginning of our century, we stand only on the threshold on which systematic botany must be raised from a comparative to an experimental science.

The experiments described in the foregoing section (§ 4) cover four generations as a rule, i. e., a selection of tricotylous plants four times repeated, and thrice repeated for those with apparently the highest values. It may, however, be allowable to suppose that stray unfavorable individuals occurred amongst the selected ones and that a selection of longer duration might possibly be crowned with success.

In order to determine this point I have, as I have stated above on page 383, continued the experiment with two species, to which I have lately added a third, through about ten generations. I devoted every possible care to the selection and carried it out on as large a scale as could be desired. The result was a genuine progress which amounted in both cases from one or two per cent to a value which reached in the seeds of some rare seed-parents, even as much as 25%. But from the character of an intermediate race with a mean value of 50% the races are still far removed, and every circumstance points to the conclusion that it is simply impossible to reach this by the method (as yet the only available one) which was employed.

The two plants used in these experiments were *Amarantus speciosus* and *Scrophularia nodosa*. The former is an annual garden plant, much in favor on account of its height and its red foliage (Fig. 83); the second is a wild perennial species which is very common in this neighborhood. It flowers in its first year and can be easily cultivated as an annual. In the previous section

(§ 4) these two species were omitted from consideration in order to describe the experiments here in their entirety.

The hope which I cherished at the beginning of these experiments has not been fulfilled, it is true; but I think that a brief notice of it will serve a useful purpose. The present form of the theory of selection would justify the expectation that a continued selection of the tricotylous individuals would result in a race which should, year after year, produce tricotyls in continually increasing quantities, until ultimately a new variety or subspecies would arise, composed solely of such individuals. This form of the theory is very accommodating. If we have regard to the law of regression (*Vol.* I, p. 83), the mean of the race always lags further behind the individuals which have been and are to be selected; so that, as a matter of fact we never attain to the type of a new and constant race. But if we neglect this law, as is now frequently done, we might expect that continual and uniform progress, which alone could account on the ground of the theory of selection for the origin of species in the vegetable and animal kingdom. And lastly we might assume an increase of variability in the chosen direction by means of selection, an hypothesis which, as I have shown in the first part (p. 9), is entirely unsupported by evidence.

The first of these hypotheses would lead us to expect a variable tricotylous race, never becoming perfectly constant, a thoroughbred race in the agricultural sense of the word. The second would lead us to expect a continuous and uninterrupted increase in tricotylous individuals resulting in a constant tricotylous subspecies. The third would point to a gradual acceleration of the process. As a matter of fact neither one nor the other has occurred. If we would speak of a thoroughbred race I only obtained a thoroughbred half race with a mean value of no more than 10 to 15% which remains dependent on the selection of seed-parents with about 25%, and is perhaps capable of some very small further improvement. The second hypothesis is so remote from facts that as yet it hardly admits of being tested; my experiments at any rate, lend no support to it. Rather might they be taken to be in favor of the third view; for the progress does actually seem to become gradually more rapid after the first few years. But then it should be remembered that selection is at first a very difficult matter, the tricotyls being still very rare, and for a large part delicate and unfit for further cultivation. In subsequent years there are hundreds of tricotyls from which the strongest may be selected; and we can even limit ourselves to the best specimens produced by the best parents and grandparents, and thus carry out a much more stringent selection. In reality the acceleration of the progress is thus brought about by a practical improvement in method and not by a biological increase in variability.

At first I entertained the hope that even if these expectations were not justified some relation between variability and mutability might perhaps exist.[1] I imagined that the capacity for producing mutations might be conditioned by external influences and therefore might itself be a variable character. The diversity amongst the mutation-coefficients of my *Oenotheras* seemed to support this view.[2] Moreover the external conditions which

---

[1] See my *Intracellular Pangenesis*.
[2] See Vol. I, p. 337.

shift the variability in the half race in the *plus* direction might perhaps be the same as those which would lead to a mutation and thereby to the sudden origin of a rich intermediate race. This would, in fact, perhaps, constitute the smallest step amongst all the possible forms of mutation.[1] If a mutation were ultimately to be induced by an improvement in the external conditions and by the choice of individuals thus modified in the desired direction,[2] it should most easily occur on the limits between the half race and the corresponding intermediate race. My hope was, therefore, that an intermediate race would suddenly arise from my improved half race and at first give about 50%, and then, by further culture, yield still higher numbers, perhaps even 80 to 90%. This hope was based on the analogous case of the origin of *Chrysanthemum segetum plenum* from *C. segetum grandiflorum*; for in this way tricotylous races behave when they happen to be found and are isolated (see § 6).

A step of this kind I have, however, not yet observed. In the case of both species I have determined the hereditary capacity of as many as 300 seed-parents in a single year, but without result. I have employed every device of culture and in *Scrophularia* I sowed the seeds of the second and sometimes those of the third year of the life of the plant; but every advance was followed by a step in the reverse direction. I believe that I have tried everything that was possible to me at the time, and I have continued to make every conceivable effort in spite of the

[1] See p. 20; with regard to premutation, Vol. I, p. 490; and with regard to varieties and subspecies, p. 64 of this volume.

[2] *Alimentation et sélection*. Volume jubilaire de la Société de Biologie, 1899; and Vol. I, p. 142 of this work where the statement is made that "selection is the choice of the best nourished individuals."

fact that the goal became ever more obviously unattainable; but nothing I did was of any avail. The half races remained half races, and the prospect of attaining an intermediate race is no greater now than it was at the beginning.[1]

I shall proceed now to the description of my two experiments.

Amarantus speciosus (Fig. 83). In 1889 I had a tricotylous and a hemi-tricotylous specimen of this fine garden plant, which usually attains a height of from 1½ to 2 meters. They flowered together, but far removed from any other specimens of the same genus.[2] I harvested the seeds separately, but only sowed those of the hemi-tricotylous plant. This had a small comb-shaped extension of the summit of the raceme as is shown in Fig. 83, and as it afterwards often occurred in this race. Its seeds gave rise to four tricotyls and one hemi-tricotyl amongst 110 seedlings, that is, a proportion of about 4.5%. The degree of inheritance therefore proved to be greater than in most of the other half races investigated (§ 4). The culture was, however, unfavorable, since only one hemi-tricotylous plant could be brought to flower, and since therefore a number of atavists had to be cultivated with it as a precautionary measure; but I only saved the seed of the hemi-tricotylous plant. This produced 6 tricotylous and 5 hemi-tricotylous seedlings amongst 250, that is, about 4.5% or the same amount as in the previous generation. This time, however, I could plant out the aberrant forms only and I managed to keep the majority of them alive. Only one, however,

[1] Mutations must, nevertheless, have external causes, and these must be found some day, but perhaps by some other means.

[2] *Amarantus speciosus* is regarded by some authors as a subspecies of *A. paniculatus*.

flowered. It was a richly branched compact tricotylous specimen, which was only a meter high. It set an abundance of seed, which produced a proportion of hemitricotyls and tricotyls, much greater than that attained

Fig. 83. *Amarantus speciosus*. Top of a plant of two meters height from the tricotylous half race.

in the previous generation. There were 89 aberrant forms amongst 700 seedlings, that is, about 13%.

A very considerable advance had therefore been made

in the fourth generation (1892) and this has been maintained since that time in spite of considerable fluctuation. Up till that time my selection had been limited by the fact that only hemi-tricotyls had survived to become seed-parents. From this point onwards I had both hemi-tricotyls and tricotyls in abundance. For the next four years I selected in these two directions, and maintained a tricotylous and hemi-tricotylous race simultaneously (1892-1896), but as no essential difference was manifested between the two I abandoned the hemi-tricotylous race, as stated in § 2, and only went on with the tricotylous one. The account which follows relates solely to this.

But before I proceed with it I wish to call attention to some facts relating to the method of culture. The seed was sown in sterilized soil in pans; the soil was not manured. As soon as the cotyledons had fully unfolded and before the first leaf was visible, the seedlings were recorded. All, or nearly all, of the dicotylous ones were destroyed and only the aberrant forms saved. Of the latter I chose what I considered to be a sufficient number of the strongest specimens, and planted them singly in pots with heavily manured soil. The best manures for this purpose are nitrogenous ones. If the number permits it, two tricotyls are put in each pot, of which the weaker is subsequently killed out. This transplantation takes place in April or May; the pots are kept under glass in the garden until June, during the nights at any rate. Then the plants are planted out into the bed at distances of from 20 to 30 centimeters, and the larger lateral branches are cut off in order that the plants may not interfere with one another. The plants are monœcious, the male and female flowers standing close together. Self-fertilization seems to be the rule; for isolated plants set as much seed as those grown in groups. The seeds do not fall out and as soon as the desired quantity is ripe the whole raceme is cut off and rubbed between the hands. One cubic centimeter per plant, and often less, is sufficient for next year's seed; as a rule I obtained about 4 to 5 cubic centimeters from each individual.

After the transplantation of the seedlings the length of the cotyledons, and of their stalks especially, increases considerably. In this way it may happen that specimens which, at their first examination, appeared to be pure tricotyls are found to have two of their cotyledons united at the base, and therefore to be deeply cleft hemi-tricotyls. I have as a rule removed such specimens when I discovered them, and not cultivated them further, although this is not actually necessary.

In the summer of 1892 I had 11 tricotylous specimens which flowered in a group as far removed as possible from the hemi-tricotylous culture already mentioned. The harvest gave a maximum of 11.9% tricotyls, but on the average only 1 to 6.5% with a mean of 3.5%. In 1893 I planted out the purely tricotylous seedlings only of a seed-parent with a hereditary value of 6.3%. I saved the seeds of 15 plants, examined between 500 and 1000 seedlings per seed-parent, and obtained values which fluctuated between 2.6% and 7.4%; mean 4.7% (1894). I selected the parent with 7.4% for the continuation of the race and only planted out the best tricotylous specimens from amongst its offspring. I had 44 tricotyls amongst 1000 seedlings to choose from; besides these there were 31 hemi-tricotyls and one tetracotyl. I repeated the evaluation of the ratio with 4000 seedlings and found 7.2%, that is to say, about the same value.

At the end of the summer of 1894 my culture consisted of the 20 best tricotyls only, and 16 of these ripened their seeds. In this generation, however, there was a considerable step back, for the individual parents varied between 0.5% and 3.7%; and one parent had not a single aberrant form amongst 200 seedlings. The mean was 1.8%. The seeds of the five best plants were again sown and 1000 seedlings from each examined. The values were now 2.6—2.8—3.2—3.2 and 3.6%. In the following years I endeavored to determine the cause of this return by a series of collateral experiments, but without success. The variability of the hereditary value in such races obviously depends in great part on causes which we do not yet understand.

I now planted out the tricotylous offspring of two plants with 3.2 and 3.6%, but in two separate groups in order to be able to confine the selection to the offspring of one of them later, if desired. These cultures consisted of 12 and 13 tricotyls, no essential difference between them being manifested. The values calculated from the batches of 300 seedlings from each seed-parent were:

SEED-PARENT	VALUES CALCULATED FROM THE OFFSPRING (Spring, 1896)											
with 3.2 %	1.3	1.7	1.7	2.0	2.3	2.3	2.7	2.7	2.7	3.0	4.0	5.5
with 3.6 %	0.7	0.7	1.0	1.3	1.7	2.0	2.7	2.7	3.0	3.0	3.7	4.7	5.3

The mean for both cases was therefore about 2.5% (actually 2.5% and 2.7% respectively). This experiment proves how little effect an enlargement of the extent of the cultures has in such an experiment in selection, for if I had only dealt with the offspring of one of the two parents, the result would obviously not have been essentially different.

In the summer of 1896 I did not go on with this culture, but tried to find out whether by planting out a considerably larger number of individuals, I could yet improve the prospects of success. For this purpose I selected the two plants of 1895 which seemed to be the best, and for which I had found the highest values in the spring of 1896, viz., 5.3% and 5.5%. In order to have a large crop to select from, I sowed 15 cubic centimeters of seed, and raised from 15,000 to 20,000 seedlings, of which the strongest tricotyls and hemi-tricotyls were planted out singly in pots, and later put out in the beds. Three sowings were made, at the end of March, at the end of April and in the middle of May, in the hope of possibly increasing thereby the variability and the prospect of a mutation. Furthermore, within the three groups, widely different positions, different degrees of remoteness of the individual plants, and different treatment in the matter of pruning, obtained. Many plants gave more than 30 cubic centimeters of seed each, but more than ten cubic centimeters was never saved.

Altogether I saved the seeds of about 450 plants, and sowed them separately. For each seed-parent 300 seedlings were recorded in 1898; and the proportion of tricotyls was calculated from these data. The result was, however, that a very great difference was seen to exist between the two grandparents of 1895; the one with the value of 5.3% proved to be a bad stock plant. Amongst its offspring, of which there were 30, the ratio was greater than 3% in ten cases only, and on the average it was 0.3—0.5%. And this in spite of the treatment, which, though varied, was the best that could be given, and in spite of the complete exclusion of atavists.

The second grandparent, with the value of 5.5%, proved as fortunate in its progeny as the former had been unfortunate. Its offspring had all been sown together, and planted out under the same average treatment on the same bed and on the same day. About 140 plants set seed abundantly.

On the average, however, this culture was not better than those of the previous years, for it only yielded a ratio of 4.5%; but the range of variability was much greater. Eight plants occurred, the hereditary coefficients of which exceeded all previous ones. Of these, six were 14 to 17%, one 21%, and one 25%. Here the possibility of a sudden advance seemed to open up.

Before I give the whole series of figures, I wish to make one further observation. If in the year 1897 I had not cultivated 450 plants, but only, let us say, one-third, I would have limited myself to the offspring of the grandparent with 5.5%, although the value is only apparently greater than the other, because the difference lies within the limits of observational error. I would then have obtained precisely the same result with only one-third of the labor. In other words, neither the selection of tricotyls as seed-parents, nor the attention paid to the hereditary values, although this excludes the poorest tricotyls in spite of the latitude of possible errors, can make the experiment independent of chance. Nothing less than carrying out the experiments on a much larger scale can effect this. But the results of the two following generations will show that even in the present very favorable case, no real or permanent advance was effected.

The values obtained, in the spring of 1898, for the 140 offspring of the best parent of 1895, which itself had a value of 5.5%, are distributed as follows: (P refers to the figures in percentages and A to the corresponding number of offspring which manifested these numbers amongst their seeds).

P	1	2	3	4	5	6	7	8	9	10	11	12	13	14	15	16	17	21	25
A	16	27	30	18	18	11	6	4	2	2	2	3	2	1	2	1	2	1	1

This series only relates to the tricotyls selected for my experiment. If I had planted out the whole progeny of the parent in question without selection, the number of low values would most certainly have been somewhat larger; and the apex of the curve would perhaps have stood over the 0 instead of over 3%. But the chief point in this series is that from 3% onwards the figures regularly and continually diminish in such a way that the two extreme variants with 21% and 25% round off the series continuously; i. e., in the ordinary manner peculiar to physiological curves. It is obvious that they do not stand in discontinuous relations to it.

In the estimation of the higher individual values a latitude of 5% must be allowed, i. e., the figure 15 denotes a real capacity of from 10 to 20%, the figure 25 one of from 20 to 30%. In the case of extreme deviations it is always well to assume that these errors may have operated in the same direction. We may only state, therefore, that by means of a cultivation through eight generations, a ratio of 20% has been attained. But whether or not this is the limit, we do not know. Also, it is uncertain whether the parent with 25% was really better than those with 15 to 21%. But it is certain that the best seed-plant, as judged by its hereditary value must be one of this group.

Therefore, at this stage in the development of our race, the process of selection has become much more certain than before and less dependent on chance. It is only necessary not to limit our choice to those which appear to be the very best, but to cultivate also some of the next best seed-parents for the continuation of the race. Of the culture of 1897 I selected five, viz., those with 16, 17, 17, 21 and 25%. In order to allow as many different external factors as possible to come into play, I sowed the following generation in two subsequent years (1898 and 1899), and in the first year in two separate lots, one in the middle of April and one in the middle of May. Altogether somewhat over 400 tricotylous plants were planted in this year, and the hereditary capacity was determined on 300 seedlings in each lot.

But in spite of every care, the result was a return to previous values, and this in the case of every one of the five parents. The mean value for all of them was only 2%; the five highest values were 9.6—10.6—10—11—11% mostly amongst the offspring of the same grandparent (E in the following table). Below I give a short review of the values found separately for each individual grandparent of 1897. The countings were made in the spring of 1899.

	A	B	C	D	E
A. Grandparents in 1897					
Their values in spring 1898	16	17	17	21	25
B. Parents sown on April 21st, 1898:					
Their values in spring 1899:					
Maximum	5.3	2.6	3.0	3.3	11.0
Mean	1.5	0.8	0.7	1.1	4.7
C. Parents sown on May 7th, 1898:					
Their values in spring 1899:					
Maximum	9.6	6.6	4.6	6.3	7.3
Mean	4.1	2.2	2.1	2.5	2.5
Mean of the two series	2.8	1.5	1.4	1.8	3.6

The two grandparents with 16 and 25% must therefor be regarded as the best; and it must be supposed that the five exceptionally high figures were brought about by peculiarly favorable circumstances which were not repeated in the following year.

The repetition of the experiment in 1899 gave practically the same result.

All in all, in the ten generations of my experiment, there occurred neither a sudden nor a gradual transition to an intermediate race.

And lastly let me summarize the whole culture in a simple table which gives only the mean and the highest value for each generation as well as the value for the seed-parent selected for the continuation of the race.

GENERA- TION	YEAR OF FLOWERING	VALUES FOUND IN THE FOLLOWING SPRING		
		MEAN	MAXIMUM	SEED-PARENT
I.	1889	—	—	4.5
II.	1890			4.5
III.	1891	—	—	13
IV.	1892	3.5	11.8	6.3
V.	1893	4.7	7.4	7.4
VI.	1894	1.8	3.6	3.6
VII.	1895	2.5	5.5	5.5
VIII. (450 Ex.)	1897	4	25	16-25
IX. (400 Ex.)	1898	2	11	—

The whole series, with the exception of the maximum figures for 1897 rather indicates a fluctuating around a constant mean value than a steady progress under the influence of selection.

Scrophularia nodosa. Of this species a series of forms occurs. Their characters have been given in the *Prodromus* of DE CANDOLLE. The leaves are broadly cordate or only rounded at the base, with little teeth of almost uniform size, or very coarsely and irregularly indented, and the fruits are large or small. The form with cordate, uniformly toothed leaves is very common in this neighborhood, and, so far as I know, the only one that occurs. The form with coarsely dissected leaves rounded at the base, and small flowers and fruits, is the one employed in my experiments (Fig. 84). In the summer of 1901 I planted out large numbers of both types in my garden, after a close examination of their characters. Both are, as far as my experience extends, quite constant for many generations, i. e., they are good subspecies.

Fig. 84. *Scrophularia nodosa.* Main stem of the tricotylous half race with ternary whorls of leaves.

My culture began in 1890 with the seeds of a tricotylous plant from our botanical garden, which, in the following year, produced, besides a number of atavists, four plants with three cotyledons and a main stem with ternary whorls up to the inflorescence (1891). From the seeds of one of these plants I again obtained in the summer of 1892 some tricotylous seedlings, the majority of which, however, became decussate afterwards. Only two of them remained ternary throughout the summer, and these were the only ones which I allowed to flower. Among their seed I counted 780 and 1000 seedlings respectively, and found the value in each case to be 1%.

In 1893, 16 tricotylous plants flowered and constituted the fourth generation of my race. Their seeds were saved separately and produced a mean of from 1 to 2%, in the best cases 2.4 to 4.1 and 5.4%. The counts were made on 700, 800 and 2000 seedlings. These three plants had, however, decussate leaves on their main stems, and this shows that the disposition of leaves on the stem is of subordinate value in selection. In the later generations I have always selected the ternary individuals where possible, but have not found that they are any better than the others.

In the summer of 1894 tricotylous seedlings of the seed-parent with 5.4% were planted out. I saved the seeds of 25 plants separately, and in the following spring examined from 300 to 500 seedlings in each crop. The values were essentially the same as in the previous year, and varied between 0.5% and 5.5% with a mean of 2%: and the five best gave 3.6—3.8—4.0—4.2 and 5.5%; the three last values are based on counts of 1000, 1500 and 2000 seedlings. It was only the offspring of these parents that were planted out in 1895, and only those with three cotyledons and with a first whorl of three leaves (Fig. 73, p. 366).

By planting them out in three groups in the following spring, a selection was made possible which related to the grandparents. Those with 4.0 to 4.2% gave offspring with 1.3 to 5% with a mean of 3%; (from 12 parents and from counts of groups of 300 seedlings). The grandparent with 5.5% proved to be considerably better. Its twelve offspring had 2 to 8%, with a mean of 4.5%. Seedlings of this group only were planted out in 1896 and only tricotylous specimens with ternary lower whorls, from parents with 6—6 and 8%. They were in all 72 strong plants each of which gave a sufficient harvest for the calculation of their hereditary values in groups of 300 seedlings. The parent with 8% gave 2 to 15%, with a mean of 6%. The two other parents 2 to 10% and 3 to 14%, with means of 6 and 7%. Obviously the difference was not sufficiently considerable to base a selection among the grandparents upon. On the other hand it was evident that the race had been improved by the process of selection which now extended over six generations; and this improvement was afterwards maintained.

For the culture of 1897 I selected the tricotylous offspring of three seed-parents with 11—11 and 15% tricotylous offspring, and planted out 100 of them, which were all strong plants with a ternary first whorl of leaves. The harvest was recorded in the spring of 1898 in the usual way. The three parents, however, which were now grandparents, again showed no essential difference in their offspring.

VALUE FOR THE GRANDPARENTS	VALUE FOR THE PARENTS			NUMBER OF PARENTS
	MIN.	MED.	MAX.	
11%	2	7	15	20
11%	3	7.5	15	24
15%	4	8.5	13	28

Here again the figures were the same as in the previous generation.

Although, as we have seen, the selection of grandparents, in this case, afforded no reliable ground from which an improvement of the race could be started, yet this must be the case whenever the hereditary capacity of the grandparent is much influenced by fortuitous circumstances which affect the selection of the seeds. Therefore the special object of such selection is mainly to exclude such cases as much as possible from the main trunk of the pedigree, by simply not planting out their seedlings.

Tricotylous Races Do Not Arise by Selection.

But when the values of several parents do not differ so greatly from one another that the differences fall outside the limits of probable error, individuals should be planted out every year from all of the best parents. This would, however, necessitate an increase of the cultures on too large a scale. Nevertheless I extended my culture as far as possible in the summer of 1898 on this principle, and saved the seeds of 300 plants separately.

The result of the 300 counts which were made on these lots of seeds, embracing 300 seedlings for each lot, corresponded with my expectations to this extent that one of the grandparents proved to be by far the best. Its hereditary value was 14%, that of its offspring had a mean of 20%, and for two individuals even 25 to 27%. Moreover, outside the limits of this group, this race also showed an increase in hereditary capacity.

The harvest of 1898 contained the seeds of 300 tricotylous plants, which in their turn had been raised from 15 parents with values from 10 to 15%. The results were as follows (expressed in percentages):

Parents	10	10	10	10	10	10	11	11	12	12	13	13	14	15	15
Offspring { Mean	8	10	10	11	14	19	10	15	11	16	10	13	20	14	14
{ Max.	8	11	20	20	27	26	19	25	20	24	21	21	27	21	22
Number of offspring	1	2	19	28	11	6	8	9	27	6	36	32	12	64	39

The mean of the whole series of counts was 12% and denoted a considerable increase in the character of the race.

Let us now compare this series of figures with that given on page 405 for *Amarantus speciosus*. That series relates to the hereditary coefficients of 140, this to those of 300 plants, which in the former case belong to the 8th, and in this one to the 9th generation. Both series, therefore are the result of a selection which extended over a sufficiently long period of time to justify the expectation that a definite result would be obtained. They exhibit, however, one striking difference. The mean of the figures for *Amarantus* is between 2—3%, and this renders it not improbable that, if the coefficients of the atavists could have been incorporated in the curve, its apex would have been at 0, that is to say, that a unilateral or so-called half curve would have been the result. But the apex of the *Scrophularia* curve is at 12% and varies amongst the individual families from 8 to 20%, indicating thereby the possibility of an isolation of a separate apex for the tricotylous specimens.

In other words, the tricotylous half race of *Amarantus* behaves, during this long period of selection, in essentially the same way as at the beginning (table p. 407), whereas *Scrophularia* behaves like other half races improved by selection. The scheme representing the influence of selection on the half race of *Ranunculus bulbosus semiplenus* (Fig. 52 on page 252 of this volume) would apply equally to this case.

The explanation of this difference in behavior is perhaps to be sought in the assumption that in *Scrophularia* tricotyly has only been indirectly improved. I am referring to the case of *Trifolium incarnatum quadrifolium* which I described above on page 239. In that case the smallest seeds gave rise to the largest number of aberrant seedlings. If the production of small seeds could be increased by selection the number of seedlings with compound primordial leaves would also be increased. It seems possible that in *Scrophularia* similar factors were at work, since the character involved need not, of course, necessarily be the size of the seeds. The repeated selection of tricotyls would not, on this assumption lead directly to the increase of this character, but to a supposed change in the structure of the seeds which would favor the anomaly. Whether or no this supposition is correct I do not know, but it is a fact that the fruits and seeds of my race have gradually become smaller; and that it is the plants producing the largest number of tricotyls which bear the smallest fruits and seeds, I find repeatedly noted in my records.

At any rate we are here in touch with a principle which may be applicable to other cases also. A selection may produce its effect on an unknown character which in its turn will affect the character actually dealt with.

With regard to the extent of my culture of 1898 it is further worth remarking that it shows that, in general, the extent of the cultures is by no means so important a factor as is usually supposed. If I had confined myself to experimenting with three or five seed-parents, as for instance in *Amarantus* in 1897 (p. 406), I should have chosen the best ones according to their hereditary coefficients; and it was exactly amongst these that the best of all occurred as the series on page 411 shows. Increased extent of the experiment deepens one's insight into the processes involved, but does not hasten the improvement of the race; although it is never advisable to confine oneself to experimenting with one single seed-parent, if this can be avoided.

The next generation, the 10th and at present the last, has repeated the progress observed in the ninth. Here again the race of *Scrophularia* behaved differently from that of *Amarantus*. I confined myself to the progeny of the plants of 1897 which exhibited the value of 14%, and amongst the offspring of which the mean attained 20%. From these I selected five specimens with 22—23

—23—25 and 27%. I planted out 165 tricotylous seedlings from amongst their offspring in pots, as soon as they proved themselves also to be ternary in their first whorls of leaves. In the following whorls about half of them reverted to the decussate arrangement. These were thrown away, and only 72 plants which remained ternary were ultimately planted out. They were fairly uniformly distributed over the crops of the five parents.

From these 72 plants, 72 values, in five groups, were calculated in the following spring (1900). The lowest values, arranged in a series corresponding with the increasing values of the five parents, now grandparents, were 9, 8, 13, 8 and 11%, the means 16, 17, 18, 17, and 19%, and the maxima 19, 22, 26, 22 and 26%. As we see the five groups did not exhibit any essential difference. The mean value of the previous generation, 12%, had now been exceeded, but the maximum remained the same.

Let us now summarize the whole experiment in the following table:

GENERATION	YEAR OF FLOWERING	VALUES FOUND IN THE FOLLOWING SPRING		
		LOWEST	MEAN	HIGHEST
I.	1890	—	—	—
II.	1891			—
III.	1892	—	—	1
IV.	1893	0–0.3	1–2	5.4
V.	1894	0.5	2	5.5
VI.	1895	2	4.5	8
VII.	1896	2–3	6–7	14–15
VIII.	1897	2–3	7–8	15
IX.	1898	2	12	25–27
X.	1899	8	16–19	26

Progress is, as we can see, a continual one, and the selection has been, although perhaps only indirectly, (p.

412), successful throughout. Nevertheless an intermediate race has not arisen, either gradually or by means of a sudden jump. The progress in the last two years was more rapid than before as the result of increased stringency of selection, without, however, affording any indication that the mean of 50% was likely soon to be reached.

Oenothera Berteriana. Besides the two cultures mentioned which were begun in the first years of my experiments in selection, I have cultivated yet a third race with the same object. This race was one in which the intercrossing of the various individuals could always be avoided. There is, however, no ground for fear that occasional unavoidable crosses in *Amarantus* and *Scrophularia* had any considerable effect on the selection process; for both species must be fertilized almost entirely with their own pollen on acount of the great number of their flowers which are open at the same time; and they are, when thus fertilized, perfectly fertile as isolated individuals show. Moreover what is spoiled by crossing is eliminated by selection.

But the evidence is more satisfactory if self-fertilization can be insured. This occurs in *Oenothera Berteriana.* Its flowers form perfectly normal fruits and seeds, when the visits of insects are excluded. I enclosed my plants in a cage of fine metal gauze. Some years I have fertilized them artificially in it; but this is quite superfluous, because when the flower withers the stigma bends downwards and thus reaches the pollen. In the two last summers the cage was shut from the beginning of the flowering period until the seed began to ripen. Nevertheless they all produced fruits with scarcely an exception. These fruits contained an abundance of seed, and a few from each plant are sufficient, and this is a great advantage as compared with the laborious process of harvesting in *Scrophularia*.

My culture began in the summer of 1896. At that time the prospect of obtaining an intermediate race in the other two races had already become very remote. From some plants in our botanical garden I saved seeds which produced 13 tricotyls, 4 hemi-tricotyls and one tetracotyl amongst 300 seedlings, i. e., 6%. This figure was considerably higher than most of the values of my half races mentioned on page 392, and about as high as the value attained by *Amarantus* and *Scrophularia* at that time. At the same time six other species of *Oenothera* were tested with reference to their production of tricotyls. I found from 0 to 1 and 2% and therefore selected the *O. Berteriana*. In the spring of 1897, tricotyls only were potted singly, and planted out in July in the beds. Their hereditary values were determined in the following year in the case of 15 plants. They fluctuated between 1 and 12% and exhibited a mean of 4%. In the following generation, 1898, I was able to plant out about 60 strong tricotyls, and since that time I have conducted the cultures on this scale with only slight modifications. 15 tricotyls from the plant with 12%, and 45 from those with 6 to 7% were planted out. The former group, however, proved to be no better than the latter. The values of these plants varied from 2 to 16%, with a mean of 4%. The best of the five other groups, however, had values varying from 6 to 16%, with a mean of 8% (from 10 separate crops). The remainder varied between 4 and 11, with a mean of 6%. I selected the three best parents whose values were 15, 16 and 16%, and I planted out the strongest tricotylous offspring of each (1899). This year 77 plants gave a sufficient harvest, the figures for the three groups (each derived from a single grandparent) did not exhibit any differences worth mentioning. The minimum was 2%, the mean 12, and the highest value 25%. Therefore a considerable advance on the preceding generation had taken place. In 1900 I had only 31 plants, bearing seed, which were cultivated in the same way as in the previous year. They constituted three groups, each from a single grandparent, but without exhibiting any differences worth consideration. The hereditary values varied between 5 and 17% and their mean was 10 to 13%. The culture of the last year (1901) embraced 40 plants, the values of which varied between 0 and 21 and had a mean of 10%.

The result, like that of the two foregoing species, may now be summarized in tabular form.

GENERA-TION	YEAR OF FLOWERING	VALUES DETERMINED IN THE FOLLOWING SPRING		
		MEAN	HIGHEST	SEED-PARENTS
I.	1896	—	—	6
II.	1897	4	12	12
III.	1898	4–8	16	15–16
IV.	1899	12	25	23–25
V.	1900	10–13	16–17	14
VI.	1901	10	21	—

The prospect of raising an intermediate race seems therefore in this experiment to be as small as in *Amarantus* and *Scrophularia*.

§ 6. THE ISOLATION OF TRICOTYLOUS INTERMEDIATE RACES.

It is just as easy to isolate an intermediate race from seed which has been bought or obtained from any other source if it happens to be already present in it, as it is difficult to raise one if it is not. Isolation can be effected in two to three generations as a rule, in fact as soon as the hereditary values of the plants raised from the original mixture of seed have been determined. Very little care and no artificial fertilization at all is necessary for this purpose. It takes place as quickly in diœcious wind-fertilized flowers, such as the hemp, as in self-fertilizing species enclosed in bags or cages.

In the choice of species and varieties one obviously has not a free hand. We must first search for cases in which the desired races happen to be present, and for this purpose sowings of considerable extent have to be made. If lots of ten to twenty grams of seed, according to the size of the individual grains, are sown, it can be seen soon after germination whether the species is likely to produce a tricotylous intermediate race or not. The occurrence of 1 to 2 tricotyls amongst about 10,000 normal seedlings does not justify this expectation, but that of one or more per thousand does as a rule.

The cultures of my half races mentioned in § 4 (p. 392) had begun in 1892 or earlier, and in the spring of 1895 there could be no doubt that I should not obtain any intermediate races from them. I resolved therefore to seek them elsewhere, and selected for this purpose about forty species and varieties of garden plants, which were chiefly annuals. The result of this sowing, which was conducted on a large scale, has already been given on page 380. Most of the sorts contained too few tricotylous seedlings. Only 8 offered the prospect of giving rise to an intermediate race, and of these I had to reject three for various reasons. The remaining five, however, fulfilled my expectations.

Before I proceed to the description of my experiments

it is desirable, in connection with what was said in the first part of this volume, to consider what we should expect to occur in the isolation of such intermediate races as may happen to be present. The intermediate race does not differ from the ordinary types or half races in any visible characters, but only in the frequency of tricotylous specimens. If the tricotylous individuals are selected for culture from crops raised from bought seed, it is by no means certain that these and these only belong to the intermediate race, if indeed such exists at all. For in addition to this, the half race is almost always present, and this, as we know, also contains some tricotyls. On the other hand the intermediate race always produces atavists and usually in considerable numbers. Provided therefore that the crop contains an intermediate race, some of the dicotyls and some of the tricotyls would belong to it; but the prospect of obtaining it from the latter will obviously be greater than from the former. For this reason I select, whenever possible, tricotyls only for transplantation. All that then remains to be done is to save their seeds separately from each plant and to determine their hereditary values in the harvest. If any of these are especially high they belong to the intermediate race, and all the rest are thrown away inasmuch as they include the half race and the doubtful cases. Ordinarily the race is by this time perfectly pure and can be improved by selection on the ordinary lines. As a rule, the means of the curves describing my intermediate races, lie between 50 and 60%, and can be brought as high as 80 to 90% by selection in occasional individuals.

In contradistinction to the unsuccessful experiments described in the previous section there can be no doubt in these successful cases that we are not dealing with the selection of the extreme variants of ordinary fluctuating variability. Even the doctrine of unilateral increase in variability as a result of selection is of no help in this case (see § 2, p. 9), for selection could hardly operate so rapidly as to produce its whole effect in a single generation. The old saying of gardeners that the first condition necessary for the production of a novelty is to possess it already, also applies to these purely experimental races (Vol. I, p. 185 and elsewhere). If the tricotylous race does not already exist it cannot, at present at any rate, be either isolated or bred.

A high percentage in tricotyls is seldom found in wild species. The highest value I have yet found occurred in *Linaria vulgaris* in the spring of 1894 in a sowing of the seed of a hemipeloric plant of the race that I was cultivating at that time (see p. 211). There were 59 tricotyls amongst the 425 seedlings, i. e., 14%. Amongst commercial seed the prospect of obtaining intermediate races seems to me to be the greatest, as I have already stated, in those sorts which are cultivated on a large scale in the field or in the garden. It is much smaller in those varieties of flowers which are only grown on a small scale every year. Moreover it seems obvious that cultivation on a large scale should favor the origin of new races.

If the intermediate race, which is being sought for, exists in some sample of seed, we may expect to find mean, better or inferior representatives of it. If the former is the case the mean character of the race, that is about 50 to 60% of tricotyly is attained at once, and this occurs in the majority of cases, as might be expected and as the table at the conclusion of this section will show (p. 439). Individuals with a higher productive

The Isolation of Tricotylous Intermediate Races. 421

capacity are rarely found at first. I came across an instance of one only once, at the beginning of my experiment with the syncotylous *Helianthus annuus* (see the following chapter). If individuals with low hereditary capacities are found, those with mean values can as a rule be easily raised from them, for they are to be regarded simply as *minus* variants of the race sought for; and will therefore, in conformity with the law of regression, revert to this value even if selection be only suspended (see above, p. 5). Experience shows, and the table already referred to will demonstrate, that one or two generations are, as a rule, sufficient for the attainment of values of 50 to 60%.

Before I proceed to a detailed description of my experiments I will give a few instances to show the course which these experiments in the isolation of tricotylous intermediate races follow.

The attainment of the mean value. On page 380 I have given, in a small table, the numbers of tricotyls which I found in some samples of seed in the spring of 1895. For some of these species these tricotyls were planted out and their seeds saved separately and sown. I obtained values from 12 to 19% in the best individuals (spring 1896), and on the rest, as a rule, much fewer. These were regarded as belonging to the half race or as of doubtful significance. The tricotyls from the best parents were now (1896) planted out, and the hereditary value for each was calculated in the following spring. Below I give a résumé of these figures in groups of 0—2, 3—7, 8—12%, with means of 1—5—10% and so forth, for convenience of comparison, and indicate, for each such reduced value, the number of individuals which exhibited it.

ISOLATION OF TRICOTYLOUS RACES.

HEREDITARY VALUES OF THE SECOND GENERATION.

	FIRST GENERATION	1	5	10	15	20	25	30	35	40	45	50	55	60	65
Clarkia pulchella	16%	0	1	1	7	1	4	7	3	4	2	1	4	2	2
Papaver Rhoeas	18%	0	0	2	8	5	4	3	1	0	0	0	1	—	—
Papaver Rhoeas	19%	0	0	1	1	8	2	1	4	4	0	1	—	—	—
Phacelia tanacetifolia	14%	0	3	3	11	3	8	6	4	2	2	2	—	—	—

It is easily seen that in such cultures the choice of specimens which belong undoubtedly to the intermediate race is made possible. All plants with a low hereditary value are simply rejected, since the possibility exists that among them may be hidden hybrids between the two races. The high values are free from this doubt and indicate the pure race we are searching for.

Thoroughbred races. Under this term those races are included in agriculture which have been considerably improved by stringent selection. In the first part of the first volume we have seen that they remain dependent on continued selection, and do not really become constant. For studying the features of such races, the tricotylous intermediate races afford most suitable material, for, after reaching the mean value of about 50%, we may select the extreme variants, and, by their culture, improve the race much farther. As instances I choose *Mercurialis annua,* a large agricultural crop, and *Silene inflata,* which I derived from seed introduced with cereals. Both cultures were started in 1892, and reached a hereditary value of 55% in the harvest of 1894. From plants with this value I raised a fourth generation, which offered me the following figures:

SELECTION OF THOROUGHBRED TRICOTYLOUS RACES.
HEREDITARY VALUES OF THE BEST TRICOTYLOUS SEED-PARENTS OF THE FOURTH GENERATION.

	THIRD GENERATION	25	30	35	40	45	50	55	60	65	70	75	80	85
Mercurialis annua	55%	—	—	—	—	1	2	2	1	2	1	2	3	1
Silene inflata	55%	1	0	2	3	4	4	4	4	7	1	1	—	—

It will be seen that regression did not accompany this selection, because the parent plants deviated too little as yet from the mean of the new race. On the other hand the high values of 75-85% were immediately reached, and this obviously indicates a considerable advance. These instances, taken together with the table which will be given on page 439, are sufficient to indicate the method by which **tricotylous** intermediate races, if they are at all present, can be isolated and improved. They will facilitate the understanding of the detailed description of my experiments. These obviously present, according to the species investigated, greater or less deviations from the instances given, but as a rule they are not essential ones. Therefore I shall give these descriptions as briefly as possible.

But before doing so I wish once more to lay emphasis on the fact that these cases are thoroughly distinct from those in which only half races are present, and from which, after several generations, often no more than 2 to 4% and only very rarely as much as 20% was reached, as maximum figures (see table on page 392).

Oenothera hirtella. Under this name I cultivate a tricotylous race which I obtained by chance from bought seed. In 1895 I was growing some samples of seed of *Oenothera Lamarckiana* and allied species which had either been bought or obtained by exchange, in order to compare them with my own strains. In the autumn I saved the seed of a number of separate specimens and sowed it in the hope of finding a tricotylous intermediate race amongst them, inasmuch as my own races and varieties offered no prospect of producing them, as has already been mentioned for a special case, that of *Oenothera rubrinervis* (p. 383). This hope was fulfilled by a single specimen, all the remaining lots of seed giving the usual very low values of from 0 to 4%. This specimen was a plant which was noticed by some striking distinctive marks. It was taller and slenderer in growth than all the other species, more than 2 meters high, with a long raceme interrupted in places by the failure of some of the lower buds. Its flowers were of the size and structure of those of *O. biennis* and, like this form, were self-fertilized within the buds before their opening. Its progeny have kept true to this type through a series of generations.

This stray plant produced from its seeds, in a lot of 300 seedlings, 7% tricotyls, and in another estimation, amongst 2430 seedlings, 8% tricotyls, of which 143 were tricotyls in the restricted sense, 59 were hemi-tricotyls, and 4 tetracotyls. Whether the parent itself had three cotyledons, I do not of course know.

Of this crop the tricotylous seedlings only, and of these only the strongest, were planted out on the 2d of April, 1896. In the middle of July, some few days before they flowered, the whole bed was covered with a great cage of fine cloth. The cloth was removed at the beginning of September and at the same time all the open flowers and buds were removed from the plants. Seeds were saved separately from 54 individuals and 300 seedlings were counted from each lot. The result was a very remarkable one. With a single exception the figures formed a close series of which the mean was 7%, whilst the great majority of figures lay between 2 and 12%. (Two parents with 0.5 and 1.0 and three with 13, 14, and 17%.) Besides these there was a single plant which stood far from the others in the series. It had produced 56% tricotyls, i. e., more than half of its seeds were tricotylous. For the sake of greater certainty this value was determined twice. The percentage value calculated from a lot of 768 seedlings was 58%; from another of 657 seedlings 54%; with a mean of 56% for 1492 seedlings.

This one plant, therefore, had a hereditary value which corresponded closely with the mean value which we should expect the intermediate race sought for to possess. Of course the experiment was continued from the seeds of this plant only. The seedlings of all the other parents, tricotyls as well as dicotyls, were thrown away.

Two courses were now open to me, either simply to maintain the race, or to endeavor to improve it by further selection. In the former case, selection would have to be avoided as much as possible, and we should have to harvest and sow the seeds of all the specimens mixed together, and in planting out the seedlings be careful not to give preference to the tricotyls. I have not done this; but as is my custom, I have harvested and recorded the seeds of each individual separately and only planted out the seedlings of the best parents, that is to say the parents with the highest hereditary values, for the continuation of the race.

In the summer of 1897, 37 tricotylous offspring of the

parent with the value of 56% flowered in a cage of metal gauze. The values, calculated in the usual way from the harvest, gave a good curve, whose mean was 72%. The lowest percentage of tricotyls was now 38%, the three highest 83—83 and 89%, that is to say, a very considerable advance.

Atavistic seedlings of the same parent were also planted out, but they naturally exhibited a somewhat smaller advance.

In the following year, 1898, I did not sow the seeds of the plant with 89%, but from various considerations those of one with 66%. The reason for this choice was that the plant had flowered early and that the harvest had been increased thereby. The parents with the higher values had flowered too late or set too little seed, and it would have been very dangerous to have continued the experiment along this line. Moreover there was no longer any particular interest in improving the race further. The mean value has been lessened thereby to about 40%, and the maximum to 74%.

Cannabis sativa. I propose to deal now with two cultures of diœcious plants, *Cannabis* and *Mercurialis*. In these cases self-fertilization is impossible; nevertheless the isolation of the intermediate races was effected as easily and almost as quickly as in *Oenothera*. Without doubt self-fertilization has in such experiments, whenever feasible, the high value which is usually assigned to it; but the experiments now to be described show that it can often be dispensed with, as well. This result is very important, because it makes isolation and selection possible in species in which an artificial fertilization of every single seed-parent would increase the labor beyond measure. Such a task I have, therefore, only undertaken with a single species, *Antirrhinum majus.*

Only some varieties of hemp seem to include tricotylous intermediate races. Amongst those which I have tested I have found such a one in the giant hemp only. In the spring of 1893 I sowed a large quantity of seed of this species, but was only able to bring 7 tricotylous plants to maturity. Unfortunately the majority of them were male, and there was but a single female plant. This, however, produced four tricotyls amongst 126 seedlings. In 1894 two of these were female and two male. The seeds of the former were saved separately and yielded 15 and 9% tricotyls amongst 400 and 600 seedlings respectively.

For the continuation of the race, only the tricotylous seedlings of the parent with 15% were planted out (1895). There were 29 plants, of which 10 bore seed. Their values were 19—31—38—40—43—47—48—50—52 and 63%. As we see they attained the mean value of the intermediate race and even exceeded it in one case, (63% amongst 316 seedlings). The tricotylous offspring of this individual alone were planted out (1896). Of these, 38 specimens set seed, and from this the values of the individual parents were calculated. I reduced them, as usual to groups with 40—45 and 50% etc., as a mean value, and found the 38 offspring of the parent with 63% to be distributed as folows:

Proportion of tricotyls	40	45	50	55	60	65	70	75	80
Number of seed-parents	5	5	3	10	5	5	2	2	1

The mean of the series is at about 55%, and the series therefore constitutes a good instance of a young, isolated race which, however, has not yet been improved to any considerable extent by selection.

In 1897 I continued the culture for the purpose of improving the race. I sowed the seeds of four plants which had 65—66—67 nd 70% tricotyls; and managed to bring 60 tricotylous plants to maturity, of which 26 set seed. Their hereditary values varied between 35 and 90%, their mean being 74%. This indicates a satisfactory advance.

Mercurialis annua. In 1892 I possessed two tricotylous plants, one a female, the other a male, which had been raised from species that had been obtained in exchange from some botanical garden. The yield was very poor; only 14 seeds germinated and they had two cotyledons each (1893). In saving their seeds I did not isolate them and found 2% tricotyls amongst the 1100 seedlings which were raised in the following spring (1894). Besides these several tricotyls had come up in the bed from seeds which had fallen out, so that I had altogether 18 female and a corresponding number of male tricotyls. This culture gave the expected result. The hereditary values calculated for the individual seed-parents were distributed regularly between 1 and 55%, the five highest being 31—34—41—52 and 55%. The intermediate race had therefore been isolated at least in these latter specimens. In 1895 I only planted out tricotylous seedlings from the one parent with 55%. Of course it was uncertain in this case whether the pollen had been produced by plants belonging to the intermediate race. But the values calculated in the spring of 1896 suggested that the race was fairly pure. I have already given this series of figures above (p. 423). The mean was at 67, the maximum had increased to 86%.

In 1896 the tricotyls of two parents, whose values were 78 and 81%, were planted out in lots of 25 and 20 respectively. The values for the two groups did not differ essentially, and were distributed between 51 and 92%, and the mean of the 25 plants (the rest had been male) was 73%. The race had, therefore, in comparison with the previous year, undergone a further improvement.

It was continued one year more in the same way (1897). 12 female and several male tricotylous offspring of the parent with 92% were planted out, and the values calculated from these were found to be distributed between 65 and 91%, with a mean of 78%.

We see, therefore, that after the figure 55% had been reached in the harvest of 1894 the mean value rose in the three following years of the experiment to 67, 73 and 78%.

Clarkia pulchella, Fig. 85. It was in the spring of 1895 that I made the extensive sowings of horticultural seeds to which I have already referred, for the purpose of isolating tricotylous intermediate races. The seeds of *Clarkia pulchella alba* produced about 1% tricotyls. 30 of these flowered, but only 18 of them produced sufficient seed. Two of them had hereditary values of 14 and 16%, the rest from 0 to 7%, with a mean of 4%. In 1896 only the tricotyls of the parent with 16% were planted out. There were 39 of these, and for all of them a value could be calculated. These values have been given on page 422, and were above 50% for eight plants. The intermediate race, therefore, was already represented by several specimens.

In 1897 I planted out only the tricotylous seedlings of the plant with 64%, and saved the seeds of 39 of them. Their hereditary values were distributed between 16 and 79% and gave a mean of 49%. The race could therefore be regarded now as perfectly pure.

Helichrysum bracteatum compositum, Fig. 86. From the same set of sowings as that which contained *Clarkia,* I planted out some tricotylous seedlings of *Helichrysum.* There was a relatively large number of them, and altogether 19 set seed. For each one the value was calcu-

Fig. 85. *Clarkia pulchella alba.* A flowering sprig.

lated separately. In the case of 15 plants these were distributed between 2 and 8% and exhibited a mean of 4%; but there were higher figures besides, viz., 12—12—16 and 41%. The latter plant was obviously a representative of the intermediate race sought for; the remainder were thrown away as doubtful, although if there had not happened to be such a favorable plant among them all, the plant with 16% would, no doubt, have offered me as good a prospect as the corresponding plant of *Clarkia*. The chosen plant was remarkable for the large proportion of tetracotyls and the low proportion of hemi-tricotyls which it produced. There were 11% of the former, and only 1% of the latter. This peculiarity has reappeared amongst its descendants, especially with regard to the tetracotyls.

In 1895 I planted out on separate beds tricotyls and tetracotyls of the parent with 41% only. Among the tricotyls 32 plants set plenty of seed. Of these one produced only 6% tricotyls, amongst the rest the values were distributed between 13 and 43%, and their mean was at 26%. The tetracotyls gave similar numbers, embracing 19 plants with from 14 to 42% and a mean of 25%; besides these there was one plant with as much as 51%. This latter fact must obviously be attributed to a fortunate chance, and we may conclude that the tetracotyls are not more likely, nor on the other hand less so, to produce tricotyls than the tricotyls themselves, but that they obviously belong to the same race, i. e., that their character is brought about by the same elementary factor. The proportion of tetracotylous seedlings in this culture was very high, but not higher than the ratio recorded for the first generation.

Fig. 86. *Helichrysum bracteatum compositum*. A flowering stem of a plant of the tricotylous intermediate race.

For the continuation of the race only seedlings of the

tetracotylous parent with 51% were planted out. Here again both the tricotylous and tetracotylous plantlets were used, but this time not in separate lots. Seeds of 37 plants were saved and sown, and they gave values which did not differ essentially from those obtained in the foregoing year. They were distributed fairly regularly between 16 and 52% and had a mean of 35%, i. e., the mean value had undergone a considerable increase.

Antirrhinum majus. It was with this plant that I made one of my first experiments in the production of tricotylous races, and the fact that the progress in this case was much slower than in later cultures, may well be ascribed to the less extensive experience which I had at that time. I am now inclined to regard the practice of selecting from an insufficient number of seed-parents, and also of inadequate manuring as some of the causes. Nevertheless I shall describe the experiment because it ultimately led to a genuine tricotylous race. On the other hand it may not be unimportant to show that, if a repetition of my tricotylous cultures is made, the attainment of the end must not always be expected to be reached in two or three generations.

I have attempted to isolate tricotylous races from two varieties of the Snap-dragon. First from the striped variety mentioned on page 120 of this volume, and figured in Plate I; but as the success attained with this after four generations (1892-1896), was not so great as in the other case, and 45% had not been exceeded, I did not proceed further with it. Therefore I shall confine my description to the latter. It was a dark red half-dwarf variety.

In 1892 I had four tricotylous plants from bought seed, and from their seeds obtained in the following spring (1893) the proportions 2—4—7 and 7% of tricotylous seedlings, counted in lots of from 300 to 500. I only planted out the seedlings of one of the plants with 7%, but the experiment was not successful, and only three tricotylous plants gave a sufficient quantity of seed. Their hereditary values were 2—8 and 8%. The seedlings of one of the best plants were planted out in 1894. In this year I had twelve tricotylous plants for which I could calculate the values. With two exceptions, they were distributed between 7 and 19%, but the exceptions attained 23 and 25%, the mean of the whole group being 13%. In 1895 the tricotylous seedlings of the two best parents were planted out in separate groups. The mean numbers of these differed considerably; but both attained a proportion of 41% as maximum. The parent with 23% had produced offspring the values of which were distributed between 7 and 31% (18 plants with a mean of 17% and one plant with 41% tricotylous grandchildren; whilst the parent with 25% gave values from 15 to 31% amongst 12 offspring, with a mean of 26%, and two specimens with 41%. In conformity with the principles of selection of the grandparents, the three plants with 41% were not regarded as analogous, and only the offspring of the parent with 25% and with a mean, calculated from its grandchildren, of 26% were planted out. The progeny of the third plant with 41% were considered as of inferior value and thrown away.

In the following spring, when the values were calculated, no essential difference could be detected between the two groups of my culture of 1896. In percentage calculated from lots of 300 seedlings these figures were as follows:

PARENT	VALUES OF THE OFFSPRING			NUMBER OF SEED-PARENTS
	LOWEST	MIDDLE	HIGHEST	
A 41%	31	45	67	16
B 41%	22	50	79	22

Since these two cultures may be regarded as typical of ordinary intermediate races, not subjected to improvement by selection, I will give the full series of figures for one of them (B). They are as follows:

```
71  79
60  60
50  51  55  55  55  56  58  59
41  42  46  47  47
35  36  39
22  25
```

Written in this way the figures show, without any further treatment, a group in the form of a curve the apex of which was between 50 and 55%.[1]

I planted out the grandchildren of the plant B only, and selected for this purpose (1897) the two parents with 71 and 79%. The two cultures consisted of tricotylous plants only, flowered on separate beds, fertilization being left to the agency of insects, however, as in previous years. The result of the determination of the values in the spring of 1898 was as follows:

PARENT	HEREDITARY VALUES OF THE OFFSPRING			NUMBER OF SEED-PARENTS
	LOWEST	MIDDLE	HIGHEST	
C 71%	34	62	74	23
D 79%	48	64	79	24

In comparison with the previous year, therefore, an advance had taken place in the mean values but not in the maximum.

[1] It is my custom to plot the harvest raised from each parent and from each grandparent in this way. When grouped in this manner the figures display the result very graphically; and immediately show whether it falls outside the limits of observational error; and whether therefore the existence of differences between the individuals is shown beyond doubt.

Let us now summarize the whole culture, which extended over six years. We obtain the following series:

	VALUES IN THE HARVESTS OF					
	1892	1893	1894	1895	1896	1897
Highest value	7	8	25	41	79	79
Mean "	5	6	13	26	50	64
Lowest "	2	2	7	15	22	48

As already stated, the intermediate race had therefore, in this case been isolated only gradually from the mixture, chiefly as the result of the originally small extent of the cultures.

Papaver Rhoeas. In 1895 I selected 21 tricotylous seedlings from a sowing of the double garden variety of this plant with mixed colors. They yielded an abundance of seed, which gave the following percentage of tricotyls, counted in groups of 300 seedlings each:

18	19	19	20			
12	12					
7	7	8				
4	4	4	4	6		
0.5	1	1	1	1	1	2

This group plainly exhibits two sections, a half curve with an apex at 1% and 4 seed-parents with the proportion of 18 to 20%. Obviously these four belong to the intermediate race sought for, whilst the rest are partly hybrids between the two races. The best tricotylous seedlings of the two parents with 18 and 19% were planted out in 1896. They flowered and set an abundance of seed. The determination of the hereditary values from lots of 300 seedlings each gave a result for one of the two parents, with 24 offspring, of 10 to 56% with a mean of 19%, and for the other from 10 to 53% with a mean of 26%. Seedlings were planted out from the two best plants of these groups and the result in the following spring, 1898, was 23 to 65%, with a mean of 40% among 17 seed-parents, and 26 to 75%, with a mean of 47%, amongst 13.

Phacelia tanacetifolia, Fig. 87. My race arose from the same set of bought seed which included *Clarkia, Helichrysum* and *Papaver*. In the summer of 1895 I had 20 tricotylous plants, raised from bought seed, in flower. Their values constituted a two-fold group like that which occurred in *Papaver*. Three seed-parents had 12—12 and 14%, but the rest had values between 1 and 10%; the three former were alone used as the basis of my race. The three cultures derived from them gave 30—6 and 9 separate harvests, which were evaluated in the spring of 1897 in the usual way. The lowest, middle and highest values for the three groups were 5—26—58, 21—28—42, and 6—14—16%. Obviously the two first grand-parents had given better offspring than the last one. I selected the former group for the continuation of the race, employing the two best plants with 54 and 58%. They gave two groups of tricotylous plants, the harvests from which exhibited a great advance on the average, but which did not differ essentially from one another. With the exception of the extremes these figures constituted a closely circumscribed group of 35 values distributed between 35 and 72%, with a mean of 57%.

The extremes were 20 and 22%, and on the other side 80—85 and 90%. The two former figures, which were probably the result of incomplete isolation, occurred in the same group. The higher figures, however, were distributed over the offspring of both grandparents.

Obviously the mean value of 57% of the intermediate

race had been attained, and therefore I did not continue the culture.

Silene inflata, Fig. 88. I obtained this race by pure chance. The stock plant was one of that series of forms which I had taken into cultivation, during the course of many years, for the purpose of finding species in a mu-

Fig. 87. *Phacelia tanacetifolia.* A flowering sprig.

Fig. 88. *Silene inflata,* A whole plant.

table state (*V*ol. I, p. 271). A single specimen which had come up from seed of weeds, accidentally imported with cereals into our harbor, was transplanted into my experimental garden. The seed from this plant produced a proportion of 3% tricotyls, and when it flowered again in 1893, one of 4%. The tricotyls of the first harvest were planted out in the summer of 1893 and eight of them yielded an ample quantity of seed. In each lot, of from 200 to 1300 seedlings, I counted the percentage of tricotyls and found most of them to be distributed between 2 and 15%, the highest numbers being 24 and 32% (these latter amongst 1300 and 1060 seedlings respectively). The mean was 11%.

Therefore the original plant obviously belonged to a tricotylous intermediate race, a fact which, however, was only proved by the behavior of its descendants.

In the spring of 1894 I only planted out the tricotylous offspring of a parent with 32%, and obtained 22 plants which set seed. From these I obtained in the following year 22 hereditary values which were distributed between 26 and 55% and reached a mean of 37%. There were three individuals with 54—55 and 55%. We see that the mean was higher than the corresponding value of the parent plant, and from this we conclude that regression did not take place in the direction of 0, but towards the other side; and this is exactly what should happen in the isolation of new races from their mixtures.

In the summer of 1895 I planted out tricotylous offspring from the two parents with 55%; I kept them in two groups and determined their values from the harvests of 31 individuals. There was no essential difference between the two groups; the numbers were distributed between 26 and 73 and their mean was 53%. The mean had, therefore, reached the value of the parent, and the race could be regarded as an intermediate race, isolated but not yet subjected to selection.

For the next generation I selected the tricotylous offspring of three plants with 66—68 and 73%. Of these plants 25 set seed, but the culture was unsuccessful in this year, owing to adverse circumstances, and only a few of the plants produced more than 200 fertile seeds. The values, therefore, were not very exact, but formed a close group ranging between 18 and 56%, with a mean of 32%. It will be seen that in spite of the selection a general retrogression had taken place.

Summary. If we look at the results of the experiments which have been described, we are struck by the fact that a certain group of hereditary values appears much more commonly than others. These are the figures ranging round 55%, and this value is usually reached at the conclusion of the second generation or, in less favorable experiments, at the conclusion of the third or fourth. With a view to emphasizing this fact I have grouped the whole series of experiments in such a way that the figures round 55% are brought together in one column:

ISOLATION OF TRICOTYLOUS INTERMEDIATE RACES.
HIGHEST HEREDITARY VALUES EXPRESSED IN PERCENTAGES FOR THE SUCCESSIVE GENERATIONS.

	FIRST GENERATION	1–50%			ABOUT 55%	60–95%		
Antirrhinum majus . .	1892	7	8	25	41	79	79	—
Mercurialis annua . .	1892	1	0	2	55	86	92	91
Silene inflata	1892	1	3	32	55	73	56	—
Cannabis sativa . . .	1893	1	3	15	52–63	80	90	—
Clarkia pulchella . . .	1895	—	1	16	64	79	—	—
Helichrysum bracteatum	1895	—	1	41	51	52	—	—
Papaver Rhoeas . . .	1895	—	1	20	56	75	—	—
Phacelia tanacetifolia .	1895	—	1	14	58	90	—	—
Oenothera hirtella . .	1895	—	1	8	56	89	—	—

In this table the first column gives the year in which the experiments began, either with the selection of tricotylous seedlings from seed which had been bought or obtained from some other source, or with plants which had been found by chance (*Oenothera, Silene*). Where necessary, this year is denoted in the table by I (first generation). The numbers which succeed each other in the row to the right of it refer to the first and the following generations. Thus for instance in *Clarkia* there were as many as 16% tricotyls in the harvest of the first generation, 64 in those of the second and 79 in the third.

The point which this table is intended to illustrate is best brought out by a comparison with the series of figures given on page 392 for the half races. In that case, a selection continued from four to six years, did not bring this value, as a rule, further than 2 to 4%, and only in exceptional cases attained 15 to 20%. In this case, on the other hand, 55% is attained in two or three generations. In the half race a continuation of the selection would presumably not have led to any considerable increase, a fact which is demonstrated by the experiments with *Amarantus* and *Scrophularia* which were continued over a longer period of time.. In this case, however, selection is as a rule very effective, inasmuch as it can increase the best representatives of the race, in a very short time, to a hereditary capacity of 80 to 90%.

Therefore there can be no doubt that entirely different factors are at work in these two cases. In the former there occurred only races with half curves on which selection has little effect. In this case, however, there occurred, besides these, the highly variable intermediate races which are extraordinarily susceptible to selection and to external conditions of life. They were easily isolated, either because one or more examples of them were already present in the original sample of seed, or because between the intermediate and the half races hybrids were met with, by the subsequent segregation of which the race was produced.

In the first experiments the isolation of the tricotylous intermediate race took from three to four generations. Later when I started the experiments on a larger scale, the number was reduced to two years. Further selection brought it up, in one year as a rule, to 70—80% and sometimes even to 90%, either immediately in the course of another year, or after two generations. Unfavorable conditions of life led to exceptions, or even to retrogression; but only *Helichrysum* and *Silene* exhibited this feature. *Cannabis, Mercurialis* and *Antirrhinum* maintained a value of 80—90% under continued selection, and it is quite probable that even 100% might have been attained in occasional individuals. The mean figures of the whole group correspond, as a rule, to these maximum values. They maintained themselves at about 55%, but are liable to be increased by selection or diminished by unfavorable conditions.

Lastly, it should be noted that the figures for these intermediate races are so far removed from those of the half races (p. 392), that all suspicion as to the possible effect of occasional errors in the choice of the samples is excluded; in other words, *that hereditary values from 1 to 5% or even of 5 to 20%, if they are maintained in spite of selection, may be regarded as characteristic of half races, whilst values of 40 to 60%, when found in separate crops, may be taken to indicate the presence of intermediate races.*

When an intermediate race is isolated from an original sample of seed, it is separated from the half race which is also present, for hardly ever is the intermediate race found pure by itself. As far as I know, at least, there is no species of plants which has so much as half of its seedlings showing three cotyledons, without being selected. If we study the process of separation by the statistical method, we find that two curves, a half curve with an apex at 0, and a bilateral curve with an apex at 50—55%, can be distinguished from one another. Sometimes in one or other of the transitional generations both curves can be more or less clearly seen side by side, constituting a so-called dimorphic curve. I have frequently observed this in these experiments, and have, in some cases referred to it.

Dimorphic curves of this kind are best obtained in the transitional generations by planting out dicotylous as well as tricotylous seedlings; for, as was mentioned before, it is very likely that many dicotyls will belong to the half race and most of the tricotyls to the intermediate race. But if the latter is once isolated, all the individuals belong to it, independently of the question whether they have two or three or cleft cotyledons. In this stage a dimorphism of this kind is no longer to be expected, unless selection is continued in two different directions.[1]

As an instance of this let me cite the case of my culture of *Mercurialis annua* in 1895 (see the table on page 439), inasmuch as this species, being diœcious, would be expected to exhibit a levelling of the differential characters. The plant with a value of 55% in 1894 could obviously have been partly cross-fertilized. Its offspring had in their seeds the following hereditary values:

---

[1] With regard to this question compare the analogous experiments with *Helianthus annuus syncotyleus* (II, §11).

A. Tricotylous offspring	{	81	82	86		
		72	74	76	78	
		66	67			
		44	48	50		
B. Dicotylous offspring	{	34	35	36	40	
		21	22	24	25	25
		18	20			

The group is therefore dimorphic, the intermediate race is not yet pure. I found the same conditions to obtain in other cases.

In *Clarkia pulchella* also, the atavists reverted when sown in 1897 alongside the tricotylous culture already mentioned and from seed of the same parent. This parent already had a hereditary value of 64%, but was the first in this race with a high figure, and therefore still belonged to the transitional period. The values are arranged in groups of 5 to 15, 15 to 25 etc., and the number of individuals which fall into each group are given under the mean percentage number, i. e., 10, 20 etc.

Proportion of tricotyls in percentage	10	20	30	40	50	60	70	80
Number of tricotylous individuals	0	2	5	8	8	12	3	1
Number of atavistic individuals	4	13	9	11	0	2	—	—

The mean for the tricotylous individuals is 49%, for the atavists 30%.

The race once isolated, the atavists obviously still have smaller values than the tricotyls; but the difference is never more than a small one, and the curve for the two races together will have only one apex. In *Ocnothera hirtella* I examined a series of values of the atavistic offspring as well as the tricotylous offspring of the same parent of 1896 which had a value of 56%. This parent had been self-fertilized, the visits of insects having been excluded; and so were all its offspring. I shall group the values in the same way as above.

Proportion of tricotyls in percentages	10	20	30	40	50	60	70	80	90
Number of tricotylous individuals	—	—	—		1	6	11	14	10	1
Number of atavistic individuals			1	0	4	3	10	11	11	6	0
Calculated for the whole progeny			1	0	4	4	19	27	32	21	1

The average value for the tricotyls is 72% and of the atavists 60%, which latter figure is of itself high enough for an intermediate race. The last row was obtained by means of a correction of the values of the tricotyls, since too small a number of these had been cultivated in comparison with the atavists. The whole curve is obviously monomorphic although much flattened.

A curve with such a great amplitude indicates a corresponding high variability and therefore also a considerable susceptibility to external influences, as we have found in several instances to be the case with intermediate races.

To this great amplitude also the fact is due that selection, however stringent, never quite eliminates the lower values. Even when the apex of the curve is between 60 and 70%, values as low as 25% may still occur. In such cases the curve has a "sweep" to the atavistic side, and thus differs most conspicuously from the curves of the half race, the shape of which might be considered as its mirrored image; but as I shall come back to this point when I deal with syncotylous races, I shall not give any instances now.

§ 7. PARTIAL VARIABILITY OF TRICOTYLY.

In experiments with tricotylous races the hereditary capacity or value is, as we have seen, the main character subjected to selection. This term indicates the number of tricotylous seedlings in the harvest. The seeds of each individual are saved and sown separately, and then recorded after germination. Whether the individual itself has two or three or merely cleft seed-leaves is a question of minor importance, and does no more than furnish a suggestion in starting the experiment. On the average, atavists and hemi-tricotyls are, without doubt, of less value in this respect than tricotyls and tetracotyls; but the differences are too small to warrant a choice of seed-parents on this ground.

The determination of the hereditary value depends, therefore, largely on the extent of the crops. The larger this is for any one individual, the more closely will the value found correspond to that which would be derived from an imaginary harvest, consisting of an infinite number of seeds. Therefore, the best plan would be to make the plants grow as vigorously as possible, and to save the seeds of all fruits from as many branches and sprigs as possible. In practice, however, this cannot be done, since it is far more important to grow as many individuals as possible in a given space. The more numerous the individuals are, the more stringent the selection becomes; and, what is far more interesting, at least in working with mixtures, the greater is the prospect of finding any desired particular sort. Unbranched, or almost unbranched, plants with terminal inflorescences only, obviously need much less room than much branched individuals; and larger harvests require relatively too much time to gather. The best plan, therefore, would be to collect only just so much seed from each individual as is necessary for sowing in the following year.

These considerations suggest the question whether different groups of fruits are alike in regard to hereditary capacity or whether they exhibit, perhaps, considerable differences. Shall we find this capacity to be different on the branches from what it is on the main stem? Will the earlier seeds give different values from those given by the later ones? Has the character of the year in which the harvest is made any such influence amongst perennial plants? Obviously these questions must be answered if the individual harvests are to be limited in the interest of the experiment.

In the last two sections of the first part of this volume we have seen that semi-latent characters manifest a certain periodicity in their distribution over the plant, and also that the choice of seeds on the plant plays some part in the process of selection. The question is, how do the tricotylous races behave in this respect?

The general rule seems to be that a bud, whether of a branch or of a flower, is more likely to reproduce an anomaly, the more vigorous it is (p. 324). Therefore with an increase in the degree of branching the expectation of the occurrence of anomalies decreases (p. 329). The first or lower fruits of an inflorescence will be stronger than the higher ones, and the fruits on the weak lateral sprigs of the primary and secondary branches may as well be thrown away.

There is no ground for supposing that the flower buds behave differently from the vegetative parts. The best instance of the phenomena in question is afforded by profusely branched specimens of the twisted race, *Dipsacus sylvestris torsus*. The torsion affects the middle portion of the main stem, but neither its upper nor lower extremity. It is repeated on the strongest branches situated on the middle of the stem, and on these only in their middle parts, and excluding the weaker ones. The stronger a stem or a branch is, the greater is the extent of the twisted part. In the branches it is always confined to single internodes, whereas the stem may frequently be entirely transformed. Branches of the second order exhibit no more than traces of the anomaly.

If we apply this instance to the distribution of tricotylous seeds on a plant, we should conclude that the lower part of the terminal inflorescence of the main stem would produce the highest proportion of tricotylous seedlings; but the flowers and fruits themselves are lateral branches, and so the question arises, how far we are justified in expecting this.

In species like *Oenothera* and *Antirrhinum*, which have a primary inflorescence rich in flowers and in seeds, I have usually limited myself to this and, where possible, to its lower and middle parts. In *Helichrysum* to the flowerheads which open first, in *Clarkia* and *Phacelia* to as many of the earlier flowering branches as would provide a sufficient quantity of seed, and so on. The question is whether these first fruits possess the same hereditary capacity as would be found from the largest number of fruits that could be gathered on the plant.

With a view to determining this point I have instituted a number of subsidiary experiments. The answer has been in the affirmative throughout. Certainly there are small differences; these, however, seldom fall outside the latitude of 5% which is the limit of observational error. I shall now present the results of these experiments in a condensed form.

A first experiment was conducted with *Oenothera hirtella*, which fertilizes itself in parchment bags without artificial aid. In the summer of 1898 I planted out seven tricotylous seedlings of one parent with a hereditary value of 66% at distances of about a meter apart, allowed them to branch freely and to develop into vigorous plants, enclosed the flowers in parchment bags and harvested and examined the seeds for each individual branch separately. Only slight differences were presented by the values obtained, and I shall therefore only give the means. I determined the values of the inflorescences of the main stem and found the average to be 38%; for the strong lateral stems produced from the axils of the rosette (see Fig. 55, Vol. I, p. 302) I found it to be 45%; for the upper branches of the stem (see Fig. 49, Vol. I, p. 282), however, 47% (calculated from 24 determinations), and for the lower branches of the stem, which in this species tend to be very much weaker, 52% (in eight counts). The distribution of the differences, therefore, was different from what would have been expected. They show that in this case, the harvest from the primary inflorescence gives a somewhat lower value than the whole harvest of the plant in question would have given. In *Dracocephalum moldavicum,* where the values are always small, I collected, in the summer of 1895, the seeds of all specimens from the main stem and lateral branches separately, but found no difference (0.4% for both). In *Amarantus speciosus* the seeds from the terminal panicle regularly gave somewhat higher values than those from the lower branches, but with very slight differences only (1892). The average calculated from 20 plants was 2.8% for the former and 1.7% for the latter.

In many of my experiments I have saved the seeds which ripened first, separately from those which ripened later: e. g., *Amarantus speciosus, Scrophularia nodosa, Mercurialis annua, Antirrhinum majus, Silene inflata,* and others. No differences of any importance were found in this way. Deviations are sometimes found in larger series, but only such as can be attributed to the unavoidable errors in the samples of seed. In *Mercurialis annua* I have for several years recorded the seedlings which came up from the first seeds that fell out on the bed. In this species, as is well known, the fruits open suddenly and scatter their seeds about. These counts gave essentially the same results as those obtained later from seeds harvested and sown by hand. I often made use of them to obtain a preliminary idea concerning the extent of the improvement to be expected.

My custom is to limit the separate harvests to the desired quantity of seed, by cutting back during the period of growth, and by stopping the saving of seed as soon as a sufficient quantity has been secured. For this purpose I have estimated this amount from the result obtained in the first generation, in the case of each species. Moreover I have frequently compared such gatherings with much more extensive ones, obtained either by not cutting the plants back or by not ceasing to save the seed before all of it was ripe. In this way *Amarantus speciosus* often gave higher values from a smaller harvest. but the differences were slight and the exceptions many. *Cannabis sativa* gave the same values from large plants bearing more than 100 cubic centimeters of seed each, as from average or weaker individuals. In such large crops, saved without limitation and amounting to 80—110 cc. of seed the average was 11%, but in crops of 20—35 cc. it was 14%. The cultures were made in 1894 and the same happened in other years. In *Oenothera rubrinervis* the value was seen to decrease as the size of the harvest increased, but only in degrees of one-tenth per cent, on the average, amongst numerous individual trials. The same occurred in *Scrophularia nodosa*.

This latter species, as well as *Silene inflata,* is a perennial form; and therefore both of them afford material for comparing the harvest of the same plant as produced in successive years. Here again I failed to find differences of any significance. I have given above the values derived in 1892 and 1893 from a single plant of *Silene*, which flowered in isolation; they were 3 and 4% for these two years. In *Scrophularia* I made a series of observations at the beginning of my cultures when the hereditary values were still small, and repeated them in the period of 1896-1899, when they had become higher (15% and more). In these years, 1897, 1898 and 1899 six plants gave the following values, the bracketed number referring to the second year: A 22 (25), B 25 (17), C 22 (17), D 23 (25), E 27 (25), F 23 (22). Obviously these figures do not justify a conclusion as to any diminution or increase in the ratio in which tricotyls are produced.

The result of all of these experiments is such as to justify my practice of limiting the individual harvest to the quantity of seed necessary for sowing.

§ 8. THE INFLUENCE OF EXTERNAL CONDITIONS ON TRICOTYLY.

Which seeds in a fruit produce aberrant seedlings? This question is at once one of the most simple and one of the most difficult presented by experimental breeding. If some day we could succeed in solving it and thereby make a control of this process possible, much light would be thrown on a whole series of phenomena connected with the origin of races.

In dealing with this question we are thrown back on egg cells and pollen grains and the problem at once becomes a double one. To these is added a third factor, namely the union of the two germ cells, which may itself be influenced by outward conditions in different ways.

I have gone into the questions only so far as was necessary for the choice of the conditions of my experiments. The course of these has, as a rule, been very regular. The results of selection seldom have been disturbed, to any great extent, by other influences. Sometimes, however, this did obviously occur; and it is exactly experiments of this kind which justify us in ascribing an important part to external conditions in the determination of the proportion of tricotylous seedlings. But when, from these facts, we proceed to analyze this influence, we do not succeed in making experiments in which the same influences have the same results.

Two instances will suffice. Under very peculiar conditions which affected the whole growth of the plant, the hereditary capacity of my race of *Amarantus speciosus* suddenly rose, in 1891, from 4.5 to 13%, without, however, maintaining that proportion even under selection (p. 407). On the other hand, in the summer of 1896 the whole culture of *Silene inflata* underwent considerable deterioration as the result of unfavorable conditions, the mean of all values falling from 53 to 32%. One year may obviously affect the plants quite differently from another, even when the treatment to which they are subjected is as uniform as possible. The effect is perhaps produced in the early stages of the plant, perhaps also in the development of the sexual cells and at the time of fertilization. Abundant starting-points for further investigations may here be found.

As a rule the likelihood of obtaining anomalies increases with the vigor of the seeds which produce them

(p. 332). It is natural to inquire, therefore, whether this vigor of the seeds can be increased by improved nutrition of the plant. Unfortunately we must, in such cases, be content with averages, and these obviously give much smaller differences than the single extreme cases. Sometimes I found that the external conditions exerted a considerable influence, but, as a rule, I could not detect any difference.

In the first place I shall refer to the effects of manuring. The prevailing opinion is that manuring with nitrates favors the development of leaves, but that phosphates favor flowers and fruits. As a nitrate manure I employed horn-meal combined with dried guano and as phosphate manure the ordinary superphosphate; but I find that the former usually produces more abundant flowering and a larger harvest of seed than the latter. In 1899 I instituted a comparative experiment with *Oenothera hirtella* by planting out the seedlings of a single self-fertilized plant. During the period of flowering each inflorescence was guarded from the visits of insects by a parchment bag. The nitrate plants grew more rapidly and luxuriantly at first than those on the phosphate bed. Moreover they began to flower somewhat earlier, and consequently set more seed; for fertilization must be stopped before all the flowerbuds have opened, since in our climate flowers which open in September have not time enough left to ripen their seed.

The harvest for each individual plant was examined separately, and from this was calculated, in the way we have so frequently described, the percentage value for each plant. In the following summary these are associated in groups of from 15-25, 26-35, 36-45; the means of the groups being 20, 30, 40 etc.

HEREDITARY VALUES IN OENOTHERA HIRTELLA IN 1899.

Percentage values	10	20	30	40	50	60	70	80	90
After manuring with nitrates	0	8	17	12	9	5	2	2	1
After manuring with phosphates	1	2	9	14	7	8	5	1	1

The number of individuals in the first experiment is 56, in the second 48; average harvest per plant, 3.5 and 2.5 cc. The seed was collected from the terminal spike only. The mean value for the nitrate plants was 37%, for those manured with phosphates, 44%. The parent of all these plants had had a value of 66%.

Manuring with superphosphate has therefore, in this case, in an otherwise uniform culture, been more favorable to the production of tricotylous seedlings than manuring with hornmeal.

I found exactly the same result in the same year with *Helichrysum bracteatum*. I planted out tricotylous seedlings only of a single parent with 11%. I determined and arranged the values as before and found:

HEREDITARY VALUES IN HELICHRYSUM BRACTEATUM IN 1899.

Percentage values	10	20	30	40	50	Mean
After manuring with nitrates	2	22	18	5	1	26%
After manuring with phosphates	1	5	20	11	2	32%

Number of individuals 48 and 39. The growth on the nitrate bed was very luxuriant; the leaves were dark-green, and the flowers abundant. On the phosphate bed the plants were yellowish-green, very little branched, and with fewer ripe flowerheads. Indeed, only 39 of the 50 plants which had been set out set sufficient seed.

Results as definite as these are not, however, always obtained, especially when a comparison is instituted not between individual plants raised from seed, but between the various parts of a single individual obtained by dividing it. I conducted such an experiment with *Oeno-*

thera (*Kneiffia*) *glauca* in 1899, but counted 5% tricotyls, both on the phosphate and on the nitrate half. I obtained the same result in *Scrophularia nodosa* by growing one half in ordinary garden soil and the other in sand, both yielding 1% (1894).

The two first experiments on the influence of manure show that an increase in the number of tricotyls is associated with a diminution of the yield. The same result can also be obtained if the harvest of the individual plants is reduced in other ways. In *Oenothera hirtella* I adopted two methods of doing this, late sowing and culture in pots throughout the whole summer. Tricotylous seedlings of a single parent (with 66%) were planted out. In one of the two groups the sowing had taken place in March, in the other at the beginning of May (1898). Some plants of the former sowing were kept in pots of 15 centimeters filled with well-manured garden soil throughout the whole summer. Fertilization took place in parchment bags and the individual crops, recorded and evaluated in the usual way, gave the following result:

HEREDITARY VALUES IN OENOTHERA HIRTELLA IN 1898.

Percentage values	10	20	30	40	50	60	70	Mean
Normal culture	1	4	4	3	5	0	1	37%
After sowing in May	0	0	10	14	11	1	0	41%
After culture in pots	0	0	3	5	7	4	0	47%

Number of plants experimented with 18, 36 and 19.

We see therefore in this case a definite though small increase in the number of tricotyls. The mean of the harvest for the normal culture was 3.5 cc., for the late culture 2.5, for the pot-culture 4.5 cc. This latter had therefore not had the expected result of making the yield smaller, and as, nevertheless, the proportion of tricotyls has increased, it is plain that the factors underlying these differences are by no means simple.

This is further shown by the fact that a repetition of the experiments does not always produce the same result. Thus in *Amarantus speciosus* (1897), and *Scrophularia nodosa* (1898), I was not able to observe any influence exerted by the various external conditions on the proportion of tricotyls, although the experiments were carried out on a large scale. I also found that if I sowed samples from the same lot of seed in two successive years, the mean ratio of tricotyls in the harvest of the two cultures was the same, (e. g., *Oenothera hirtella,* 1898 and 1899).

Moreover, unfavorable conditions can sometimes, by diminishing the yield, lead to an increase in the percentage values. This is shown by an experiment which I made in 1898 with *Antirrhinum majus*. I covered half of my culture with a cage made of fine metal gauze painted black, after the plants had developed a stem of 10 centimeters above the cotyledons. Within the cage it was considerably darker than outside, and the plants grew very weakly, produced only few branches, and developed but a small number of flowers and fruits on the terminal spikes. Fertilization was artificial, and all the plants were guarded from the visits of insects. The yield in the cage was 0.5 cc. per plant, and outside 1 cc. per plant. In the former case as many fruits as possible were gathered but in the latter the upper flowers of the spike were not fertilized. The material used in this experiment consisted exclusively of the offspring of a single parent of 1897 which had had a value of 14%, but tricotylous and dicotylous seedlings were both used. each group being planted half within the cage and half outside. I collected the seeds and determined the values in the same way as usual, and obtained the figures which appear in the following summary:

HEREDITARY VALUES IN ANTIRRHINUM MAJUS.

Percentage values	1–4	10	20	30	40	Mean
Dicotyls in the shade	0	4	2	2	2	20%
Dicotyls in the sun	5	8	4	1	4	14%
Tricotyls in the shade	0	3	2	3	4	35%
Tricotyls in the sun	1	14	9	—	—	14%

Number of seed-parents 10, 21, 12 and 24.

We see that in both cases the proportion of tricotyls was considerably increased by cultivation in the shade.

The experiments which we have passed in review in this section prove that the proportion of tricotyls in the offspring may be considerably modified by external influences, especially in eversporting varieties, but within certain limits. The differences are such as cannot be neglected in comparative experiments and cultures, and show the importance of carrying out the latter under as constant conditions as possible. In general it must be assumed that a treatment, favorable to the individual flowers and seeds, will increase the production of tricotylous seedlings, but the question as to the nature of this relation in the individual cases must, for the present, remain unanswered.

II. SYNCOTYLOUS RACES.

§ 9. HEMI-SYNCOTYLY, SYNCOTYLY, AMPHI-SYNCOTYLY.

Alongside seedlings with split or double cotyledons there occur others, the seed-leaves of which have fused so as to form one single organ. They are, however, much rarer, but not so rare that they cannot be found every year, at least in some sowing or other. Tricotyls may be found in the proportion say, of one in a thousand, but syncotyls, as a rule, only one in tens of thousands of seedlings.

In the sowings of 1895 which we have already mentioned (p. 380), I found, amongst more than 250,000 seedlings of 40 species, only 10 syncotyls as against 150 tricotyls. Where they occur more abundantly this is due to inheritance, for as soon as the syncotyls are allowed to flower in isolation and their seeds are saved separately, it is found that they inherit their character in almost the same degree as do tricotyls.

Instances of syncotyls were afforded me by *Aster tenellus, Clarkia elegans, C. pulchella, Cerinthe gymnandra, Chrysanthemum Myconis, Helichrysum bracteatum, Phacelia tanacetifolia, Silene hirsuta, Anagallis grandiflora, Epilobium hirsutum, Hesperis matronalis, Pentstemon gentianoides, Robinia Pseud-Acacia,* and many other forms;[1] but always in small proportions. A few

---

[1] Further instances are given by H. B. GRUFFY, *Irregularity of Some Cotyledons,* Science Gossip, N. S., Vol. II, 1895, p. 171.

species, however, produced them more abundantly. Among these were the following which produced the proportions of syncotyls from the seeds of single individuals, indicated by the attached figures: *Oenothera* (*Kneiffia*) *glauca* 16% (Figs. 89 and 90), *Picris hieracioides* 8%, *Valeriana alba* 3%. *Dahlia variabilis* and *Sycios angulata* were also found to be rich in syncotylous seedlings,

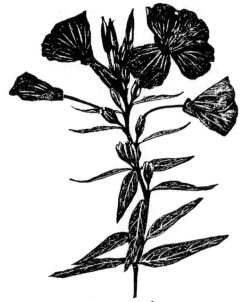

Fig. 89. *Oenothera glauca*. A plant which produces an abundance of syncotylous seedlings.

and in *Centranthus macrosiphon* I found 37% of them amongst the seeds of a single syncotylous plant.[1]

By syncotyls we mean, in the strict sense of the term, those individuals whose seed-leaves are completely, or almost completely, fused together along one side. Smaller degrees of fusion are called hemi-syncotyls, and they constitute an uninterrupted series between normal di-

[1] *Eine Methode, Zwangsdrehungen aufzusuchen,* Ber. d. d. bot. Ges., Vol. XII, 1894, p. 25.

cotyls and types in which the fusion has extended to the very tip. The lesser degrees of fusion are usually rarer than the greater, and, therefore, in order to get a fairly complete series of transitional forms, cultures from the seeds of selected individuals are required. It is often very difficult, especially in forms with stalked cotyledons, to distinguish between the lowest grades of fusion and pure dicotyls. If the crops are recorded before the cotyledons and their peduncles have grown to their full extent, it may easily happen that occasional seedlings,

Fig. 90. *Oenothera glauca.* Syncotylous seedlings. A and B, ordinary syncotyls; C and D, pitcher-shaped syncotyls or amphi-syncotyls.

which, after further growth, will prove to b hemi-syncotylous, are mistaken for dicotyls.

The hemi-syncotyls constitute the minus variants of the series, the corresponding plus variants being the amphi-syncotyls of double or two-sided syncotyls (amphicotyls). In these latter cases the cotyledons are fused on both sides, often more on one side than the other, sometimes equally high on both sides. The cotyledons constitute a pitcher (Fig. 90 *C* and *D*, and Fig. 91) which is either cylindrical or urn-shaped or disc-shaped, according to the breadth of the cotyledons on the normal plant. I found instances of these remarkable and pretty structures almost every year in *Helianthus annuus syncotyleus* and in larger numbers in *Mercurialis annua* and *Centranthus macrosiphon*,[1] and also occasionally in *Antirrhinum majus, Sinapis alba*, etc.

The fusion of these cotyledons sometimes results in a pressure being exerted on the plumule and interfering with its growth, and we often find that syncotyls grow slower than normal seedlings, at least at first, and that

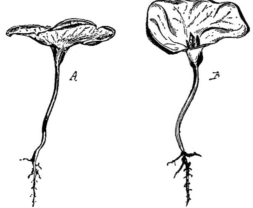

Fig. 91. *Raphanus Raphanistrum*. A cotyl pitcher expanded in the form of a disc. A, from the side; B, the same, seen obliquely from above and showing the plumule.

amphi-syncotyls are especially backward. The extent of this influence cannot, as yet, be measured; but the fact that syncotyls are so much rarer than tricotyls may perhaps be explained by this check on their growth. Sometimes the plumule breaks sideways through the cotyl-pitcher by splitting its lower part (*Centranthus macrosiphon, Mercurialis annua*). Sometimes it succeeds, although very late, in elongating in the normal direction. In *Helianthus annuus* I have sometimes operated upon such

[1] Figured in *Ber. d. d. bot. Ges.*, Vol. XII, Pl. II, Figs. 3 and 4.

pitchers by splitting them up along one side with a sharp knife, as soon as I saw that the plant would not grow without this help. By this means I succeeded in bringing the plants to flower, but they were weak, thin-stemmed individuals.

The growth of the plumule is often entirely suspended, both in ordinary unilateral syncotyls and in pitcher-like structures. In such cases the seedlings live much longer than usual, the cotyledons grow to a great size, often twice the normal, but finally the plant dies without producing a stem. This occurred frequently in *Helianthus annuus* and *Dahlia variabilis,* and less often

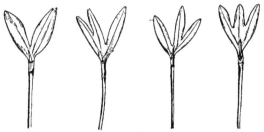

Fig. 92. *Polygonum Convolvulus.* Instances of tri-syncotylous seedlings with various degrees of splitting and fusion.

in *Penstemon gentianoides* and *Cannabis sativa.* In the two latter species one or two leaves were sometimes developed, but after that the terminal bud ceased to grow, whilst those leaves attained an abnormal size and thickness (Fig. 93). In such cases the question arises whether the inhibition of the growth may not, both here and elsewhere, have some other cause also.

Syncotyly may be combined with tricotyly, and since, as we have seen, the fluctuating variability of both presents a long series of forms, the multiformity will be much greater still in the series of the so-called tri-syncotyls. Thus, for instance, in *Polygonum Convolvulus*

(Fig. 92), and also in *Chenopodium album, Thrincia hirta*, etc.

Between syncotyly and disturbances in the normal arrangement of leaves in the later life of a plant, there

Fig. 93. *Cannabis sativa.* A, a seedling which unfolded its cotyledons in April, but up till June had formed no plumule; B, another with two abnormally large and thick first leaves, the terminal bud of which underwent no further growth. Both are from a culture, in the remaining seedlings of which the length of the stem had become 10–20 centimeters during the same time (1894).

Fig. 94. *Fagus sylvatica.* Syncotylous seedling with much elongated epicotylous internode and abnormal development of the plumule.

is a close connection, as there was in the case of tricotyly. Frequently this disturbance is only a small one (Fig. 94); but in species with a decussate arrangement of the leaves or at any rate of the first leaves, there often follows on the syncotylous organ a node with only one leaf, which then usually stands opposite to this structure. I have frequently observed such cases in *Mercurialis annua* (Fig. 95) and *Helianthus annuus*.[1] Further disturbances may then follow, such as unequal size of the two leaves

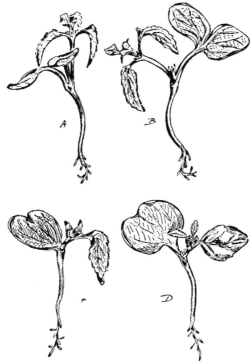

Fig. 95. *Mercurialis annua*. Syncotylous seedlings. A and B, hemi-syncotyls with normal arrangement of leaves; C and D, completely syncotylous; the first leaf is single and opposite the cotyledon.

of a pair, leaves with two apexes, and in the same cultures such even occur not infrequently immediately above the cotyledons.

Not infrequently spiral torsion and fasciation occur

[1] *Over de erfelykheid van Synfisen.* Bot. Jaarboek, Gent, 1895, Plate IV.

after syncotyly.[1] I found the former case, for instance, in *Anagallis grandiflora, Collinsia heterophylla, C. grandiflora, C. bicolor, Scabiosa atropurpurea, Centranthus macrosiphon,* and very abundantly in *Mercurialis annua*

Fig. 96. *Mercurialis annua.* A lateral branch with spiral torsion. All the leaves of the twisted region are situated on one side of the stem.

(Fig. 96). Fasciations also occurred in the latter and several other species, especially in *Helianthus annuus.* Moreover, pitcher formation seems often to be associated with syncotyly, either on the same individuals or amongst

[1] *Ber. d. d. bot. Ges.,* Vol. XII, *loc. cit.,*

other individuals of the same culture (Fig. 97). As instances I mention *Mercurialis annua, Anagallis grandiflora, Antirrhinum majus, Fagus sylvatica, Polygonum Fagopyrum, Spinacia oleracea* and *Raphanus Raphanistrum*.[1]

Syncotyly is a heritable anomaly; if we isolate the stray syncotyls we will find the phenomenon repeated

Fig. 97. *Coriandrum sativum*. A, a flowering sprig; B, a hemi-syncotylous seedling; C, a dicotylous seedling from the same culture, whose first leaf is transformed into a pitcher (1894).

amongst their offspring; as for instance in *Valeriana alba* (1892, 3%; 1893, 6%). Moreover the seeds of non-syncotylous individuals from the same crop may reproduce the syncotyly; as for instance in *Amarantus speciosus, Polygonum Convolvulus, Scrophularia nodosa*, etc.[2]

[1] *Kruidkundig Jaarboek*, loc. cit., pp. 172-177.
[2] *Kruidkundig Jaarboek*, loc. cit., p. 159.

In the spring of 1894 I planted out into separate beds some syncotyls and amphi-syncotyls from my tricotylous intermediate race of *Mercurialis annua*, which produces a large number of syncotyls every year. There were three female plants and some male ones. Saved separately, their seeds produced 4, 19 and 24% syncotyls, amongst which hemi-syncotyls and amphicotyls appeared in considerable numbers. There is little doubt that I would have succeeded in isolating the syncotylous intermediate races from *Mercurialis* (with 24%) as well as from *Centranthus* (with 37%), and also perhaps from *Picris* (8%) and *Valeriana* (6%); but the difficulties, due mainly to the retarded growth of the plumule, have led me to limit my experiments in this direction to a single species, *Helianthus annuus*.

§ 10. HELIANTHUS ANNUUS SYNCOTYLEUS.[1]

As early as 1887 I found a syncotylous race of the ordinary sun-flower and isolated it immediately. Since then I have grown this race for ten generations. It has always been extremely variable, both in the extent of the fusion of the cotyledons and in the percentage number of syncotylous seedlings. On the other hand it has proved perfectly constant and immutable.

For instance, every effort to bring it to a uniform condition, free from atavists, or, on the other hand, to reduce it to the corresponding half race, has failed, in spite of the most stringent selection. Therefore we have here as good an instance of a constant type, which, in its visible properties merges into other races, i. e., is transgressively variable, as was furnished by the best

---

[1] *Over de erfelykheid van Synfisen.* Kruidkundig Jaarboek, 1895. pp. 136-142 and Plate IV. See also the same journal, 1894.

tricotylous strains. Whether from this a pure, almost unvariable syncotylous race can be raised, which would no longer be an intermediate race, but show to this the same relation as that between the *aurea* varieties and variegated plants (see p. 21), is a point on which we are at present completely in the dark.

As already stated, I found my race in the year 1887. At that time I had sowed a large quantity of seed of *Helianthus annuus,* obtained in exchange from different botanical gardens, and found 18 syncotyls amongst 500 seedlings. I planted out these only. They flowered together, but the seeds of each individual were saved, sown and examined separately. In 17 lots the proportions of syncotylous seedlings were distributed between 1 and 15%, and in the case of only one of them it was 19%. The latter plant had had its cotyledons fused up to their upper margin and was selected as the basis of my race. The offspring of the other plants were not grown, and moreover from this selected individual only syncotylous seedlings were chosen for further cultivation.

These flowered together and were left to be pollinated by insects. This is apparently necessary in this species, or at any rate in my race of it, because isolated individuals fertilized with their own pollen, either artificially or by humble bees, set no seed. In the later generations the seed has always been collected separately from each plant, and in the following spring sown with a label bearing the number of the parent, and examined. In this way I obtained hereditary values for each single plant.[1] The seeds are large and few plants produce as many as 300 seedlings, so that the values are not so

---

[1] This rule holds without exception, and therefore I shall not mention it in the description of the experiments which are to follow.

accurate as those calculated from more extensive crops. Nevertheless they fall into definite groups, so that this source of error may be neglected.

In the harvest of 1888 it was at once evident that the race had been obtained in its pure state by the selection carried out in the foregoing year. Twelve plants set an abundance of seed; 9 of them had from 30 to 55% syncotyls, whilst the remaining three parents had 76, 81 and 89%, in samples of 121, 275 and 128 seedlings. Of course the plant with 89% was chosen for the continuation of the race.

It was important to find out whether, if these three high values had not been found, the remaining parents would have given the same result. For this purpose I made a subsidiary culture in 1890, and chose for this purpose the syncotylous seedlings of a parent with 51%. They produced 23 seed-parents of which the values were distributed evenly between 24 and 91% with a mean of 55%. The four highest numbers were 77, 79, 84 and 91%. The result can therefore be reached also by an indirect way, as the following scheme shows at the first glance.

ORIGIN OF THE SYNCOTYLOUS EVERSPORTING VARIETY OF HELIANTHUS ANNUUS

HARVEST OF	1887	1888	1889
Hereditary values of the best parents in percentages	19——	⎰ 89———	—81
		⎱ 51———	—91

In order to consider this fact in combination with the figures found for the tricotylous races, we must compare it with the table on page 439. Into this table the experiment with *Helianthus* could be inserted straight away. On the one hand it confirms the expectation, which that table justified, of finding a value of 55% in the second generation; on the other it shows that this can sometimes be passed over, inasmuch as in a fortunate case the high value of 80-90% may be directly attained; notwithstanding that this never appeared until after the 55% stage had been reached in the species dealt with in that table. When I repeated the experiment described for my tricotylous race of *Helichrysum bracteatum,* I had to pass through several generations before reaching the hereditary value of about 50%, and this also shows that in different cases the same value may be reached by means of different numbers of generations. We may therefore conclude that the isolation of a syncotylous race takes place according to the same scheme as that which has been found for the tricotylous intermediate races.

The general statements on page 17 of this volume and elsewhere relating to the isolation of eversporting varieties, and the apparent increase in variability, by which we explained the isolation of the five-leaved red clover, find an experimental confirmation in the present instance. The new race departs rapidly from the type of its species, but only because it is approaching its own new type.

Into the group of figures given on the preceding page I have inserted a value for 1889 relating to the first-mentioned line of my race. It would take me too long to describe the subsequent generations singly, but, as is very important in the culture of such forms, they were treated in the same way ever year. I shall therefore give the pedigree of the whole culture on the following page.

In this summary I have written at the top the percentage number of syncotyls in the crops in such a way that the single numbers have been associated into small groups; thus 48 signifies the values from 46 to 50%; 53 the values from 51 to 55%, etc. The numbers which occur in a vertical line below these figures give the number of individuals for each generation with the hereditary value indicated by the upper figures. The barb in the bracket, below one of the values for a generation, denotes the character of the parent plant, that is to say, that plant whose offspring alone constituted this generation.

HELIANTHUS ANNUUS SYNCOTYLEUS.

PEDIGREE OF THE SYNCOTYLOUS EVERSPORTING VARIETY.

Under Selection of the Individuals with the Highest Hereditary Value.

	HEREDITARY VALUES IN PERCENTAGES.												
	19	25–40	48	53	58	63	68	74	78	83	88	93	98
9th generation 1896					1	1	3	8	4	1	2		
8th generation 1895						1	2	5	2	4	7	3	
7th generation 1894			1	1	0	1	1	1	1	2	0	1	
6th generation 1892								1	2	3	8	9	5
5th generation 1891									1	2	2	11	6
4th generation 1890						1	0	1	2	4	8	15	1
3d generation 1889		2	1	0	0	1	1	0	0	2	1		
2d generation 1888		2	2	5	0	0	0	0	1	1	1		
1st generation 1887	1												

The pedigree shows that from an original plant with

a value of 19% a race with values ranging between 50% and 89% arose immediately, and that this value was fairly constantly maintained under continued selection. Some low values (25-50%) still occurred in the first two generations which followed the original plant of 1887, but they have not occurred since. The highest values attained in the various years deviate but little; the deviations lying well within the probable error of observation, or due to occasional differences between the climates of the succeeding years. The hereditary values of the parents selected each year were as follows:

HARVEST OF	1887	1888	1889	1890	1891	1892	1894	1895	1896
Values of the parent plants in percentages	19	89	81	94	97	98	86	93	82

In some years the parent was not the individual which possessed the highest value. This was due to the fact that when the differences in the values are small it may be necessary to pay attention to other circumstances also and especially to reject weak individuals, or such as flower too late, or set little seed, or happen to be fasciated.

The choice of a parent must further be left to chance in groups of equally good plants with practically the same hereditary values; because sunflowers are too large to permit of diminishing the elements of chance by doubling the principle of the selection by the application of the method of selection by grandparents (see p. 385); but, as it happened, I was mostly fortunate in my choice, with the exception of the last year in which the mean sank from 85% to about 68%.

My race is not to be regarded as a normal eversporting variety but as an improved breed within it. Instead of selecting parents with 80-90% and over I could have used the plants with a mean value (say 50-55%) for the continuation of the culture. But then I would have obtained a non-selected strain, and the figures for the first two generations following on the original plant, clearly indicate that this mean would have been about 50-55%, which is also, as we know, the mean value for tricotylous intermediate races.

The figures given in the pedigree do not present a complete picture of the whole improved race, for in each year the hereditary value of the best offspring alone was determined. Dicotylous seedlings and those with a low degree of fusion in the seed-leaves were excluded from further cultivation, as were also weak plants. If I had not applied this selection, the mean values would obviously have turned out somewhat lower; but the difference would not have been a very considerable one, as the next two sections will show.

The chance of obtaining a pure syncotylous progeny, i. e., a crop without dicotylous seedlings, may appear to be very great in this experiment. In 13 individual crops 96% and over was reached. But appearances are deceptive. Only once did I have a perfectly pure crop (100%) and that even at the outset of my experiment, in 1890; but this plant had produced only 105 fertile seeds; and of course we must acknowledge the possibility of some stray atavistic seedlings occurring amongst them if the harvest had been larger. In other words, selection leads the race as close as possible to the highest degree of purity, without, however, enabling it to reach it. Moreover, the table shows at a glance, that the progeny of the plant with 100% would probably have fallen back from this high value, in the case of many of its offspring at any rate.

If all the seeds of a single head are planted out in groups according to their degree of syncotyly, very little differences will be seen in the results. Atavists and seedlings, in which the fusion extends less than half way up, give a progeny the value of which is, on the average, somewhat lower, but only if we fix our attention on the mean values, and not if we compare the single individuals. Many an atavist or hemi-syncotyl has a higher value than most of the average syncotyls, and amongst these latter, the question whether the fusion is so complete that the apex of the double leaf exhibits no invagination, has no effect on the hereditary values of its offspring, so far as I have been able to determine. In the same way the disturbances in the disposition of the leaves, which so often follow on syncotyly, are of no value as selective characters (§ 9); but it is not necessary to describe the experiments which prove this point.

Hemi-syncotylous seedlings are always rarer, and often much rarer than the true syncotyls; amongst these latter on the other hand, the highest degrees of symphysis are more abundant than the lower ones, which have an obvious invagination of the apex. I have often recorded separately the various degrees of syncotyly in my seedlings. In this way figures are obtained which give curves with two peaks, such as have been found for other anomalies, especially for fasciation.[1] The atavists constitute one peak and the most complete syncotyls the other. From the former the curve drops rapidly to mount again gradnally with the increasing degrees of symphysis; whilst on the other side of the apex of the syncotyls there is another rapid drop to the very rare cases of amphicotyly,

[1] *Sur les courbes galtoniennes des monstruosités*, Bull. Scientif. de la France et de la Belgique. Publié par A. GIARD, Vol. XXVII, p. 396, April, 1896. See especially the curve on page 397.

in which the seed-leaves are fused together on both sides (Fig. 90, C and D). Fig. 98 represents such a curve from the harvest of 1889, which includes 2439 seedlings. The absolute numbers of the seedlings which possess the various degrees of symphysis are given below the figure. The groups are obviously arbitrary, but it is not likely that a grouping according to a different scheme would have any essential effect on the form of the curve, for the seedlings with fused peduncles only would always constitute a minimum between the two peaks.

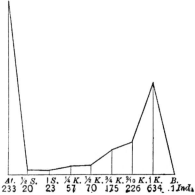

Fig. 98. *Helianthus annuus syncotyleus*. Curve representing the degree of syncotyly in the seedlings of the harvest of 1889. *At*, atavists; $\frac{1}{2}S$, $1S$, peduncles fused along half their extent or entirely; $\frac{1}{4}K$, $\frac{1}{2}K$, $\frac{3}{4}K$, $\frac{9}{10}K$, $1K$, seed-leaves fused over $\frac{1}{4}$ of their extent and so forth; *B*, pitcher-cotyls or amphicotyls.

If we examine the series of figures of our pedigree statistically, we see that they are, as a rule, asymmetrical, i. e., the highest values are closer to the one end than to the other. In all the years with the exception of the first and the last, the highest peak is shifted in the direction of selection. In order to examine this effect more closely I instituted an experiment in the summer of 1890 on a larger scale alongside the one dealt with on page 470. For this purpose I planted out about 60 syncotyls from a parent with a value of 81%; 55 of these gave a sufficient harvest. The values have been arranged in groups in the same way as in the pedigree on page 470, and the sizes of the groups thus obtained, are graphically displayed in Fig. 99 and given in a lower row of numbers. The asymmetrical form of the curve is seen at a glance, although the selection which preceded this generation was limited to two years only.

If we compare this curve with the groups of figures representing the tricotylous half races, we see that it is not simply its mirror-image. It is true that both have their highest point over or close to the end of the base line. But in the case of the half race the longer side of

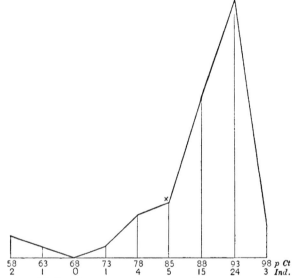

Fig. 99. *Helianthus annuus syncotyleus.* Curve of the hereditary values of 55 syncotylous offspring of a parent with 81% (Culture 1890). Given as a type of a curve resulting from selection.

the curve extends seldom so far as 25%, (*Scrophularia*) and ordinarily much less, and this in spite of the selection in the plus direction (see the table on p. 392). In the case of *Helianthus*, however, the longer side extends almost to the middle or about 60%, and this in spite of the selection of the best syncotyls with the highest values. If the selection, in this case as in the other, had been directed towards the middle of the base line, the sweep of the curve would obviously have been a still more gradual one. We see, therefore, here as elsewhere, that the intermediate race or eversporting variety is far more susceptible to selection than the half race.

§ 11. IMPROVEMENT OF A HEMI-SYNCOTYLOUS RACE.

Like tricotyly, I regard syncotyly as a single elementary character the external manifestation of which is subject to fluctuating variability. Completely fused cotyledons, which are the type occurring most abundantly (see Fig. 98), constitute the normal or typical structure, whilst the lower degrees of fusion and the bilateral unions are the *minus* and *plus* variants of the same series.

One of the most important problems in the theory of mutation is to bring together the various expressions of the same elementary character in each particular case. The delimitation of such groups is often clear at once, but often it can only be reached by the actual observation of the process of mutation. Still, the question is obviously a fundamental one, for our whole conception of affinities must rest on it, both in the question of species and in that of hybridization. We must therefore seek for methods which will lead to a solution of this problem.

With regard to the present case we may start from the following discussion. If the hemi-syncotyls and amphi-syncotyls were representatives of distinct elementary characters, we should expect to be able to isolate the corresponding races. The amphicotyls are too rare and too difficult to cultivate, on account of the disturbance in the growth of the plumule, to offer much prospect of success. Hemi-syncotyls, on the other hand, are well adapted to such an experiment in isolation. Our experience with the tricotylous races teaches us what to expect; for, obviously, either a half race or an intermediate race must arise, if hemi-syncotyly is at all capable of separate existence.

But if such an isolation cannot be effected, i. e., if the hemi-syncotyls are only *minus* variants of the syncotylous intermediate race, the selection of hemi-syncotyls will obviously do no more than maintain this latter race, and only modify it slightly in the *minus* direction. We should then expect to obtain a strain which should not differ essentially from a true syncotylous intermediate race, except by a slight shifting of its mean value. In this case the number of the hemi-syncotyls will be somewhat increased by selection, but not, however, to the exclusion of the syncotyls.

From this discussion it is clear that we can furnish the experimental proof that the hemi-syncotyls are *minus* variants of the syncotyls by an appropriate experiment in selection. For this purpose I started in 1890 with a lateral branch of the pedigree on page 470, by selecting every year only hemi-syncotylous plants as seed-parents. I have continued this experiment in the same way for seven generations, and the result was, as we shall see, a confirmation of the above conclusion.

This experiment was conducted in another garden from that in which I cultivated the syncotylous race, but otherwise carried out in exactly the same way. In the crops the hemi-syncotyls and the true syncotyls were always recorded separately, so that for each seed-parent two values were obtained. The individuals to be planted out were at first chosen from two parents, but later only from one; the selection being made according to the proportion of hemi-syncotyls among the seedlings, i. e., that parent was chosen which produced the highest percentage of hemi-syncotyls, no regard being paid to the number of syncotyls produced.

I shall now give the values of these selected parents for several generations in tabular form:

HEMI-SYNCOTYLOUS RACE.

HEREDITARY VALUE OF THE SEED-PARENTS SELECTED FOR THE CONTINUATION OF THIS RACE.

From the harvest of	Proportion in percentages of			Degree of symphysis of cotyledons
	Hemi-syncotyls	Syncotyls	Totals	
1889	5	76	81	$\frac{1}{2}$
1890	40	36	76	$\frac{1}{4}$
1891	15	15	30	$\frac{1}{2}$
1892	20	20	40	$\frac{1}{2}$
1893	28	27	55	$\frac{1}{2}$
1894	31	55	86	$\frac{1}{2}$
1895	34	35	69	$\frac{1}{2}$
1896	37	30	67	—

In this table the degree of symphysis in the cotyledons is given for each parent selected. This was, where possible one-half, i. e., the cotyledons were fused along half the distance between the upper end of their petioles and their tips. In the earlier years this was more or less a matter of chance, and in the harvest of 1890 it was by no means the plant with the highest degree of fusion that gave the best hereditary value. Later, I could limit myself to planting out those seedlings in which the fusion had only extended half-way and so I only determined the values from these.

The first row of the table (1889) refers to that individual of the syncotylous race (p. 470), whose offspring constituted the basis for this special culture. The selec-

Improvement of a Hemi-Syncotylous Race. 479

tion and the planting out of hemi-syncotylous seedlings suddenly increased the proportion of the latter and correspondingly diminished that of the syncotyls; but only for a time. In the harvest of 1891 both have become low, and from that time onwards selection has gradually increased, with but slight deviations, both the number of hemi-syncotyls and that of syncotyls. In the last harvest (1896) the numbers are given for the plant with the largest proportion of hemi-syncotyls. The average number of the hemi-syncotyls of the whole crop was, however, 29%, and that of the syncotyls 31%.

The result of this experiment of seven years was, therefore, that by the double selection of seed-parents, which produced the largest number of hemi-syncotyls, and of hemi-syncotylous individuals for seed-parents, we do not even approach a pure hemi-syncotylous race. In spite of the inevitable fluctuation of the numbers, the syncotyls remain in about the same proportion as the half type. If we think how many degrees of symphysis this latter embraces as opposed to the uniform group of the true syncotyls, the number of the latter actually found assumes even greater importance.

In the pedigree on page 480 I have given a summary of the whole course of this culture. In this I have entered only the totals of hemi-syncotyls and syncotyls; but as half of these values belonged to hemi-syncotyls, the pedigree would not have been different if these alone had been entered, except of course that the percentage figures at the top of the table would have to be halved. As usual, these figures relate to means of small groups, which in this case embrace values between 1 and 9, 10 and 19, 30 and 39, etc. The figures which are given for the several generations indicate the number of individuals whose hereditary value is represented vertically above them. The barbs in the bracket show the parents of the previous generation which had been selected and correspond therefore to the totals in the table on page 478.

The chief point which this pedigree brings out, is that the selection of the hemi-syncotyls does not modify the

HEMI-SYNCOTYLOUS RACE.
PEDIGREE OF THE TOTALS OF HEMI-SYNCOTYLOUS AND SYNCOTYLOUS SEEDLINGS.

	5	15	25	35	45	55	65	75	85	95
7th generation 1896				1	3	2	10	3		
6th generation 1895						1	6	7	3	
5th generation 1894				3	4	3	1	3	3	
4th generation 1893			1	5	2	3	2	2		
3d generation 1892	3	4	3	5	2					
2d generation 1891				3	5	4	5	4	5	1
1st generation 1890						1	1	7	8	4
Syncotylous race, 1889								1		

race either in one direction or in the other. If we omit the crop of 1890 the several generations fluctuate round a mean value which does not deviate greatly from 55% and thereby agrees sufficiently closely with the average of tricotylous individuals in tricotylous intermediate races. In other words, *the selection of hemi-syncotyls gives a syncotylous race of average value*; and if we compare the several generations of this race with a true syncotylous race (see the table on page 470), we only find in most of them an oscillation within much wider limits.

The hemi-syncotyls are, therefore, only variants of the syncotylous race.

§ 12. ATAVISTIC RACES.

Although the atavistic specimens of *Helianthus annuus syncotyleus* are exactly similar to the normal seedlings of the ordinary sunflower, they nevertheless belong to the syncotylous race and do not depart from it. With regard to their visible characters they are aberrant forms; with regard to their progeny, however, simply extreme *minus* variants; but the latter is true only on the average, and not for the particular atavists whose values often approach those of the best syncotyls, and not infrequently exceed the mean of the race.

If the intermediate race were not pure we should, of course, still be able to select the corresponding half race from it. But after it has once been purified by selection, this is no longer possible. A syncotylous intermediate race can no more give rise to a half race than a tricotylous half race can be transformed by selection into an intermediate race (see *Amarantus* and *Scrophularia*, pp. 398 and 407). In the years 1890, 1891, 1892 and 1894, I made extensive cultures of atavists derived from my syncotylous race and also from my hemi-syncotylous race, and determined their hereditary values for one or two generations. In these experiments curves were obtained which were not asymmetrical, with a peak at one end, but were flattened curves, extending over the whole length of the base line. Thus, for instance, I obtained from a dozen atavists from the seeds of the seed-parent of 1890, selected for the syncotylous race, the following percentage numbers of syncotyls:

19 48 54 56 62 68 69 73 80 84 88 96

If we compare this series with the pedigree on page 470 and especially with the figures given there for 1890 and later years, it will be seen that it is almost only the first two figures 19 and 48% which fall outside the group of values of the syncotylous offspring of the selected seed-parents.

A selection of atavists as seed-parents continued through several generations, and a selection amongst these of specimens which produce the smallest percentage of syncotyls, will obviously reduce the mean value of such a race, but not to such an extent as to justify the expectation that a continuation of the process will lead to the origin of a half race. I started such an experiment in the summer of 1894, from the seeds of the harvest of 1892. I selected a specimen from my syncotylous race, (p. 470), which had produced 92% syncotyls and only 8% atavists, and planted out the latter only. Since that time I have cultivated the race continuously and on an isolated spot and have planted out every year only the seedlings from those parents which had produced the largest number of atavists amongst their offspring. From these I always selected the pure dicotylous seedlings only.

I determined the hereditary value for each example and combined these in small groups in the usual way. At the top of the pedigree on page 483 will be found the mean values of these groups ($5 = 1$—9, $15 = 10$—19 etc.), and, vertically below these figures the corresponding number of individuals. The figures at the top give the percentage numbers of syncotyls. The pedigree can,

Atavistic Races.

therefore, be directly compared with those on pages 470 and 480. The brackets in each case indicate the selected seed-parent.

HELIANTHUS ANNUUS SYNCOTYLEUS.
ATAVISTIC RACE.

	5	15	25	35	45	55	65	75	85	95
4th generation 1897	0	1	2	3	4	7	9	2		
3d generation 1896				7	5	8	1	2		
2d generation 1895							1	0	5	12
1st generation 1894								2	1	3
Syncotylous race, 1892										1

We see that the *minus* selection has actually had its effect in that direction. The extremes and the mean have regularly decreased. The experiments correspond exactly

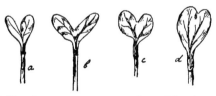

Fig. 100. *Helianthus annuus syncotyleus.* Hemi-syncotylous seedlings with various degrees of fusion of the cotyledons. A and B small, C and D larger degrees of symphysis.

with that with maize, which was represented in Fig. 18 on page 73 of the first volume, except that in this case the change took place in the reverse direction. On the other hand, the signification of the pedigree is, in this case, quite different from that of the main syncotylous race (p. 470). In that case we found a sudden transition from 19% to about 98% in the first generation, and since then only fluctuation around the obtained value. In this instance, however, there was a regular retrogression which has reduced the mean value from 90% to about 50 to 55%. This value corresponds to that of an inter-

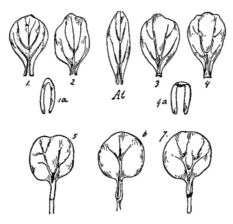

Fig. 101. *Helianthus annuus syncotyleus.* 1-7, fused cotyledons, each figure of a single plant; *At,* atavistic normal seed-leaf; 1 and 2, simple opposed syncotyls of which 1a represents a transverse section; 3-4, doubly folded syncotyls with 3 peaks; their transverse section in 4a; 5, with very slight invagination of the apex; 6 and 7, without invagination; 6, with one vein and 7, with two.

mediate race without selection, especially of tricotyls, and is about the same as that which was reached by the hemisyncotylous culture (p. 480). There can, however, be no doubt that a further selection in the *minus* direction would have reduced the proportion of syncotyls in my race to a figure much below this mean.

We thus see that *after a minus selection extending over four generations, the race still contains individuals with 65 to 75% syncotylous offspring,* and could therefore be brought back in one or two generations to the highest point attained by the original race. It is therefore far removed even from the slightest semblance of a half race.

How far, in the case before us, selection could lead in the course of years can, of course, only be surmised. But the sunflower is not a suitable species for the continuation of these experiments, on account of the risk of occasional pollen grains of the ordinary sort being brought by insects from distant gardens; for under these conditions selection would promptly extract a half race which would be the result of hybridization and not of selection, and artificial fertilization in the sunflower is beset with very great difficulties.

§ 13. THE INFLUENCE OF EXTERNAL CONDITIONS ON HEREDITARY VALUES.

In these cultures my practice has been, first to record the seedlings in the pans and then to plant them out into pots with well manured soil, from which they were transplanted to the beds in June. Here the plants stood at distances of over half a meter apart. As soon as axillary buds became visible they were removed, and only the terminal head was allowed to flower. Fertilization was left to insects, for my sunflowers do not set seed without crossing. As soon as the seed is ripe, I cut off the whole head in order to collect and clean the seed.

The question presents itself, does this method of culture exert any considerable influence on the proportions of syncotyls? In order to provide an answer I have instituted a number of subsidiary cultures in different years, preferably with seedlings which did not differ much from those of the main culture. These have, however, practically always exhibited no more than slight deviations; and where these happened to be exceptionally large, the cause of the deviations has remained obscure. As a rule, we may assume that favorable conditions increase the

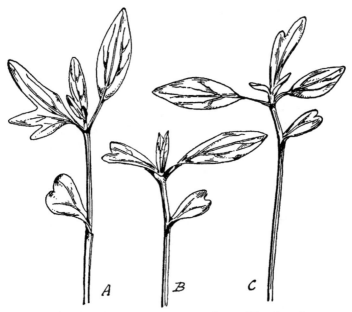

Fig. 102. *Helianthus annuus syncotyleus.* The first leaves of syncotylous seedlings. A, a leaf of the first pair with three peaks and inserted above the syncotyl; B, a leaf with two peaks placed opposite the syncotyl; C, a leaf with two peaks above the first pair of leaves.

hereditary values; but only to an inconsiderable extent, and especially so within the relatively narrow limits of the scale on which our plant can be cultivated in the garden.

In 1892, and at other times, I also collected the seeds from the flowers on lateral branches; both the flower-heads and their fruits are smaller here than at the top of the main stem, and they produce a correspondingly smaller number of syncotyls, but only in the proportion of 87 to 80%, on an average for 12 plants. If the seeds of single inflorescences are harvested in three lots in such a way that the outer, inner and middle areas are separated, the former generally contain slightly more syncotyls, for the fruits in the middle of the head tend to be somewhat weaker. From each of these three groups I have allowed the best seedlings to flower and bear fruit, but I found no essential difference in their hereditary values. It appears that improved conditions during the early stages of the plant, have some small influence in a positive direction, but crowding in later life and partial removal of the leaves during the flowering period had little noticeable effect (1891). Furthermore a culture on good sandy soil, instead of garden soil, made no difference in the values (1892).

Fig. 103. *Helianthus annuus syncotyleus*. A and B, seedlings whose plumule has not developed during the course of several weeks; the syncotyl is abnormally enlarged. C, Amphi-syncotylous plant, also with inhibited development of the stem.

Striking exceptions, however, sometimes occurred; for instance in the year 1891, when three plants in my main culture became affected with *Peziza* a short time after flowering, and died, not however without ripening some of their seed. These three produced the smallest values of the whole group. (76, 84 and 85% as against 86 to 99%). But perhaps they were the weakest individuals and contracted the disease for this reason.

III. THE INCONSTANCY OF FASCIATED RACES.

§ 14. THE INHERITANCE OF FASCIATIONS.

Fasciations are amongst the commonest anomalies which occur in the vegetable kingdom.[1] Until about ten years ago the prevailing opinion concerning them, as indeed in regard to monstrosities in general, was that they were not heritable but owed their origin to external influences only. The coxcomb, *Celosia cristata*, was considered an exception to this rule. It was, however, well known that the phenomenon occurred more frequently among certain species than among others; but the conception that some plants possessed a greater tendency to the production of such anomalies than others was taken for a sufficient explanation of this fact.

But since I have succeeded, in the case of a series of apparently fortuitous fasciations, in establishing, by isolation and selection, races in which the deviation is repeated regularly and in a considerable number of individuals, it has become evident that we are concerned here with heritable qualities which are handed on from one generation to another in certain strains of individuals, and which really differ from the characters of ordinary varieties only in the fact that they are always accompanied by reversions. It never happens that every branch

[1] See A. GALLARDO, *Fasciación, Proliferación y Sinantia.* Anales del Museo Nacional de Buenos Aires, Vol. VI, pp. 37-45.

on a sufficiently branched individual is fasciated, any more than that all the individuals in a large crop produce the anomaly without exception. The *Celosia cristata,* which comes closest to perfection in this respect, is only an apparent exception to the rule.

Fasciations, therefore, afford valuable material for the study of inconstant characters. Moreover they are

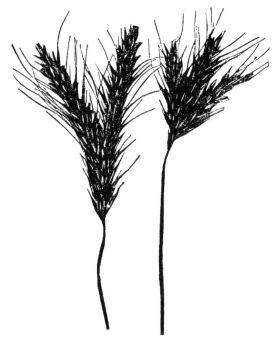

Fig. 104. Ears of rye with two and three tops, such as are sometimes found in mowing the fields (1891).

known everywhere, can be procured by everybody, and are fairly easy to cultivate; and the successful cultures give rise to beautiful instances of fasciation in a third or even in a greater proportion of the individuals. There is a complete series of transitional forms between the atavists and the most abnormal specimens, inasmuch as the broadening of the axis can be either very slight or very considerable, and all the intermediate stages can often be observed on the lateral branches of a single plant.

The most important result which can be deduced from the experiments which follow is the discovery *that the atavists or non-fasciated individuals of the race can only be regarded as reversions in the morphological sense; but that physiologically considered, that is, as transmitters*

Fig. 105. *Ranunculus bulbosus.* A fasciated stem which has grown up from a broadened rosette of radical leaves. The terminal flower is also broadened and split. Hilversum, 1894.

of fasciations, they are scarcely inferior to the best fasciated specimens of the race. The character in question is only latent in them, or rather only temporarily invisible; perhaps simply not developed by reason of the absence of some necessary external factors.

The atavists do not, so to speak, depart from the race, as do those of *Oenothera scintillans*; the race constitutes a uniform group of individuals and only differs from constant races or true varieties by the extraordinarily high degree of variability of its distinguishing character.

This degree of variability relates not only to the degree of expansion of the axis, but also to the manner in which the anomaly is manifested. First we have to

Fig. 106. *Viola tricolor maxima,* the garden pansy. A forked flower stalk arising from the axil of a double leaf (a, a); s and s, the outer stipules; s, the inner unsplit stipules of this leaf. The continuation of the main axis has been bent down laterally, (b).

distinguish between the split branches and the fasciated branches *sensu stricto*. Split ears of the rye are sometimes found in the fields (Fig. 104) and then preserved by the country folk. The axis of these ears may be divided once or oftener, the parts above the division being perfectly normal themselves. Or again the haulm below the ear may be split and produce two ordinary ears side by side. Amongst other plants also forkings of this kind are by no means rare, but as a rule they occur associated with typical fasciations.

The latter are of the same breadth from below, that is to say in the strict sense ribbon-shaped, or they begin with a cylindrical form below, and gradually flatten out towards the top. The latter case is the normal one; from it the former has been derived. This may occur either on an axis of two or more years of age, or on the lateral branches of stems which are themselves fasciated. Axes of two or more years of age begin by being circular in transverse section, and can, in the same summer, obtain a maximum breadth which they then retain in later years. This is observed especially in stems which arise from a rosette of radical leaves developed in the preceding year (*Crepis biennis, Aster Tripolium, Picris hieracioides, Primula japonica, Ranunculus bulbosus* [Fig. 105] etc.). In trees and shrubs and especially in firs (*Abies excelsa*) a fasciation that has once appeared, frequently reappears for several years in succession.[1] The lateral branches of fasciated axes often have an expanded base and then grow on without further increase in breadth.

Like the base of a lateral twig, a leaf on a fasciation can undergo expansion in its youth. In this way there arise broader, more or less deeply split, leaves, not infrequently even with similarly expanded or split axillary twigs (Fig. 106); or again the germs may be split and parted quite early, and two or more leaves arise in this way instead of a single one. Especially when the leaves

---

[1] C. DE CANDOLLE, *Fasciation chez un Sapin,* Archiv. Sc. phys. et nat., 1889, Vol. XXI, p. 95. Pl. II; and *Over de erfelykheid der fasciatiën,* Botan. Jaarboek, 1894. Pl. XI.

are arranged in whorls, this kind of multiplication looks very striking (Fig. 107).

If we examine the growing point of a fasciated shoot, the abnormal conformation can already be distinguished.

Fig. 107. The Madder, *Rubia tinctorum.* Fasciated stems with an increased number of leaves in the whorls, found in a madder field near Ouwerkerk and broken off at the rhizome (1890).

Where the fasciations are very broad this was to be expected, as for instance in the case of *Sedum reflexum cristatum* figured in the first volume (Figs. 34-35, pp.

182-183). Expanded combs of this kind are often seen at the top of inflorescences, especially in *Veronica longi-*

Fig. 108. *Veronica longifolia.* Inflorescence with the tip expanded in the form of a comb.

folia (Fig. 108), *Amarantus speciosus* (Fig. 83, p. 399), *Oenothera Lamarckiana* and *O. brevistylis*, etc.

These phenomena have been more closely investigated by NESTLER who regularly found inside the terminal bud of a fasciated branch a line of fasciation instead of a point, in other words a growing comb instead of a growing cone.[1] The leaves arising from this latter are produced in very large numbers and in an abnormal arrangement, and in consequence of this, the disposition of the leaves on fasciated stems tends to be extremely irregular; but although this subject is obviously very important in its bearing on the whole question of phyllotaxy, it still awaits a thorough investigation. In this respect it would be particularly valuable to make a study of fasciations in conifers, for our knowledge of the normal structure of their cone of vegetation is much more extensive than it is in regard to that of angiosperms.[2] Moreover there is no lack of material, for *Cryptomeria japonica monstrosa* and the fasciated varieties of several other species are on the market (Fig. 109).

Besides the cases of ordinary fasciation, and of those in which the surface may be bent by unequal secondary growth, there are some cases of most peculiar conformation which hitherto have been very little investigated. It is not known whether these are the expressions of the same internal character or whether they must be referred to special factors. Their morphological structure is in favor of the latter view; but the former is supported by the fact that they have hitherto always been observed in conjunction with ordinary fasciations, i. e., in species which are particularly liable to them. Examples are afforded by the multi-radiate and annulate fasciations.

---

[1] A. NESTLER, *Untersuchungen über Fasciationen*, Oesterr. botan. Zeitschrift, 1894, No. 9 ff., with 2 plates.

H. DINGLER, *Zum Scheitelwachsthum der Gymnospermen*, Ber. d. d. bot. Ges., Vol. IV, 1886, p. 18.

496 The Inconstancy of Fasciated Races.

In the ring fasciations the vegetative cone is transformed into an annular wall which ultimately develops into a smaller or larger funnel. I found these remarkable structures repeatedly in my cultures of *Veronica longifolia*, where they remained quite small, scarcely reaching a centimeter.[1] On the other hand I have observed a funnel-like structure in *Peperomia maculosa* of more than a decimeter in length.[2] Some of the best known instances of ring fasciations occur in *Taraxacum officinale*; these have frequently been described and I have myself often had the opportunity of investigating them.[3] Within a thick, tube-like flower-stalk, often as many as 10 to 20 slender stalks arise, each in the axil of a leaf and each

Fig. 109. *Crytomeria japonica monstrosa*. A commercial variety very rich in fasciations. *a*, laterally expanded tip of a branch producing a normal twig at *d* by means of splitting; *b* and *c*, further fasciations of this branch.

---

[1] A. NESTLER, *Ueber Ringfasciation*, Sitzungsber. d. k. Acad. d. Wiss., Wien, Vol. CIII, Part I, 1894, Plates I-II.

[2] *Sur un spadice tubuleux du Peperomia maculosa*, Archives Néerlandaises d. sc. ex. et nat., Vol. XXIV, p. 258, Pl. XII. Afterwards the anomaly occurred again on the same plant (1892).

[3] MICHELIS, *Botan. Zeitung*, 1873, p. 334; 1885, p. 440. Further literature will be found in NESTLER'S paper already cited. For *Helian-*

The Inheritance of Fasciations. 497

bearing a more or less normal inflorescence. From the seeds of such an individual, I have for many years cultivated a large bed containing over 100 plants which, however, only produced ordinary fasciations, but no ring-fasciations.

Sometimes I have found radiate fasciations in my cultures of *Amarantus speciosus* alongside of the more

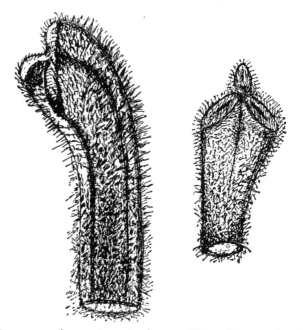

Fig. 110. *Amarantus speciosus.* Tri-radiate peaks of inflorescences.

common, ordinary fasciations.[1] The tip of the inflorescence was not flattened, but tri- or sometimes quadri-radiate (Fig. 110). In *Digitalis lutea* I have observed a similar case, and in *Celosia cristata* I found a good

thus annuus, see PAUL RICHTER, *Ber d. d. bot. Ges.*, 1890, Vol. VIII, p. 231, Pl. XVI.

[1] *Over de erfelykheid van fasciatiën*, Botanisch Jaarboek, Gent, 1894, p. 90.

instance of a quadri-radiate apex on a branched individual (1893). Tri-radiate fasciated heads have often been found in Composites; for instance in *Chrysanthemum Leucanthemum, Helianthus annuus* and *Erigeron bellidiflorus*. In the latter species these have occurred with tolerable frequency in my garden. All these cases are greatly in need of a closer investigation.

For the production of a fasciation the presence of the internal factor is not of itself sufficient. Favorable conditions of life are also quite necessary. The stronger a plant or a branch is, the more liable is it to expand and flatten out. This is best seen in those biennial or perennial plants which occasionally have the power of flowering in the first year. If they do this, either only small fasciations, or none at all, are developed, whilst it is amongst the specimens which remain in the rosette stage during the first year and do not develop their stem till the second, i. e., after they have undergone considerable increase in strength, that the most numerous and finest fasciations will be found. Thus, for instance, I obtained through the kindness of Professor Lagerheim from Stockholm seeds of a fasciated plant of *Hieracium umbellatum*, and in the summer of 1901 I had from these a bed with nearly a hundred plants without a trace of fasciation. Some plants, however, did not make a stem that year but, after they had survived the winter, produced in the following spring some beautiful expanded stems with comb-shaped inflorescences at the top. The same occurred in my cultures of *Aster Tripolium, Picris hieracioides, Oenothera Lamarckiana* and others. The first species, when grown as an annual, developed tall stems which remained fairly cylindrical in the lower part, and then began to flatten, without, however, attaining a greater breadth than about 2.5 centimeters; but in large cultures there were sometimes as many as 60 or 70% of such broadened shoots. When grown as a biennial, however, the hearts of the rosettes gradually expand in the first autumn or during the winter, and from these, stems are produced which sometimes attain a very considerable breadth. Thus, for instance, in the summer of 1895 I measured some of from 3-6 centimeters. *Picris hieracioides* seldom produces fasciations in the first year, and when it does they are not broad; whilst the stems produced in the second year from the broadened rosettes of radical leaves ordinarily afford some of the finest instances of this anomaly.[1]

Just as the age of the various individuals of a culture has a great effect on the production and development of the fasciations, so also has the time of the year at which the seed is sown. Many biennial or perennial plants which quickly manifest the character of the race when sown early, remain apparently normal if the sowing was made late, and they cannot grow out to sufficient strength before winter. My fasciated races of *Crepis biennis* and *Taraxacum officinale* are very instructive in this connection. Sowings of *Crepis,* made in April and May, gave from 30-40% of fasciated individuals. Sowings made at the end of July produced 20% only, and those made in September none at all. Similarly, *Taraxacum officinale,* when sown in spring, produced 13 to 27% of fasciations, whilst a sowing made in August did not produce even so much as a single flattened flower-stalk.

[1] *Sur la culture des monstruosités,* Compt. rend., Paris, Jan. 1899; *Sur la culture des fasciations des espèces annuelles et bisannuelles,* Revue gén. d. Bot., 1899, Vol. XI, p. 136; and *Ueber die Abhängigkeit der Fasciation vom Alter bei zweijährigen Pflanzen,* Botanisches Centralblatt, 1899, Vol. LXXVII.

In races of this kind all conditions of the environment are of importance. By crowding, the proportion of fasciated individuals can be reduced, for instance, from

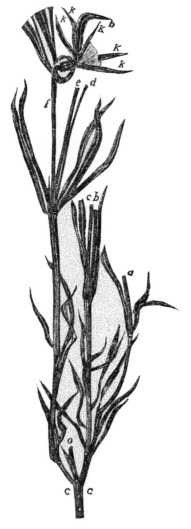

Fig. 111. *Agrostemma Githago.* A plant the main axis of which had been cut off at *o*. The cotyledons have dropped off at C and C'. Their axial twigs are fasciated, and instead of producing decussate leaves, bear multiple whorls. They split occasionally at the nodes, forming forks with branches at *a, b, c, d, e* and *f.* Just above *f* the calyx of the flower is spiral (*k k*), and fused with the uppermost foliage leaf (*b*), by which the flower stalk is seen to be twisted. (1892.)

40 to 5%. Even if the number of fasciations is calculated per area of the bed, a small number of specimens give a prospect of obtaining more fasciations than a

larger number planted on the same area, but too close together. The number of fasciations can be diminished by cultivation on sterile sand and increased by the addition of manure. The most numerous and the finest instances of the anomaly are produced by cultivation in pots with heavy manuring and by subsequent transplantation to the beds; and so forth.[1]

Further, a considerable effect can be exerted on the production of fasciations, as upon that of monstrosities in general, by pruning and by the selection of buds which accompanies this process. Thus, GOEBEL says in his *Organographie*[2] that fasciations can be artificially produced by diverting the "sap" with great intensity into a lateral bud which would otherwise have obtained only a small quantity of it. It is for this reason that fasciations are particularly abundant on adventitious branches and shoots from cut stems. Even in annual plants such as *Phaseolus multiflorus* and *Vicia Faba,* fasciation can be induced by cutting off the plant above the cotyledons. A plant which is peculiarly suitable for the demonstration of this method is *Agrostemma Githago,* which in my cultures always produced an abundance of anomalies, but did so with greater certainty when I had removed the main stem, just above the cotyledons or the first pair of leaves. The axillary buds, which, as a rule, do not develop, grew out under this treatment and frequently became fasciated (Fig. 111).

The phenomena of periodicity are also associated with the effects of nutrition. Lateral branches of fasciated shoots are usually of normal structure; but broadened ones not infrequently occur amongst them. If they do,

[1] *Botan. Centralblatt,* 1899, *loc. cit.*
[2] Vol. I, p. 234.

they manifest a certain order in their disposition since they are usually found in that region where other normal and abnormal characters also attain the maximum of their development.[1] This phenomenon which can easily be demonstrated in the fasciated race of *Tetragonia expansa,* is, however, in need of more thorough investigation.

But the chief point seems to me to be, as GOEBEL pointed out in his work to which we have already referred, that the latent factor for the production of anomalies must be present in all those parts of the plant in which external influences are able to induce fasciations or anomalies. If this is absent no amount of manipulation is of any avail. It seems, however, that the factor for the production of fasciations is pretty generally distributed throughout the vegetable kingdom; though it is not so general but that there are certain groups in which it does not occur. It is a curious fact that fasciations are much commoner amongst dicotyledons than amongst monocotyledons, although some very well-known instances are presented by the latter (*Asparagus, Lilium, Fritillaria, Orchis*).

§ 15. HALF RACES WITH HERITABLE FASCIATION.

It is only by conducting special breeding experiments that we can discover for certain whether fasciation is heritable in a given species, and to what extent. If, however, such cultures are made, we shall soon find that these anomalies fall into two categories which are perfectly analogous to the half races and eversporting vari-

---

[1] See T. TAMMES, *Ueber die Periodicität morphologischer Erscheinungen bei den Pflanzen* Kon. Acad. v. Wet., Amsterdam, 1903.

eties distinguished in the first part of this volume (p. 18). In the former case the anomalies are rare and their frequency can not be increased by selection to any considerable extent. In the second case the fasciations occur even in the field in obviously larger numbers; and it is only necessary to isolate the examples in question in order to be in immediate possession of a race producing fasciations abundantly. These experiments are perfectly analogous to those which we have described in the case of other anomalies and especially to those relating to tricotylous and syncotylous cultures (see page 343). Nevertheless, in the case of fasciations, we have by no means so stringent a character to select by, as is presented by the percentage hereditary values, calculated from the seedlings of tricotylous races, and therefore the subsequent development of the race after the initial isolation is a matter of much greater difficulty.

Postponing the consideration of the eversporting varieties or intermediate races to the next section, let us here attempt to obtain some insight into the races in which the anomaly occurs more rarely. Unfortunately, in many instances the available data are not yet sufficient to enable us to decide with perfect certainty to which of the two types a given case belongs.

Fasciations occur in so many commercial varieties that they are accessible to everybody; sometimes even the variety owes its name to the frequency of this character, as, for instance, the Sword-elder (*Sambucus nigra fasciata*); or it may be an almost constant attribute of the cultivated sorts, as in the Japanese spindle-tree (*Evonymus japonica*, Fig. 112). When the varieties are either largely or exclusively multiplied in the vegetative way, it is possible that the anomaly, although only heritable to a slight degree, may be manifested frequently; as, for instance, in *Lilium speciosum album corymbiflorum* and in the so-called sword-shaped variety of *Fritillaria imperialis*. The well-known monstrous species of Cactus should also be referred to here (*C. peruvianus monstrosus*). Experimental estimations of the constancy

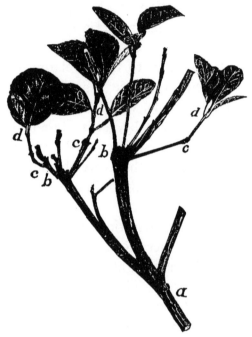

Fig. 112. *Evonymus japonica*. A fasciated much-split twig of this common garden plant. A, B, growth in 1898; C, D, in 1900. Photographed, Aug. 1900, from life.

of these abnormalities from seed do not yet seem to have been made.

The same general conditions obtain amongst many species of trees and shrubs, of which any collector can easily obtain a fasciated sprig. As instances from my own collection I mention *Fraxinus excelsior, Alnus glutinosa, Crataegus nigra, Azalea indica, Robinia Pseud-*

Acacia, Salix purpurea, Salix alba, Spiraea callosa atropurpurea. Moreover I have obtained, by the kindness of Prof. W. JOHANNSEN of Copenhagen, beautiful broad

Fig. 113. *Helianthus tuberosus.* A fasciated stem which is split into two almost equal fork-branches *b* and *c*, at the node marked *a*. From the fork there arise two leaves *d* the mid-ribs of which have fused along their dorsal side.[1]

fasciations of the underground stems or runners of *Spiraea sorbifolia* from the nursery of Mr. ZEINER LASSEN in Helsingör. Instances of the same fasciations have

[1] I have also observed this remarkable occurrence of fusing by the backs in leaves in the forks of split twigs, in *Robinia Pseud-Acacia* and *Evonymus japonica* (PRINGSHEIM, *Jahrb. f. wiss. Bot.*, XXIII, p. 81) and also in *Collinsia heterophylla* (1892), *Epilobium hirsutum* (1892), *Echium vulgare* (1892), *Chrysanthemum segetum* (1892), *Agrostemma Githago* (1892 and 1894), *Acer Pseudo-Platanus* (1891), *Crepis biennis* (1893), *Amarantus speciosus* (1894), *Mercurialis annua* (1894) and *Lamium purpureum* (1895).

been described and figured by CASPARY and therefore appear to be fairly abundant in this species.[1]

In perennial weeds fasciations also occur freely, and here we may often observe that the phenomenon is repeated more or less regularly in successive years on the same specimens or groups of specimens. For instance, we have in the botanical gardens at Amsterdam a plant of *Sonchus palustris* on which I first observed a splendid tall broadened stem in 1890. Since then the plant has produced almost every year one or several such structures, sometimes 2 meters in height, sometimes not more than 1 meter, and attaining a breadth of 6 centimeters by a thickness of 1. These stems are cylindrical at the level of the ground, but flatten out gradually upwards.

Fig. 114. *Plantago lanceolata.* Ears which have split one, two or three times. Cultures of 1894 and 1895.

Similar instances were afforded by *Aconitum Napellus,* and *Helianthus tuberosus* (Fig. 113) in our garden, by *Justicia superba* in the greenhouse, and by *Agrimonia Eupatorium* and *Chrysanthemum Leucanthemum* in the field. I frequently observed fasciations in annual and biennial species, and usually in the course of several years in the same locality; for instance in *Raphanus*

[1] R. CASPARY, *Eine gebänderte Wurzel von Spiraea sorbifolia L.,* Schriften d. Physik. Oec. Ges. Königsberg, 1878, XIX, p. 149, Plate IV. As a matter of fact, however, it was not a root; see PENZIG, *Teratologie,* I, p. 421.

Half Races with Heritable Fasciation. 507

Raphanistrum, Pedicularis palustris and *Oenothera biennis*. In my cultures of *Amarantus speciosus, Helianthus annuus* and *Oenothera Lamarckiana*, the anomaly was reproduced almost every year through the course of ten years.[1]

One of the best known instances is afforded by the sugar-beet, fasciated specimens of which can be found almost every year in the fields. We frequently find amongst them quite long, broad and wholly flat stems. In spite of the obvious fact that they are not selected as seed-parents they recur regularly, and this fact is sufficient to demonstrate the heritability of the anomaly.

I have further to mention *Plantago lanceolata*, the variety *ramosa* of which I have described in detail in the first part of this volume (page 148) and which I have cultivated every year since 1889. It sometimes produces split ears (Fig. 114), especially amongst the inflorescences which are not branched at their base, i. e., the atavistic ones. In this

Fig. 115. *Artemisia Absynthium*. A fasciated branch which has been heavily bent in consequence.

---

[1] Further details relating to this subject and more instances of the phenomenon will be found in *Over de erfelykheid der fasciatiën*, Botanisch Jaarboek Gent, 1894, p. 72.

race the anomaly is obviously in the latent condition and only to a slight extent heritable.

N. MEZZANA records an instance of a fasciated stem of *Cucurbita Pepo*, the upper part of which gradually became broader over about a meter of its length, and was thickly set with leaves and flowers. The phenomenon was observed on a number of specimens which had been raised from seed of the same fruit and MEZZANA concludes from this fact that the anomaly is inherited.[1] The fact that I have frequently observed such fasciations in my own cultures of *Cucurbita* supports this conclusion. In *Artemisia Absynthium* also, fasciations are sometimes very common as I observed in 1883 (Fig. 115), 1887, 1888, 1889 and 1890, and the phenomenon was repeated from seed in 1889 and 1891.[2] The remarkable forms which the fasciated branches of this species so frequently assume offer a profitable subject for future inquiry.

§ 16. EVERSPORTING VARIETIES WITH HERITABLE FASCIATION.

Some wild species produce, in certain districts at least, a much higher proportion of fasciated examples than others do. According to my experience, such cases suggest the occurrence of heritable races, the individuals of which are mixed with those of the normal species or occasionally occur by themselves alone. So far as I am aware, such races do not consist exclusively of fasciated plants, but partly of these and partly of normal ones. Without cultivation the latter cannot be distinguished from the normal plants of the species in question, and

---

[1] N. MEZZANA, *Sopra un caso di fasciazione nel fusto di Cucurbita Pepo*, Bull. d. Soc. Bot. Italiana, Florence, 1899, pp. 268-273.

[2] *Botan. Jaarb. Gent*, 1894, p. 97.

therefore we can not directly see whether both races or only the first grow in the particular locality; but their great rarity points to the mixed condition.

The heritable races which have hitherto been found and isolated in this way, behave like eversporting varieties inasmuch as each generation consists both of fasciated individuals and of atavists, even under conditions of the most stringent selection. Moreover the proportion of these two types appears to be pretty constant, at least under similar conditions of life. As a rule, there are about 40% fasciated individuals and 60% atavists. Higher percentages of the former occur only under favorable circumstances, whilst the proportion of the latter very easily increases under unsuitable conditions of culture, in spite of selection.

The first instance that I shall describe was afforded by *Crepis biennis,* an exclusively biennial plant, fasciated stems of which have been frequently observed in various localities in Holland. The starting point of my culture consisted in two fasciations, which I found in May 1886, in a meadow near Hilversum, amongst hundreds of normal plants of *Crepis*. The broadening of their stems was small and limited to the top. I collected ripe seed in this meadow in June, but from normal plants only. Whether all or only some of these belonged to the eversporting race I was in search of could, of course, not be determined then.

This seed furnished in the following year about one hundred plants, of which three were already fasciated in the rosette stage, whilst in the following year nine more of them developed more or less flattened stems or branches. The total proportion, therefore, was 12%. In order to make perfectly sure, I retained only the first mentioned specimens as seed-parents and destroyed the rest before they flowered.

These three plants therefore formed the beginning of my race. Denoting the wild specimens of 1886 as the first generation, the second grew in 1887 and 1888, the third in 1889 and 1890, and so on. This third one consisted of 120 plants, of which 48 or about 40% already exhibited a comb-shaped linear growing point in the heart of their rosettes in the winter. This comb was in some specimens as much as six centimeters long. I selected the three finest fasciations as seed-parents and removed the rest before flowering. The fourth generation raised from this seed produced a slightly lower proportion of fasciations, containing, as it did, only 30% of them. In the fifth generation a further reduction took place, viz., to 24%.[1] The sixth generation (1895-1896) was very rich in fasciations, and in ten plants out of the 40, produced a growing comb of from 4 to 7 centimeters. Unfortunately the remaining fasciations of less degree mostly died in the winter before they were recorded, so that an exact percentage value can not be given. Nevertheless it was obvious that the character of the race had been displayed oftener than in the previous generation. In the following, viz., the seventh generation, I also recorded only the expanded rosettes before the winter, and found ten of them amongst 49 plants, that is about 20%. The eighth generation was not sown till 1902, and only on a very small scale. The combs became visible during the winter 1902-3.

If we summarize the results which we have described, we see that the seed collected in the field, without selection, gave about 12% fasciated offspring, whilst the seeds

[1] *Botanisch Jaarboek, Gent,* 1894. p. 80, and 1897, p. 66.

of the best cultivated individuals gave values between 10 and 20%, in the course of the five generations which followed. These would probably have been higher if the cultures had been larger and the external conditions more favorable, and especially if a closer search had been made for smaller fasciations on the lateral branches. We may therefore regard the constitution of this race as fairly constant under normal conditions, and put it on the average about 30-40%.

No doubt, this figure is somewhat lower than the normal value of tricotylous intermediate races which we described in the second chapter of this part. In that case the value was about 55%; but this difference does not seem to be of great significance, especially when we remember that tricotyly is already determined in the seedlings, whilst a long period of time elapses between the ripening of the seed of the fasciated plants and the manifestation of its character, during which period all sorts of external influences may be affecting the result.

A second difference between tricotylous and fasciated races also demands brief notice. In the former, selection soon led to a transgression of the original normal value; and values of 70 to 90% were often reached or even surpassed without much trouble. In the fasciated races, on the other hand, it is very difficult to raise the value above about 40%. The explanation of this seems to me to be as follows: In my tricotylous races a twofold selection took place, inasmuch as, first, the tricotylous individuals were selected for subsequent cultivation, and, secondly, a selection was based on the proportion of tricotylous individuals amongst their seeds. By the selection of the parents with the highest hereditary values, this value itself was seen to increase in the following generation. In the fasciated races, on the other hand, we are obliged to limit our selection to the best representatives of the anomaly; but there is no further reason to suppose that these also possess the highest hereditary values. Thus, an essential part of the selective process as applied to the tricotyls is omitted in this case. This is mainly due to the impossibility of calculating the hereditary values in the seed pans, and the fact that these would need cultivation on a very large scale in the garden. In order to calculate the hereditary values for only 20 seed-parents from lots of only 100 offspring each— and even this would hardly give reliable results—80 square meters of the garden would have to be devoted to *Crepis*, and this can scarcely be done in an ordinary garden. It is to be hoped that institutions will soon be erected where such determinations can be carried out.[1]

Besides *Crepis biennis* I discovered one or two other species behaving in the same way and succeeded in raising eversporting varieties from them.[2] The first to be mentioned is *Aster Tripolium*, of which I obtained a splendid fasciated example with ripe fruits in the autumn of 1900, from this neighborhood. At first I grew the plant as an annual and reached only a low proportion of fasciated individuals as a result of this. The figure was 7% for the fourth generation. In the fifth generation, however, in the summer of 1894, the plants were subjected to better treatment, and more than half of them produced fasciated stems, amongst which many were more than 3-4 centimeters broad. I shall deal with

My experimental garden contains 75 beds of about 4 square meters each.

[2] *Botanisch Jaarboek, Gent*, 1894, and *Bull. Scientifique de la France et de la Belgique*, publié par A. GIARD, XXVII, 1896, p. 402.

Geranium molle fasciatum in the next section. Of this race, one-third consists, as a rule, of individuals with fasciated branches. In 1895 I was growing its sixth generation. I have also cultivated six generations of *Taraxacum officinale fasciatum*. This species, as a rule, produces 30%, and sometimes more, of fasciated individuals. Beautiful instances of fasciation have been furnished almost every year, since 1885, by *Tetragonia expansa* in the botanical garden in Amsterdam, and the proportion of these was, in the fourth and fifth generations after isolation, slightly over 50%.

The same general behavior was observed in my fasciated races of *Thrincia hirta, Veronica longifolia, Hesperis matronalis, Picris hieracioides* etc.

From these data we may draw the general conclusion that such races, after having been isolated and subjected to good treatment, and by the selection of the finest instances of fasciations as seed-parents, consist of a little less than one-half of fasciated individuals, and of a little more of apparently normal, atavistic, plants. This proportion, however, depends to a large extent on external conditions. By means of suitable cultivation it can be considerably increased, but on cessation of this care it soon sinks to quite low values.

Many of the known instances of fasciations probably behave in the same way. For instance KÖRNICKE has grown for many years a perfectly constant race of a fasciated pea (*Pisum sativum*) in Poppelsdorf, and RIMPAU has informed me that he cultivated this fasciated race from seeds during several years in good garden soil and found it constant. The result of sowing the seed of *Sedum reflexum cristatum* (Vol. I, p. 183), in this garden, was the reappearance of the abnormality in large numbers. To this group, also *Asparagus officinalis* and several other species seem to belong.

§ 17. THE SIGNIFICANCE OF THE ATAVISTS.

As we have seen in the first section of this chapter (page 490) a proper understanding of what is meant by atavism is a necessary basis for the discussion of our appreciation of the inheritance of fasciations, and of anomalies in general. Here, the atavists are not individuals which step out of the race; on the contrary, they are to be regarded merely as specimens in which, from some external cause or other, the anomaly is not manifested during their lifetime. In the selection for the continuation of the race they are, of course, not usually preferred, but as a matter of fact they may serve just as well for this purpose as the fasciated individuals themselves.

Extensive investigations are still to be carried out before a complete and proper understanding of the principles which underlie these phenomena can be attained. The knowledge however, which we already possess, appears to me to be sufficient to demonstrate the correctness of the generalization just enunciated. In the first place I mentioned the remarkable fact that the anomaly can remain latent during a whole generation without disappearing forever or even becoming noticeably diminished. Sometimes indeed two or several generations can be skipped in this way. Let me give some instances as proof of this.[1] In the fall of 1887 I collected some seed of *Tetragonia expansa* from fruits on very broad stems and obtained, in the following years, 1888-1890, three further

[1] *Botanisch Jaarboek, Gent,* 1894.

generations which produced a greater or less number of fasciations. The seeds of the finest fasciations of 1890, however, produced nothing but normal plants in 1891 which did not exhibit the anomaly, even on a single lateral branch. They were weak plants and it looked as if the anomaly were lost once and for all; but seeds of these plants produced in the following year, 1892, fourteen plants, of which seven were fasciated. Six of them had 1—1—2—2—3—3 broadened stems, and one plant had as much as four large fasciations. Moreover the lateral branches were so much affected by the anomaly that I found about one-third of them to be modified in this way. Since that time the anomaly has remained constant in this strain. In the third generation of my race of *Amarantus speciosus* (1891) the fasciations were also absent, but returned in the fourth and fifth generations in 30 and 50% of the individuals. In *Helianthus annuus* they were also absent from the third generation (1889), whilst the fourth contained about 20% of fasciated individuals, and the anomaly has since remained constant. In the maize I observed fasciated ears in a cultivated race in the years 1888, 1889, 1892 and 1893, but not in the generation of 1891, between these. From the seed of a very broad stem of *Picris hieracioides* (1887) I raised three generations under unfavorable conditions, and they did not produce a trace of the anomaly on many hundreds of branches and stems. It was not until the fourth generation that the anomaly reappeared, although only to a slight extent. Besides this strain I have cultivated a race of biennial individuals, and these have presented fine instances of fasciations in greater or less abundance in every generation.[1]

[1] *Revue générale de botanique*, 1899, Vol. XI. p. 136.

In the summer of 1895 I isolated some of the atavists of my race of *Crepis biennis* described above, before they flowered.[1] The seeds were sown in the following year and produced over 350 plants. About 20% of these exhibited the comb-like structure in the center of the rosettes, and this line sometimes reached a length of five centimeters. Thus we see that the seeds of the atavists may produce fasciations in about the same quantities as do the selected fasciated individuals. Inasmuch as the monstrosity frequently lowers the strength of the plant, it might perhaps even be advisable to choose the seeds of the atavists or of individuals which are only fasciated on their lateral branches.

A further proof of the view that the atavists are only fasciated individuals with their character in a latent condition is afforded by experiments on the effect of thorough manuring. For the individuals which under normal favorable circumstances behave as atavists can be induced by it to a relatively considerable development of fasciations. In 1895 I made such an experiment with my strain of *Crepis biennis fasciata* which then contained some 20 to 40% of fasciated individuals every year. I manured a group of 41 plants with an abundance of horn-meal. At the time of ripening only six of these lacked the anomaly, i. e., 85% instead of 40% were fasciated. The plants stood fairly close together, at distances of about 20 centimeters apart. If I had given them enough room I should probably have succeeded in inducing the anomaly in every one of them.[2]

The fasciated commercial race of *Celosia cristata*,

---

[1] *Botanisch Jaarboek, Gent*, 1897, p. 66.

[2] *Botanisch Jaarboek*, 1897, p. 66; and *Bull. Scientif., loc. cit.*, Vol. XXVII, p. 413.

The Significance of the Atavists. 517

generally known as the coxcomb, is peculiarly well suited for an investigation of atavism. The great combs produced by this favorite garden plant are merely the selected well treated and highly nourished individuals, i e., the extreme *plus* variants of the race.[1] All the rest are thrown away in their early stages by the gardener; but if we wish to know how the race really behaves, we must

Fig. 116. *Celosia cristata.* An almost entire plant under poor treatment. Top comb-shaped but small; most of the lateral branches were also broadened at their tips.

make sowings ourselves and plant out all the individuals without selection, and cultivate them further. It is well known that in this way we obtain numerous plants with branched stems and with a much lower degree of the anomaly (Figs. 116 and 117). By the selection of these *minus* variants for further cultivation one might expect,

[1] For the mode of cultivation suitable for this variety see MÖLLER's *Deutsche Gartenzeitung*, 1892, p. 200.

as Solms-Laubach suggests, to ultimately obtain the unfasciated original form of this highly modified plant.[1] But according to the view laid down in this book, a mutation would be necessary for this; and, so far as we know, these appear only very seldom and fortuitously, unless we happen to meet with a plant in a mutational period.

During the years 1893-1897 I endeavored to obtain a race devoid of fasciations, but in vain; and inasmuch as the general rule is that favorable circumstances increase the production of the anomalies, and as correspondingly the worst nourished and weakest individuals have always borne the smallest fasciations, I was ultimately obliged to give up the experiment, because my strain gradually became very much debilitated without, however, producing the desired result.

Fig. 117. *Celosia cristata.* Top of a plant about ½ meter high; the stem was much branched and its inflorescences were ear-shaped but mostly with comb-shaped tops.

In 1893 I raised about fifty plants from bought seed, but did not make any detailed record of them. Most of them produced combs in every degree of development,[2] but usually of small size, 5 — 10 centimeters broad or less.

---

[1] H. Solms-Laubach, *Bot. Zeitung*, 1900, p. 42.

[2] The plane of the fasciation is the median plane of the cotyledons; and this is also the case in *Crepis biennis*. This fact might be used as a starting point for an inquiry into the ontogeny of fasciations.

Six plants had a terminal spike at the top of the main stem with a small comb; in six others this comb was absent, although small fasciations occurred here and there amongst the lateral branches. These six plants were selected as seed-parents, and from their seeds I obtained the second generation (1894), in which about half the individuals bore terminal spikes without a comb. There were 41 of these. The remaining 53 had combs ranging from 1-3 centimeters broad and were thrown away. Of the selected atavists, several produced lateral combs later, and each one of these plants was destroyed as soon as I discovered the anomaly, until at the end of the harvest period there were left only five plants which, though profusely branched, manifested no trace of fasciation. In the following year there were 29 plants with small combs and 6 without any at all; these latter were again isolated as seed-parents. In 1896 I had 38 individuals, all of which, without exception, produced combs, the length of which ranged between 2 and 8.5 centimeters with a mean of 4.5. Only one plant had no comb at the apex of the stem; but had a much flattened lateral branch instead. This, and the plants with the smallest terminal combs, were selected as seed-parents. Again, in the following year (1897), no progress was made, inasmuch as only a single weakly individual failed to produce a comb. Therefore I gave up the experiment and I conclude that complete atavists are very rare in *Celosia cristata* and that even under repeated selection in the *minus* direction they will only be obtained in very small quantities. At present at least there seems not to be any prospect of obtaining a pure atavistic strain.

The experiments I have described show that, as a rule, there is no sharp limit between the fasciated individuals and the atavists. This fact can be illustrated by statistical examination of the material, provided it can be made sufficiently large by cultivation. As an example of this I may again cite my race of *Crepis biennis fasciata*.[1]

In order to obtain a pure curve, I sowed the seeds of a single broadly fasciated individual of the third generation of my race (1890) in March 1894. The plant had flowered together with two other fasciated individuals of the same ancestry, but this little group had been completely isolated. The seeds were sown in pans in the greenhouse and afterwards planted out at suitable distances in the bed. Whilst still in the rosette stage many of them produced in the first year a comb-shaped heart. Others did not exhibit fasciation until the stem began to develop in the second year. Still others had not produced a single comb at the time when all their branches were fully developed. The external conditions had been as favorable as possible, so that I obtained a relatively high proportion of fasciated individuals. When I examined the plants in June of the second year, I obtained the following result:

```
Stems without fasciation  . . . . . .   33
  "    with slight fasciation at the top  .    9
  "    fasciated along their whole length  108
                                    Total  150
```

The breadths of the stems of these latter 108 plants were distributed as follows:

Cm.	2	3	4	5	6	7	8	9	10	11	12	13	14	19
Ind.	9	9	4	11	11	11	13	15	11	6	3	3	1	1

The numbers are plotted in Fig. 118 in the form of a curve. In this curve 0 indicates the group of 33 ata-

[1] *Sur les courbes Galtoniennes des monstruosités,* **Bull. Scientif.,** publié par A. GIARD, XXVII, 1896, p. 396.

vists, 1 the nine plants with cylindrical stems and slight broadening at the top; whilst 2-20 denote the breadth of the stems in centimeters.

The curve is seen to have two peaks, one of which is formed by the atavists (*a*), the other by the fasciations of mean breadth, i. e., of about nine centimeters. Expressed in words, the result is that transitions between

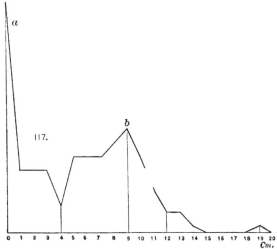

Fig. 118. *Crepis biennis fasciata.* A curve representing the breadth of the main stems of all the individuals of my culture of 1895. The numbers under the base line indicate the breadth of the stems in centimeters. 0, stem round; 1, stems only flattened at the upper end; 2-20, stems flattened along their whole length. The height of the ordinates gives the number of individuals. Total number of individuals 150. *a*, the peak of the atavists; *b*, that of the fasciated individuals.[1]

the fasciated individuals and the atavists do occur, but are relatively rare. The race produces by preference the two pure types, and the same thing is seen to happen in chance fasciations and in many other monstrosities. I call to mind the relative rarity of hemi-tricotyls and hemi-syncotyls, both in the wild state and in the tricotylous

[1] *Bull. Scientif., loc. cit.,* Vol. XXVII, p. 397.

and syncotylous races described in the first two chapters of this part.

Similar results may be obtained with other fasciated

Fig. 119. *Geranium molle fasciatum.* Fasciated branches with broadened and split fruits, *a, b; c.*

races. Thus I obtained in the fifth generation of my fasciated race of *Aster Tripolium,* under biennial culture:

	WITHOUT FASCIATION		WITH FASCIATION					
Cm.	—		1	2	3	4	5	6
Individuals	16		2	6	8	1	1	1

The curve representing these 35 plants would have a distinct peak representing the atavists and another corresponding to the fasciated individuals of the average breadth of 3 centimeters.[1]

In *Geranium molle fasciatum* the variation of this character proved, after a statistical examination of the material to be represented by a many-peaked curve. This

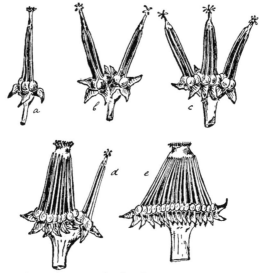

Fig. 120. *Geranium molle fasciatum.* a, fruit containing 6 single fruits; b, split into one group of 4 and one of 5; c, split into three groups containing 5—7 and 5 single fruits; d, one group has 16 and the other 5 divisions of fruit and stigmas; e, fruit with 33 divisions and stigmas (1895).

race[2] is remarkable from the fact that the stems have, as is well known, a sympodial structure (Fig. 119), Therefore the anomaly affects, as a rule, one member of the sympodium only; but sometimes it extends to two or more adjacent ones. These are again succeeded by atavistic members. Each part terminates in a flower.

[1] *Revue générale de botanique*, 1899, Vol. XI, p. 143.
[2] *Botanisch Jaarb. Gent*, 1894, p. 81; and 1897, p. 67.

In the fasciated parts these flowers are broadened and usually more or less divided, the fruit sometimes forming a flat structure (Fig. 120 e), with or without one or two lateral fruits in the same flower (Fig. 120 c and d). Often these latter are pentamerous. Lastly the whole fruit can be split into two or three nearly equal parts

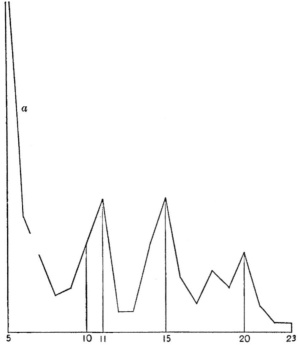

Fig. 121. *Geranium molle fasciatum.* Curve representing the number of sections of the fruit in the individual flowers of the sixth generation, June 1895; a, number of normal flowers far above 100; number of flowers with 6 to 23 stigmas, 120.

(Fig 120 b). In these various types of splitting there seems to be a tendency to the production of whorls of five, and the lateral flowers nearly always present this number.

My race began with a specimen found wild in 1888 and in the third and fourth generations produced 25 to

30% individuals with fasciations. In the two following generations they were much more abundant. In the sixth there were 65% fasciated specimens, in a culture of 220 plants, and these afforded me sufficient material for a statistical examination. For this purpose I collected, shortly before the ripening of the seed, all the aberrant flowers from a certain number of plants, and counted the number of the divisions of the fruit or that of the stigmas of 120 individual flowers taken at random. The figures which I obtained were as follows:

NUMBER OF
Stigmas	6	7	8	9	10	11	12	13	14	15	16	17	18	19	20	21	22	23
Flowers	13	9	4	5	10	15	2	2	10	15	6	3	7	5	9	3	1	1

I did not count the normal flowers, but their number far exceeded that of the whole group of the fasciated ones. In the curve in Fig. 121, in which the above series of figures is graphically displayed, the apex representing the atavists is therefore only formally indicated. The secondary peaks fall at 11 (10), 15 and 20 pistils, and the normal pentamery of the flowers is thus clearly repeated in these multiple figures.[1]

Besides illustrating this curious fact, the curve shows that low grades of fasciation are relatively rare and that atavistic and normal flowers constitute two distinct groups, although connected by intermediate stages.

Let us now summarize the conclusions we have come to in regard to fasciated eversporting varieties.

1. The races always consist of fasciated individuals and atavists.

2. The proportion of the former varies greatly, often amounting to only 40% or less, but not infrequently

---

[1] A further inquiry into this point is, in my opinion, urgently called for.

to more (*Geranium* and *Crepis* with 65% and 85%; *Celosia cristata*).

3. The fasciated individuals and the atavists are connected by transitional forms, which are, however, rare; and the statistical curves representing them have therefore two apices.

4. These proportions are to a large extent dependent on external conditions of life, which can transform atavists into fasciated plants and *vice versa*. This transformation obviously takes place during the plastic period in youth, before the character in question is actually developed.

5. The atavists, as well as the selected individuals, produce fasciated offspring, and often in proportions very little lower than those in which the selected individuals produce them. They may therefore be used for the perpetuation of the race.

6. Between the broadened specimens and the atavists there is no essential or fundamental difference, in spite of the great difference in their external forms.

IV. HERITABLE SPIRAL TORSIONS.
(Plate VI.)

§ 18. THE SPIRAL DISPOSITION OF THE LEAVES.

In the case of spiral torsion the difference between normal and abnormal individuals is far more striking than in that of fasciations. *Valeriana officinalis* is one of the best known and the most frequently figured instances (Fig. 122). Here the whole stem, instead of growing to a height of more than a meter, can be reduced to about a decimeter, and becomes more or less funnel-shaped. Low down the leaves are disposed spirally, but higher up, the spiral gradually becomes steeper, until, in the expanded upper part of the funnel all the leaves are directed to one side like a fan. The terminal inflorescence surpasses the highest lateral flowering branches very little or not at all.

By no means every stem of the spirally twisted plant manifests the anomaly. On the contrary, very few of them do as a rule. Since 1889, I have had a specimen which has gradually increased by runners and now covers an area of several square meters in the botanical garden of Amsterdam. It produces spiral torsions every year, but they are rare, and, as a rule, there are not more than two or three among several hundreds of normal stems.

The same rarity is seen in the inheritance of the anomaly, both in *Valeriana* and other species; as a rule, only a small proportion of the plants repeat the anomaly, even when the seeds have been saved on the most highly modified stems. Races which produce the anomaly abundautly have been very rare hitherto. They will be described in the following section (§ 19).

Fig. 122. *Valeriana officinalis.* An entire plant the stem of which attained the height of 10 centimeters only. Photographed from life in June 1900. Disposition of the leaves spiral below, unilateral above.

In consequence of this rarity the belief prevailed, up till about 10 years ago, that these torsions were not heritable. The general view was that they were induced by special external influences which operated immediately upon the individual every time the anomaly arose. At that time the experiments which I had made with a twisted

The Spiral Disposition of the Leaves.

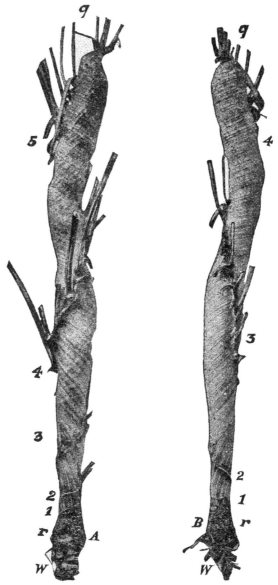

Fig. 123. *Dipsacus sylvestris torsus.* Two spirally twisted main stems, *w*, root; *r*, stem with the scars of the radical leaves and at 1—5 the windings of the spiral; *g*, stalk of the terminal inflorescence; A, leaf-spiral left hand, stem right hand; B, leaf spiral right hand, stem left. Stems thickened and hollow.

race of *Dipsacus sylvestris torsus,* during over thirteen years, together with a long series of further observations on the inheritance of this anomaly in other plants, have proved that this character is as heritable as other anomalies are. Plate VI gives a view of a culture of this race, reproduced from a photograph of one of my beds.

Real spiral torsion only occurs in those species which normally have a decussate or whorled disposition of the leaves. It consists in the substitution of a spiral arrangement for this. The leaves arise from an unbroken spiral, along which they are attached to one another more or less closely to their bases (Fig. 123). This close spiral is sometimes interrupted and normal internodes are intercalated in the twisted part. Not infrequently the torsion is limited to a greater or a lesser part of the stem (compare, e. g., below *Dianthus,* Fig. 129). Indeed no single stem is completely abnormal from the very beginning.

As might be expected the fusion of the base of the peduncles into a continuous band results in an inhibition of the longitudinal growth of the stem. The internodes cannot elongate normally, and as they strive to extend, they partly unwind the leaf spirals. In consequence the spiral becomes steeper and not infrequently unwinds so much in the upper parts of the stem, i. e., those parts which normally grow in length, as to become a straight line. When this occurs the leaves and their axillary buds arise in a longitudinal series on one side of the stem (Fig. 122). Obviously this can only be reached by the stem twisting itself in the opposite direction to that of the leaf-spiral (Fig. 123). Inside the twisted stem, if it is hollow, the diaphragms which normally occur at the nodes, do not exist as such, but are united together to a continuous screw-like band which corresponds exactly to the leaf-spiral on the outer side.

Twisted stems look as if they were inflated (Figs. 122 and 123), and are much thicker than the normal stems of the same species. Longitudinal growth has, so to speak, been changed into a tangential growth, as the course of the otherwise vertical ribs clearly shows in our figures. The longer the particular internodes on the normal individuals are, the broader are the corresponding parts on the twisted ones. In this way the funnel shape of the twisted *Valerian*, as well as other specific and local differences, are easily explained.

From this we see that a right-hand torsion of the stem (mounting in the direction of the movement of the hands of a clock) must be associated with a left-hand leaf-spiral and *vice versa* (Fig. 123).

The explanation here given was first suggested by BRAUN, and later demonstrated by KLEBAHN, by the microscopical examination of the top of the stem of a twisted *Galium*.[1] It can now easily be confirmed by every one on the material afforded by my heritable races.[2]

In *Dipsacus sylvestris torsus* the spiral arrangement of the leaves can be detected towards the end of the first summer in the heart of the rosette of radical leaves, with the naked eye, and without any damage to the plant. After germination and in the earlier stages the leaves are decussate (Fig. 124 A) in all the plants with very rare exceptions; it is not until later that this arrangement

---

[1] AL. BRAUN, *Monatsber. d. k. Akad. d. Wiss.*, Berlin, 1854, p 440. See *Bot. Zeitung*, 1873, p. 31; H. KLEBAHN, Ber. d. d. bot. Ges., Vol. VI, p. 346. See also *Ueber die Erblichkeit der Zwangsdrehungen*, same journal, Vol. VII, p. **291**.

[2] For the literature of the subject see *Monographie der Zwangsdrehungen*, Jahrb. f. wiss. Bot., Vol. XXIII, 1891.

is changed into the spiral one, and in different individuals this occurs at varying ages. If a transverse section of the plant is then examined under high power, at the level of the growing point, the spiral arrangement of the leaves can easily be seen. Taken later, i. e., during the elongation of the stem, such transverse sections have still more or less the same appearance (Fig. 124 B). The outer leaves of this figure have been cut through their lower parts where they are fused laterally with one another; their left-hand spiral can easily be followed

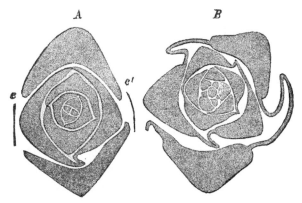

Fig. 124. *Dipsacus sylvestris torsus*. A, a transverse section through a seedling a little above the growing point showing the normal decussate arrangement of the leaves; c, c', the cotyledons; B, a transverse section through the still young point of the stem of a twisted individual with spiral arrangement of leaves.

in the figure. The subsequent leaves were still very young and were therefore cut across their upper free parts, but are nevertheless obviously arranged in a spiral. The three youngest leaves do not seem to form part of the spiral, but a trimerous whorl, such as very often occurs in the upper part of the stem of twisted specimens of *Dipsacus* If the angle of divergence of the leaves is measured it usually corresponds pretty closely to one of the ordinary types of leaf arrangement, (e. g., ⅜).

In the rosettes of radical leaves, where the internodes do not lengthen, the spiral disposition of the leaves does not involve any further disturbance. The leaves simply grow out and retain their original position. But when, in the second year, the young internodes begin to elongate, this cannot happen equally on all sides of the stem, because the line of attachment of the leaves acts as a check. In consequence of this the stem must twist and unwind the leaf-spiral, the angle of divergence between successive leaves becoming gradually smaller. The number of windings decreases, and on the other hand, the numbers of leaves on a single section of the spiral (i. e., from a given point on the stem to another vertically above it) increases, as our Fig. 123 clearly shows.

Spiral torsion is, therefore, a mechanical result of the loss of a single character, the decussate arrangement of the leaves. Once this is lost, the ancestral spiral disposition steps in, but now accompanied by peculiarities in the structure of the basal parts of the leaves, which on normal plants never occur independent of an arrangement in whorls, since they can only, so to speak, agree with the normal structure of plants with this arrangement. Therefore, in my opinion spiral torsion is due to a retrogressive transformation of the decussate arrangement of leaves.

That it is a mechanical result can be proved by experiment; all that is necessary is to remove the cause of the twisting at an early stage, by cutting through the leaf spiral. If this is done carefully, the general growth of the plant is not interfered with, but the torsion will be locally inhibited, or, more strictly speaking, does not appear at the place operated on. Thus a straight internode will become intercalated in the middle of an otherwise twisted stem. This can easily be seen in Fig. 125 at *a,* in the gaping wound and in the vertically ascending longitudinal ribs. Sometimes nature itself makes a simi-

Fig. 125. *Dipsacus sylvestris torsus.* The two leaves *a* and *b* of the twisted stem have been separated from one another in earliest youth by a cut; between them is seen the gaping split. The one leaf, *b,* is situated now about 2 centimeters higher than the other, *a.* Above *a* the stem has remained straight for more than 2 centimeters. Some of the lower leaves were removed before photographing in order to show the important parts more clearly.

lar experiment, and longitudinal growth tears the leafspiral in one or two places, without, however, making so deep a wound as results from an experiment. The result of such a natural experiment is a straight internode of normal length, for instance of a decimeter or more, in the middle of a twisted stem.

Like other anomalies spiral torsions are to a very large extent dependent on external conditions. Under unfavorable treatment the anomaly may be almost entirely absent, even from cultures from seeds which otherwise would give rise to a third or more of individuals with fine spiral torsions. The experiments which I have made on this point with *Dipsacus sylvestris torsus* and which have been confirmed by the results of my experiments with other species, seem to justify the following conclusion.[1]

The more favorable the conditions of life and the more vigorous therefore the growth, the greater will be the proportion of beautifully twisted plants in a given culture, and the more complete will be the torsions which are produced.

The most important of these conditions is the space given to the individual plants. They ought not to be shaded by one another nor touch one another. More than 20-25 plants should never be grown on a square meter. Grown thus they nevertheless stand in close contact in autumn, and it is obviously still better that no more than 10-15 plants should occupy a square meter.

Fewer torsions are produced on a given area when the plants are crowded than when the distances between them are greater. No useful purpose is therefore served by increasing the number of plants on the same bed. If nevertheless this is done the twisted individuals will be found almost exclusively along the edges of the bed.

The time of sowing the seed is a point of considerable

---

[1] *On Biastrepsis in Its Relation to Cultivation*, Annals of Botany, Vol. XIII, 1899, p. 395.

importance, inasmuch as it determines the length of the life of the plant up to the moment of the production of the stem. The longer this period continues under favorable circumstances the greater is the likelihood of the leaves becoming spirally arranged.

Sowings, made in summer or early autumn, which gave rise to stems in the next year, reduced the prospect of obtaining torsions almost to nil. On the other hand, autumn sowings which do not give rise to stems until the summer after the next, contain very large numbers of twisted specimens; so that if the seeds are sown late in the autumn the proportion of plants with a spiral arrangement of their leaves is even greater than amongst plants raised from seeds sown in the spring.

Little effect is produced upon the result of a culture by sowing in March or in April or even in the beginning of May. Also it does not matter much whether the sowing is carried out in the greenhouse in pans and the seedlings planted out later into the beds, or whether the seeds are sown where they are to grow. For various reasons I have for many years preferred the former method, as it is more convenient and safer, especially in dry springs.

A good loose soil with a strong manure rich in nitrates seems to be an essential condition. On unmanured sandy soil even the best seeds do not produce twisted individuals, and on hard or barren soil the proportion is considerably diminished.

It is possible to confine the cycle of life of *Dipsacus sylvestris torsus* within the limits of a year by sowing the seeds under favorable circumstances immediately after they are ripe. In this way a generation can be grown every year, and an annual twisted race might possibly be raised by selection. At present, however, the annual habit and the torsion are mutually exclusive, the anomaly being represented on the stems of such plants either not at all or as faint indications only.

§ 19. RARE SPIRAL TORSIONS.

From time to time spiral torsions are also found on wild and cultivated plants under conditions which make it impossible to make any other observations on the inheritance of the anomaly than that they occur relatively frequently on the several branches of the same plant or in more or less numerous examples in the same locality; or recur during the course of several years. They may be found in dozens in *Weigelia amabilis,* and are also well known in several species of *Galium.* In *Galium verum* and *G. Aparine* I have collected them in this neighborhood. *Equisetum* is also a well-known example, which deserves special mention as belonging to the vascular cryptograms as well as on account of its peculiar leaf-whorls. Our figure 126 is photographed from a stem which Dr. TH. WEEVERS found near Nymegen in the summer of 1900. Here it grew among several other instances of torsion in the same species. *Casuarina* also sometimes forms such anomalies on its branches; for instance, several occurred in 1897 in the botanical garden of Amsterdam (Fig. 127 a).

In the first chapter of this part we have seen how, as a result of the correlation between various abnormal types of leaf arrangement, the selection of tricotylous seedlings often leads to the discovery of spiral torsions in species from which otherwise they can be obtained only very rarely. As instances of such species I may

538   *Heritable Spiral Torsions.*

recall *Dracocephalum moldavicum* (Fig. 74, p. 369), *Asperula azurea, Centranthus macrosiphon* and especially *Mercurialis annua*. In the latter species these malformations appeared almost every year and often in consider-

Fig. 126. *Equisetum Telmateja.* Spiral torsion of an erect stem.

able numbers, in my tricotylous and syncotylous races (Fig. 96, p. 464).

Unless direct experiments in isolation have been made it is impossible in the majority of cases to be certain whether the race we are dealing with is a half or an intermediate race. Nevertheless the rarity of the anomaly in repeated sowings strongly indicates the former alternative. This was certainly the case in *Lupinus luteus*,

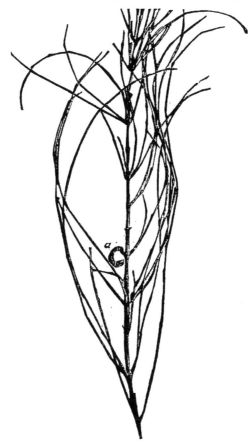

Fig. 127. *Casuarina quadrivalvis*. A branch with a lateral torsion, *a*. From the botanical garden in Amsterdam, 1897.

Silene noctiflora and other species which occasionally produced spiral torsions in my cultures. In *Valeriana officinalis* (Fig. 122, p. 528), which has already been mentioned, the proportion of anomalous individuals raised from seeds gathered from twisted stems was very small, and in *Rubia tinctorum,* of which I have had several beautiful instances of spiral torsions from the fields in Holland, they are never by any means common, although

Fig. 128. *Agrostemma Githago.* The calyx is open at one side and fused with the uppermost pair of leaves and transformed into a spiral. Culture of 1896.

they occur nearly every year.[1] In *Agrostemma Githago* I have often observed torsions in a race which I have cultivated during ten generations since 1888, but always

[1] *Eenige gevallen van Klemdraai by de Meekrap (Rubia tinctorum).* Botan. Jaarb. Gent, 1891, p. 74 and Plate IV.

in small numbers. Sometimes these were confined to the vegetative parts, but sometimes the calyx, together with the upper pair of leaves, was transformed into an open spiral (Fig. 128).

In order to determine the intensity of inheritance of this anomaly in such cases, I have conducted an experiment with *Dipsacus laciniatus*. This plant was of peculiar interest to me because it belonged to a genus which has been known for a long time to produce fine spiral torsions and to which also the *Dipsacus sylvestris* belongs, from which I have been able to isolate an eversporting variety which produces this anomaly in abundance. In *Dipsacus laciniatus*, on the other hand, the attempt to effect a similar isolation was unsuccessful. In this case an intermediate race was not present, but only a half race; for in spite of extensive cultures for two generations no more than slight local torsions were obtained.

A plant of *Dipsacus laciniatus* which I had seen in full bloom in the botanical garden at Groningen formed the starting point of this experiment. It was over two meters high with a perfectly straight stem which, however, had a little group of spirally arranged leaves at one of the upper nodes and here exhibited local twisting. Prof. J. W. MOLL kindly sent me seeds of this plant. From these I grew in the two following years 400 plants on a bed of about 16 square meters, that is about 25 per square meter, which may be regarded as giving ample room. The sowing took place in the beginning of April. In the summer of 1897 all the stems became straight; not a single one was twisted, as in *Dipsacus sylvestris* (Fig. 123, p. 529). Moreover, the arrangement of the leaves was decussate in all of them. Slight abnormalities in this arrangement certainly occurred, such as split leaves, trifoliate upper whorls, fusion of two pairs of leaves by a so-called line of tearing extending over the internode, a phenomenon which shows that the leaves in their early stages were fused in a spiral manner instead of in whorls. Besides these, there were eight plants, i. e., about 2%, which had a distinct local torsion at one of the upper nodes of the same development as in the parent plant. Their leaf spiral involved from 3-6 leaves. Only these plants were allowed to flower, and there was no other *Dipsacus laciniatus* in the garden. Five of them produced a sufficient quantity of seed.

The second generation of this culture occupied the years 1888 and 1889 and consisted of 435 plants which produced stems and were allowed considerably more space than their predecessors, 12 specimens being grown to each square meter. Nevertheless the result was the same as in the foregoing generation. All the plants formed erect and tall but straight stems, from 2-2½ meters high. Amongst them there were five specimens which had a somewhat larger local twisted part than occurred in the preceding generation, inasmuch as in these cases from 7-12 leaves were combined into an uninterrupted spiral. They arose from three of the five parents. Less pronounced torsions (at the nodes) which affected from 2-3 or rarely from 4-6 leaves, occurred fairly commonly in this year. I counted the percentages of 5, 5, 13, 13, and 28% of such individuals per parent.

A general advance had therefore taken place, brought about partly by selection and partly by the ample room given to the plants; but whilst the isolation of twisted specimens of *Dipsacus sylvestris torsus* resulted immediately in a proportion of 34% of fine, almost complete malformations, as soon as the culture was given sufficient room, nothing but small local twistings occurred in this case, in spite of highly favorable conditions of growth. From this I concluded that an eversporting variety analogous to my *Dipsacus sylvestris torsus* cannot be raised from these seeds.

§ 20. SPIRALLY TWISTED RACES.

The often mentioned race of *Dipsacus sylvestris torsus* (Figs. 123-125, pp. 529-534, and Plate VI) has now completed its ninth generation. It consists of about 40% of individuals with fine torsions and about 60% of atavists with, as a rule, decussate, but occasionally ternary, arrangement of the leaves. The atavists and the twisted plants with incomplete or interrupted spirals have always been destroyed before flowering. The completely twisted individuals, however, were left to be fertilized by one another through the agency of insects.

This very stringent process of selection has had the result of maintaining the race at a fairly constant level. During the first two generations I did not know the proper conditions and sowed the seed much too thick, and consequently only obtained twisted individuals at the edges of the beds, and even here in small numbers. But with the improvement of these conditions the proportion of twisted stems at once mounted to 34%, and the hereditary value has since remained about the same, although subject to fluctuation due to the greater or less favorable climatic conditions obtaining in successive years. But no advance could be discovered.

In order to exhibit graphically the result of this culture, which extended over 18 years, I shall give here a brief summary of the nine successive generations.

GENERATION	NUMBER OF PLANTS	PLANTS PER SQUARE METER	PROPORTION OF TWISTED MAIN STEMS
1st. 1884–1885	—	—	—
2d. 1886–1887	1643	50	0.1
3d. 1888–1889	1616	35	4
4th. 1891–1892	107	25	34
5th. 1893–1894	45	22	10 20[1]
6th. 1895–1896	33	8	42
7th. 1897–1898	70	16	46
8th. 1899–1900	1295	22	32
9th. 1901–1902	492	22	41

The mean of the six generations was therefore about 35%.

We see from this table that the continued isolation and selection of the finest torsions maintained the race at its level, but did not improve it to any considerable extent.

Two points about this result should be noted. In the first place the percentage figures given do not denote the proportion of twisted individuals, but relate solely to those with torsions on the main stem. In some experiments, however, instead of pulling up the atavists, I only removed all their inflorescences before the flowers opened, and by this means I found that several, and often many of them, had the power of developing more or less fine local torsions on their lateral branches. In 1902 I made as accurate an account of these as possible, and found 71 of them or 14%. Of twisted individuals there were, therefore, 41+14=55% altogether, i. e., somewhat more than half the individuals produced torsions. It seems of some interest to note that this percentage is the same

[1] The percentage figures for the 5th, 6th and 7th generations are hardly decisive on account of the small size of the cultures.

as has been observed elsewhere in intermediate races, and especially in the tricotyls (see page 439).

Secondly, I call the attention of the reader to the remark which I made in the previous chapter (§ 16, p. 511) with regard to selection. Even as in the case of fasciation, a double selection, by the visible characters of the seed-parents and by their hereditary value, cannot here be carried out under the ordinary conditions of the experiments. The most valuable character to select by, the hereditary coefficient, fails, and in consequence of this the improvement of the race, which in the case of the tricotyls so soon resulted in a proportion of from 70-90%, obviously cannot be carried to its full height here. Nothing less than a lucky chance or the conducting of the experiment on a much larger scale can bring this about. In the eighth generation I compared the hereditary coefficients of ten parents. They ranged between 10 and 55%; but as only from 100-140 offspring of each parent could be compared, the percentage numbers did not seem to me to be of much importance from the point of view of selection. In the 9th generation I saved the seeds from over 100 very finely twisted individuals, in the hope of still being able to carry out an experiment in selection by hereditary coefficients.

I shall now give a detailed description of the whole experiment.[1] The starting point of my race was formed by two individuals with a twisted main stem which flowered in 1885 in a culture sown in 1884 in my garden. All the remaining plants were destroyed before they flowered.

In 1886 I obtained a second generation from their seeds. As I have already mentioned, I was not at that

[1] *Annals of Botany*, Vol. XIII, No. LI, Sept. 1899, p. 401.

time familiar with the most favorable conditions of this culture, and obtained only two twisted individuals amongst nearly 1650, probably for this reason alone. These two flowered in isolation and set an abundance of seed. The third generation (1888-1889) raised from seeds of the second, produced 67 twisted stems in a culture of about the same extent, i. e., about 4%. Here again the seed-parents flowered in isolation.

The fourth generation was sown partly in 1890, and produced as many as 10% twisted individuals which however could not be used for the continuation of the race. The rest of the seed was sown in 1891 with a better knowledge of the conditions, whereby the proportion of twisted individuals rose to 34%, a value which the later generations on the average maintained, without surpassing it to any considerable extent.

The improvement in the method of culture consisted essentially in providing the young plants from their very earliest stages with more room. In the two previous generations about 50 specimens were grown per square meter In this generation, however, the number was reduced to about 25 by the removal of the superfluous ones in June, as soon as the plants began to touch one another.

The seed had been sown in May in the beds. At the beginning of October I noticed, amongst about 100 individuals, 6 with a spiral arrangement of leaves in the heart of a rosette. In the beginning of November I was able to remove more than half of the plants as undoubted atavists, and when, towards the end of May 1892, the stems grew up vigorously, they were finally examined and recorded. The result, together with that obtained in November, i. e., for the whole culture, is as follows:

Spirally Twisted Races.

```
Twisted stems . . . . . . . . . . . .  37  =34%
Stems with ternary whorls . . . . . . .  12  =11%
Atavists with decussate arrangement of leaves  58  =55%
                                        ___
                                  Total 107
```

It is very important to notice that the number of twisted stems increased as the result of the greater distances between the plants, not only relatively (i e., per 100 individuals), but absolutely (i. e., per square meter). In the third generation there were 50 plants per square meter, of which 4% (1-7%, from 1-4 specimens) were spirally twisted. In the fourth generation, however, there were 37 twisted individuals on 4 square meters or 9 per square meter.

I selected the seven plants with the finest torsions on this bed as seed-parents. They all, moreover, presented local torsions on several branches. I isolated them before they flowered.

The fifth generation (1893-1894) was not so successful. Only 25% with twisted stems were produced. The seed to produce this generation had not as hitherto been sown on the beds but in seed-pans in the greenhouse attached to my laboratory. This method has since proved more convenient and safer, and has been employed in the two generations which follow.

The seed saved in September 1892 was sown in the middle of March 1893. In the middle of May the best seedlings were planted separately in 10 centimeter pots with good, richly manured garden soil, and planted in the bed in the middle of May, at the same distance apart as were the plants in the preceding generation (22 plants per square meter). In the following year (1894), all the stems developed and were recorded. The result was as follows:

	A	B	% A	% B
Twisted main stems	5	2	20	10
Ternary main stems	1	1	4	5
Decussate main stems	19	17	76	85
Total	25	20		

A and B are two groups which had been produced from seeds separately harvested from two seed-parents of 1892.

The extent of the culture is, as we see, too small to allow of an exact determination of percentage values.

Seed was saved in the autumn of 1894 from the four best plants which had been isolated from the rest before they flowered.

The sixth generation (1895-1896) again produced a better result, viz., 42% twisted main stems, a result due in part at least to the increase in the distance between the plants in the culture, which was in other respects the same as in the previous year. The seed of 1894 was sown in the middle of March in pans in the greenhouse. The seedlings were transplanted to pots in April and into the bed in May Altogether there were only 33 plants on 4 square meters, that is about 8 per square meter. In October I found 14 rosettes with a spiral arrangement of their inner leaves, whilst seven were ternary and 12 decussate, that is 42% spiral, 21% ternary, and 36% decussate. This result was confirmed in May and the decussate and ternary individuals were thrown away. The six individuals with the finest torsions were selected as seed-parents and isolated before flowering.

The seventh generation (1897-1898) was grown in practically the same way. Seeds of 1896 were sown in the greenhouse in May 1897, were transferred to pots later, and to the beds at the beginning of July at the rate of 16 individuals per square meter. Altogether there were 70 plants.

A record made at the end of May 1898 gave the following result:

Twisted stems	32	=46%
Ternary stems	21	=30%
Decussate stems	17	=24%
Total	70	

In the eighth generation the plants were treated in exactly the same way as in the preceding one. As already mentioned, the offspring of 10 distinct parents were compared. Moreover for each parent the seeds of the terminal inflorescence were compared with those of the inflorescences on the main branches. For the latter determination I selected from each parent 4-8 of the strongest branches which arose from that part of the middle of the stem which had undergone most torsion. They corresponded, therefore, with the period of maximum activity of the stem,[1] and confirmed this by the fact that small local torsions appeared on them. From the seed of the primary inflorescences I raised 645 plants with 31%, from those of the secondary inflorescences 650 individuals with 34% twisted main stems. There was therefore no difference of any significance.

The ninth generation was composed partly of offspring of twisted individuals and partly of those of atavists. The method of culture employed was the same as before; but I shall revert to the result of it in the next section.

Besides *Dipsacus sylvestris* I have grown two other constant races with an abundance of fine torsions which must

[1] See T. TAMMES, *Die Periodicität morphologischer Erscheinungen bei den Pflanzen,* Kon. Akad. v. Wet., Amsterdam, 1903.

be regarded as intermediate races. The first is *Dianthus barbatus torsus*. I obtained a fine twisted branch of this form in the autumn of 1894 from Mr. J. ENSINK in Ruurlo. The torsion was similar to that represented in Fig. 129, but the fruits were ripe and full of seed. I could not, however, sow the seed until the spring of 1897.

Fig. 129. *Dianthus barbatus torsus.* Twisted Sweet William The stem has a decussate arrangement of leaves below and spiral above. June 1900.

This species does not flower till the second year, 10-20 shoots being produced on each individual. I made a culture of about 300 individuals, which were planted out at distances of 20 centimeters apart. Plants began to flower in June of the second year (1898); the great majority of their stems were normally decussate; whilst several had ternary whorls along the whole length, or in other cases, only in the upper parts of the stems. Others again were twisted and manifested a spiral arrangement of the leaves. The twisting, too, seemed to prefer the upper half of the stem, just underneath the inflorescence. Sometimes however the whole stem, or a great part of it, was twisted or the torsion extended into the inflorescence. The maximum development of the anomaly largely corresponded, so far as I could determine, with the period of maximum growth, as in *Dipsacus*. Altogether I counted amongst somewhat over 4600 normal stems, 53 more or less distinctly twisted ones in all degrees of development, 33 torsions extending over a greater part of the shoot and

Fig. 139. The dark-eyed Viscaria (*Viscaria oculata*) with spiral torsion of the main stem. *a*, the terminal flower of this main stem; the flowerstalk, first bent downwards by the torsion, has subsequently turned upwards again. *b*, *c*, the normal straight portion below the twist; *d*, a local interruption of the torsion (1900).

20 over smaller portions. This means a proportion of about 1% of the total number of the stems. The 33 best torsions occurred on ten plants of which five had 12, 6, 5, 3 and 2 twisted shoots, and the remaining five, one such anomaly each. The 20 smaller abnormalities were distributed at random over the plants in such a way that about 10% of the individuals had developed the anomaly. It is worth remarking that the most abnormal individuals occurred chiefly on the most sunny edge of the bed.

The selected individuals mentioned were not allowed to flower before all the remaining plants had been removed. In the spring of 1899 I only sowed the seeds of the plant with 12 torsions. I planted out 180 individuals, at the same distances apart and also in other respects under exactly similar conditions as those which obtained in the preceding generation. They flowered in the summer of 1900. At the beginning of the flowering period I had 2246 stems, in 1246 of which the arrangement of the leaves was decussate, whilst in 414 it was ternary. Further, there were 227 stems which were twisted over more than half their length and 359 which were twisted in the upper part only, i. e., 26% twisted shoots as against 1% in the preceding generation, on the average 3 torsions per plant. According to a rough estimate well over half the individuals had produced one or more twisted stems.

Two generations, therefore, had sufficed for the isolation of a genuine intermediate race from an original insect-fertilized sample of seed.

The other race referred to above occurred in *Viscaria oculata* (*Lychnis Coeli-rosa*), a favorite garden plant. The plant is annual and its culture easy. In twisting, its stems become much shortened, and the plants become low, but they flowered freely (Fig. 130). In 1897 in a culture grown for another purpose, I found a twisted plant like the one figured and saved its seed separately; its flowers had not been protected.

From this seed I raised in 1898 a culture of 300 plants of which 259 were normal, whilst 40 exhibited torsions in the main stem and 28 others in one or several of the lateral branches, i. e., a proportion of 21% twisted individuals. I only allowed the 21 finest specimens of these to flower and set seed; and of these I only sowed the seeds of the one finest twisted plant. In the following year, 1899, I raised from it 385 individuals of which 137 or about 35% showed torsions. Here also, as in the case of *Dipsacus* and *Dianthus,* many individuals exhibited a ternary arrangement in the whorls. I counted about 100 of these, which therefore constituted about one-quarter of the whole culture.

In 1900 I grew the fourth generation, but on a smaller scale. Torsions were as abundant as before, and amongst their number was the plant represented in Fig. 130.

There is little doubt that similar intermediate races could be raised by an isolation of the spirally twisted individuals of several other species. And the best chance would obviously be given by those which frequently give rise to this anomaly without selection. Thus, for instance, *Gypsophila paniculata, Urtica urens* (of which I have already cultivated two generations with success), and perhaps also *Scabiosa atropurpurea.* On the other hand, as stated in the previous section, my sowings of the seeds of spirally twisted examples of *Valeriana officinalis, Saponaria officinalis, Galium Aparine*[1] and others, have offered no such prospects.

[1] *Bydragen tot de leer van den klemdraai.* Botanisch Jaarboek, Gent, IV, 1892, p. 154, Pl. XV.

The existence of constant intermediate races or ever-sporting varieties exhibiting spiral torsion seems to me to be conclusively demonstrated by the experiments described. They were found by chance and then easily extracted by isolation, and induced to produce a proportion of 30-40% twisted individuals. This result could be made permanent by subsequent cultivation, but could not be increased by a continuation of the selection.

§ 21. THE SIGNIFICANCE OF THE ATAVISTS.

In spite of every attention and in spite of repeated and careful selection spirally twisted races will continue to produce numerous atavists. These are either decussate individuals of the normal structure of the species, or they have ternary leaf whorls, as is also often seen in twisted individuals above the torsions.

I propose now to examine more closely the nature of, and especially the mode of, inheritance in these atavists, as I have done in the case of the fasciations. The essential point about them is that the atavists do not, as we may express it, depart from the twisted race; but can be used almost as safely for its continuation, as the twisted individuals.[1]

A fact of the highest importance is that there is no sharp limit between the atavists and the twisted individuals. Of course the difference between the tall erect stems and the short stunted anomalous ones is most striking, as our Plate VI shows. It is quite clear that this is not a case of the fluctuating variation of a single

---

[1] In Stocks the double specimens are sterile, as is well known, and the variety has to be propagated by means of the seeds of the single ones. Under cultivation in the field as a rule about 50% of "single" atavists are produced.

character, but that two antagonistic factors are at work, the one excluding the other, although never completely. Even the stems with the most pronounced torsions produce branches, most of which revert to the decussate arrangement of leaves. It never occurs that this character is completely excluded from the whole plant. Conversely, as we have already seen, atavistic individuals with perfectly erect main stems and with a decussate or ternary arrangement of the leaves frequently exhibit torsions in their lateral branches. In 1887 I cut half the atavists of my culture of *Dipsacus sylvestris torsus* close down to the ground; they shot out from the base of the stem. In this way I obtained about 2000 branches of the second and third order. Amongst them 235 had a slight but quite definite torsion and 26 had a small many-leaved spiral. In the third generation I repeated the experiment with the same result; and moreover observed torsions on the lateral branches of some atavists which had been allowed to remain on the beds until they were just about to flower.

Other abnormalities in the arrangement of leaves also betray the real nature of the atavists. First, there are the individuals with ternary whorls. Such whorls do not occur in the early stages of the plant, and tricotylous seedlings are even very rare. At first the arrangement of the leaves is always decussate, and it is not until late summer or autumn, at the time when other specimens begin to produce their leaves in a spiral, that the decussate arrangement gives place to a ternary one. But when this has once appeared it usually remains on the stem up to the terminal flowerhead. Such plants look quite normal, and especially their leaves do not produce those forkings of the mid-rib which are so common in the

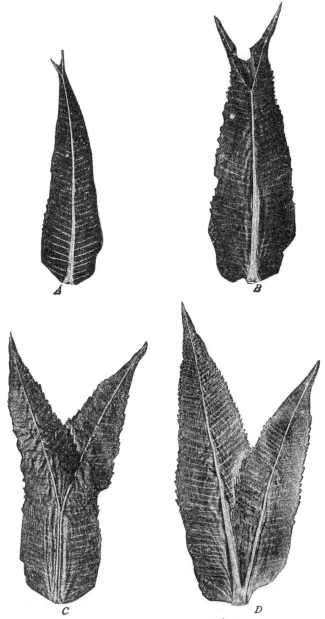

Fig. 131. *Dipsacus sylvestris torsus*. Split leaves from the decussate atavists of the eighth generation. A-D, increasing degrees of splitting.

decussate individuals. On the lateral branches of these whorled stems, on the other hand, these splittings, as well as local torsions and other anomalies are by no means rare.

The relation of the ternary whorl to the spiral arrangement demands closer investigation. Perhaps the former is to be regarded as a lower stage of the anomaly; and this view is supported by the fact that highly twisted individuals which have one or more straight internodes above the torsion, usually exhibit ternary whorls between them.

The decussate individuals often produce leaves with split midribs (Fig. 131 A B), and in all degrees of splitting from leaves with two tips to leaves split down to the base.[1] Sometimes they even produce one or two ternary whorls in the upper parts of the plant. The range of variation in these splittings has been dealt with by DELPINO,[2] and material for a complete demonstration of these forms may be furnished by every generation of my race.

In the third generation I left three atavists on the bed until shortly before they flowered. They all bore some split leaves in the upper part of their stem; in the ninth generation I observed the same anomaly on the main stems of 172 of the 200 decussate atavists, that is to say, in about 80% of the whole group. Several of the remainder exhibited the anomaly on the lateral branches.

From the axil of a split leaf there usually arises, according to my observations, a single shoot; but sometimes two of them, or a single broad flattened one with two

---

[1] *Ber. d. d. bot. Ges.*, Vol. VII, 1889, p. 296.
[2] F. DELPINO, *Teoria generale della Fillotassi*. Atti della R. Universita di Genova, IV, Parte II, 1883.

inflorescences at the top. In this respect also, my results confirm those of DELPINO.

External conditions exert a great influence on these secondary anomalies just as they do on the main torsions. The more favorable the conditions, the rarer are the individuals, all of whose branches and leaves are normal. *Obviously this fact suggests that the factor for the malformation must be present in all of them.*

Brief mention should also be made of the so-called local torsions. They occur occasionally in all twisted races. Fig. 132 represents an instance of them in *Valeriana officinalis*. It flowered in the same year and on the same bed as the completely twisted stem shown in Fig. 122 (p. 528).

In *Dipsacus laciniatus* the malformation was confined to these, and in *Dipsacus sylvestris torsus* I observed them, under very special conditions of culture, on erect and otherwise decussate stems; but on the lateral branches of twisted individuals, especially on the strongest ones, such as those which arise from the axils of the radical leaves or from the middle of the twisted stems, they are always seen to be most profusely produced.

As in the case of fasciations, forms intermediate between these local torsions and the normal arrangement of the leaves, are relatively rare. This is not true, however, if the ternary whorls, split leaves, and the local torsions on the lateral branches are included, but only if we confine our attention to the torsions on the main stems. Unfortunately the exact measurement of the part bearing leaves in a spiral is a matter of considerable difficulty; inasmuch as the spiral begins inside the rosette at a time when the oldest leaves have already rotted off and disappeared. Nevertheless I have recorded the beginning of the spiral accurately in some cases by marking and counting the pairs of leaves, as they made their appearance, from the moment of germination. But I have not yet plotted a curve from data collected in this extremely laborious way. On the other hand the number of straight internodes above the torsion can easily be

Fig 132. *Valeriana officinalis.* A stem with a local torsion, from the same culture as Fig. 122. (June 1900.)

counted; and their number obviously varies inversely with the extent of the twisted part.

In 1900, when my eighth generation consisted of 1295 flowering plants, I recorded the number of straight internodes above the torsion in every twisted stem. Omitting the long stalk of the terminal inflorescence, which is always present, I obtained the following series.

	PLANTS WITHOUT TORSION	WITH 0–6 STRAIGHT INTERNODES ABOVE THE TORSION						
Straight internodes		6	5	4	3	2	1	0
Individuals	900	2	3	1	2	40	200	148

The curve constructed from these figures is obviously one with two peaks, and essentially the same as the corresponding curve for fasciations seen in Fig. 118 (p.521). Torsions with two, with one, and without any straight internodes above the torsions are far the commonest. Smaller torsions only occurred in nine cases among 1295, that is less than 1% of all the individuals, or in about 2% of the twisted individuals. I repeated the same examinations in 1902; the intermediate forms were somewhat more numerous and reached a proportion of about 7% in 492 individuals. The form of the curve, was, however, not essentially modified.

In order to determine the hereditary coefficients of the atavists I instituted an experiment in the eighth and ninth generation of my race. In July 1900, I had some highly twisted plants, some completely decussate ones, and some with ternary whorls only, on the same bed. Before the flowering period I reduced the inflorescences to the required number and afterwards insured pure fertilization in the following way. All the inflorescences were enclosed in parchment bags, and the bags were taken off the individuals falling into one group, for several hours one day out of three for each group. The humble bees flying about could thus fertilize only twisted individuals on some days, on others only decussate, and on still others only ternary ones. This practice was continued until all the flowers were gone. The seeds were saved separately from each plant. In the spring of 1901 I sowed the seeds of two decussate, three ternary and four twisted seed-parents and recorded the main stems produced by each parent in 1902, just before the flowering period; but since the numbers from the individual seed-parents of each group did not differ to any considerable extent amongst themselves, I shall give the total result only.

SEED-PARENTS	OFFSPRING, IN PERCENTAGES.			NUMBER
	DECUSSATE	TERNARY	TWISTED	
Decussate	48	8	44	201
Ternary	39	24	37	136
Twisted	45	14	41	155
			Total	492

We see that the offspring of the atavists produced just as large a proportion of twisted stems as the offspring of the twisted individuals. On the other hand the choice of decussate or of ternary atavists seems to have exerted some influence in the direction of ternary offspring.

The twisted stems of *Dipsacus* are partly right-hand and partly left-hand spirals; and about equal numbers of the two sorts are always found.[1]

A very curious question thus presents itself, viz., whether this equilibrium can be disturbed by selection and whether the balance can be upset in a given direction.[2] According to the view laid down in this book, that it is not true that with selection all things are possible, we might expect that this would not be a fixable character and that selection would have no influence on

[1] *Annals of Botany,* loc. cit., p. 404.

[2] R. M. YERKES, *Variation in the Fiddler Crab, Gelasimus pugilator,* Proceed. Amer. Ac. of Arts and Sciences, 1901, Vol. XXXVI, No. 24, p. 417. On page 441 this author states that right- and left-handed animals occur in approximately equal numbers.

it. For this reason I allowed plants with a right-hand spiral only to flower in the seventh generation; and in the following one also, the spiral of all the chosen seed-parents was right-handed. The result was as follows:

	LEAF-SPIRAL	
	RIGHT-HAND	LEFT-HAND
8th generation	205	215 individuals
9th generation	40	24 individuals

These constant but eversporting twisted varieties offer favorable material for attempting to transfer the malformation to related species by means of hybridization. At present, however, only a single attempt of this kind has been made, viz., one which was communicated to me by Prof. LE MONNIER of Nancy.[1]

He sent me two twisted stems of *Dipsacus fullonum* which exhibited a torsion as pronounced and as complete as the best instances of my race (Fig. 123 *a* and *b*, p. 529); and which owed this malformation to a cross between the species in question and my *Dipsacus sylvestris torsus*, which Mr. LE MONNIER had cultivated on a large scale for many years.

In 1896 my race flowered at the same time as the normal *D. fullonum* in the botanical garden at Nancy, at a distance of 100 meters apart but in great quantities of individuals. Pollen could easily be transferred by insects from one bed to the other. From the seeds of these *Dipsacus fullonum* there arose besides numerous normal plants three with twisted stems; one of them had upwardly directed bracts in the involucre, a character of *D. sylvestris* which distinguishes it from *D. fullonum*. The investigation of this important question has, however, not been continued.

[1] *Journ. Roy. Hortic. Soc.*, 1900, Vol. XXIV, p. 69.

The facts described in this and the preceding sections seem to me to furnish sufficient proof that the twisted eversporting varieties behave in exactly the same way as the corresponding fasciated races; and that, in both cases, atavism is merely a morphological phenomenon and not a real deviation from the race.

PART III.

THE RELATIONS OF THE MUTATION THEORY TO OTHER BRANCHES OF INQUIRY.

I. THE CONCEPTION OF SPECIES ACCORDING TO THE THEORY OF MUTATION.

§ 1. SYSTEMATIC BIOLOGY AND THE THEORY OF MUTATION.

In every case in which we were able to obtain a deeper insight into the nature of the hereditary character of an organism by direct observation or experiment we have found this character to be of a compound nature. No plant transmits its peculiarities to its offspring as an inseparable whole, as has been the general view until now. On the contrary we have described a long series of phenomena in which a single character or a smaller or larger group of them can be separated from the rest and behave in an entirely different way. When new species or varieties originate, it is not the whole nature of the organism that is changed; on the contrary everything remains in a state of rest except at one or two points, and it is only to the changes of these points that all the improvement is due. In hybridization the two types which sexually unite are always alike in the vast majority of their characters, and the differences between them are limited to a few definite units, which, in the simplest cases, can be dealt with numerically.

The *analysis of organisms,* therefore, leads us to the hypothesis of units, which are in many respects analogous to the molecules of the chemist. They are, however, of a much more complicated structure and have arisen in a historical way. They cannot be isolated and then subjected to experiment like chemical bodies; we can only investigate them by studying the relation of closely allied species and varieties, i. e., forms in which a definite unit, or several of them, are present in one plant and absent in the other. For this reason our investigations are, for the present at any rate, confined to the units which have arisen most recently.

But even as it is the business of comparative science in general, first, to apply the conclusions derived directly from the facts, to cases that have not been themselves observed and then to extend them gradually further and further, it is our duty and our right to test the applicability of our conclusions as thoroughly and as widely as possible.

Therefore we have now to face the question whether the theory of the origin of species by mutation and the theory that hereditary characters are composed of elementary units are in harmony with the theoretical conceptions to which systematic science on the one hand and embryology on the other have given rise. If it can be shown that the mutation theory satisfies the demands of these sciences better than the present form of the theory of selection, its justification as a theory of the nature of inheritance will, in my opinion, be placed on a sure foundation.

For this reason I shall devote the last part of this work to general considerations of this kind. In doing so I leave the safe ground of facts and venture into a region in which I can no longer mainly depend on my own experience. But experimental inquiry must derive its problems from this more general aspect of the questions; and it does not by any means seem superfluous to ask ourselves from time to time what has already been achieved and what is to be done in the immediate future. In doing so I shall have to keep a close rein on myself and, whenever possible, conform to the opinions of recognized authorities, limiting my own views to such points as may throw light on the relation between those opinions and the theory of mutation. I shall try to avoid questions of minor importance or concerning uncertain or subordinate points; the literature of the subject has long since grown so prodigiously that it is no longer possible to keep pace with it.

My only object is to demonstrate the applicability of the theory of mutation to the main conclusions of the doctrine of evolution; and only to do this briefly and sometimes no more than in outline. New theories and new hypotheses I shall not have to introduce, the more so, as I am convinced that the doctrine of mutation will lead everywhere to a simplification and a clearer conception of the problems. The prospect of recognition of a theory rests on the one hand on its empirical foundation, and on the other on its suggestiveness and the number of facts which it explains. Therefore the consideration of this latter point will now be my task. I shall devote the several chapters of this part to the various problems involved.

§ 2. PROGRESSIVE, RETROGRESSIVE AND DEGRESSIVE MUTATIONS.

Progress in organic nature consists essentially in increase of differentiation. The peculiarities which go to make up the individual character of the species become more numerous. Every more highly organized being has, as a rule, more of them than its ancestors of long ago had. In applying this principle to the doctrine of elementary characters we see at once that the number of these units must increase with increasing differentiation; or, conversely, that the degree of differentiation is ultimately determined by the number of elementary characters. Whenever a new unit is added to those already existing, differentiation advances a step forward. If it were possible to count the units we should have a measure of the degree of differentiation of all organisms.

Obviously the individual steps are only small ones, at the present time at least; and any single one of them can hardly effect a noticeable increase in differentiation. At any rate we have at present no means of so exactly measuring the degree of differentiation, since we cannot estimate the possible influence of a single unit on a complex built up of thousands of them. Only groups of units produce clear and obvious differences in the degree of organization; but within the limits of a small genus or of a multiform collective species the several types seem to us to be almost always equivalent.

The individual steps into which, according to this view, the process of gradual differentiation can be analyzed, we propose to call mutations; and since they constitute an advance, progressive mutations. Each of them contributes a new character to the complex of hereditary qualities already present.

Such a new character need not, however, become visible as soon as it arises, since we are not dealing solely with external qualities but with the internal factors to which they owe their appearance. Even as the germ contains large numbers of qualities awaiting development, so we can imagine that a new character remains latent for some time after its first origin, its phylogenetic birth, if I may so express it, and does not become active until after the lapse of a lesser or greater period of time.

According to this view, every progressive mutation is fundamentally a double process, and consists in the production of a new internal factor and in its activation. Both may sometimes happen at the same time, but this is not necessary. It is therefore desirable to apply separate names to the two processes; the internal one I have called premutation, the externally visible one a mutation *sensu stricto*.

The premutation is therefore of a hypothetical, the mutation, however, of an empirical nature.

It further follows from this that an internal factor does not of itself lead to the origin of an external change. As in ontogeny so also in phylogeny an elementary character can be sometimes active, but at other times latent or inactive. If a new character emerges from its original latent conditions and becomes active, we call the process a progressive mutation; conversely we can denote its return from the active to the latent condition as a retrogressive mutation.

The experience of gardeners and of the systematists with the smaller species and varieties teaches that retrogressive mutations of this kind are common phenomena. Almost any character may disappear. This applies not only to the superficial characters, such as color, hairs, or thorns, but also to those deeper ones which affect the inner organization of the plant, such as the decussate arrangement of leaves, and even the symmetry of the organism. Spiral torsion and peloria show how profound an alteration in the appearance of a plant or in the structure

572 Species According to the Theory of Mutation.

of a flower may be brought about even by only one of these elementary characters becoming inactive.

Retrogressive mutations give the impression of something being lost; some character or other disappears from the picture. But everything seems to point to the con-

Fig. 133. *Castanea vesca.* Abnormally leaved catkins. A, with two leaves; B and C, with one leaf each. C has also a lateral twig, *a* male, *b* female flowers; *c* normal leaf. Apeldoorn, 1896. Collected by Dr. P. F. ABBINK-SPAINK.

Fig. 134. *Mercurialis annua.* A sprig of the male plant with stray fruits on the long thin ears. On the female plants the fruits are inserted on short stalks in the axils of the leaves.

clusion that, in the vast majority of cases at least, this loss is only an external one; and that the factor remains in the inner organization of the plant, in an inactive state. This view is especially supported by those cases in which a systematic character which has become latent is occasionally manifested as an anomaly; as for instance the appearance of female flowers on male specimens of diœcions species (Fig. 134), or of leaves on normally leafless inflorescences (Fig. 133).

Two races, which only differ in the latency or activity of a single factor, therefore possess the same number of elementary units in their internal organization. Obviously the relation between them is different from that between two races, of which one has arisen from the other by the formation of a new factor; in which case there is a difference of one, in the number of units, between the two.

But before we examine this relation more closely we must face the question whether the active and the inactive states are the only ones in which an internal factor can occur. Theoretically this is obviously not necessarily the case, for we can easily imagine various degrees of activity between the two extremes, and as a matter of fact, experience shows that these intermediate stages do actually occur. We have described them above (p. 20) as semi-latent; and have given the name of middle races to those which possess such semi-latent characters. Of these there are two types, which we frequently meet both in nature and in our cultures, half races and intermediate races, or eversporting varieties. In both of them the semi-latent quality is associated in such a way with some active character that the two cannot be manifested at the same time. They exclude one another, if we may so express it, and so constitute a mutually vicarious pair. Trifoliate and quinquefoliate leaves of clover, tricotylous seedlings or split cotyledons and dicotylous ones, normal and peloric flowers, cylindrical and fasciated stems, ordinary and petaloid stamens, constitute such pairs. The same leaf cannot be at once trifoliate and quinquefoliate and so on; in a word, an organ cannot be both normal and abnormal.

These vicarious pairs of characters are the sources of a great variability inasmuch as the anomaly can appear in all degrees of development. In such cases the individuals of a group are not ranged round a mean in respect of their external qualities, as with ordinary fluctuating or oscillating variability, but between two types which are often widely separated and more or less antithetic to one another. They have the appearance of being inconstant; and races and varieties of this kind are usually so described, but this is only true in the sense that the range of forms which they present is a very wide one; and, moreover, is ditypic or dimorphic. But it would not be true in the sense that any individual could transgress the boundaries of this range and found a new race. In this sense the so-called inconstant races are just as stable as the best constant species and varieties.

The difference between half and middle races lies solely in the difference between the mutual relation of the members of the vicarious pair in the two cases. If, under ordinary conditions, and in the absence of selection, one of them predominates over the other to a very large extent, the race is, so to speak, unilateral and is called a half race (e. g., Fig. 135). But if, under similar circumstances, neither of them predominates but an equilibrium is maintained, we have an intermediate race (e. g., Fig. 27 of the first volume, page 138). In the case of tricotyls and syncotyls the half race rarely contains more than a very few anomalous individuals, in the absence of selection; whilst the intermediate race consists as a rule about half of normal, and the other half of tricotylous or syncotylous individuals. The seeds, however, of both types give the same proportion of tricotyls or syncotyls even when self-fertilization has been insured.

If we wish to elaborate terminologies still further, the term semi-latent may be limited to the anomalous character of the half race, and the character of the middle race may be described as semi-active (see p. 21). We can then distinguish four conditions of one and the same factor; the active, the latent, the semi-active and the semi-latent. This classification may suffice for the time; and at any rate it can be said that it is in accord, so far as my experience goes, with the facts known at present.

Such a factor cannot be transferred at will from one of these conditions to the other, either by selection or by any other means, at any rate in the present state of our knowledge. Such a transition is only effected by combinations of causes of which we still know nothing, or as we say by chance. Moreover the transitions, so far as we can observe them, are not slow or gradual, but take place suddenly. The new race appears on the scene at once and unexpectedly, as in the case of the peloric *Linaria*. Sudden transitions of this kind are exactly what we call mutations; and to distinguish them from the progressive and retrogressive types, we may refer to them as *degressive mutations*.

Fig. 135. *Papaver commutatum polycephalum.* The same anomaly which occurs in *P. somniferum* as a middle race (Vol. 1, Fig. 27, p. 138) occurs here as a half race, manifesting the character very rarely and only to a small extent.

Every mutation therefore consists fundamentally in

the transposition of an internal character; from being latent it becomes active; from semi-latent, semi-active; and so on. If new factors are becoming active for the first time after having been latent through a shorter or longer series of ancestors, we speak of *progressive mutations*. If the active characters again become latent, the process is a *retrogressive* one. In all other cases it is *degressive*.

The phenomena of hybridization find a ready explanation in the principles derived, in the first part of this volume, from our consideration of the origin of species and varieties. There are two main types of crosses, the bi-sexual or Mendelian and the uni-sexual. The former conform to the laws of segregation, they lead to various combinations of elementary characters, and thus can lead to the origin of as many new races as the number of possible combinations indicates. These races are constant; the hybrids, however, always exhibit segregation in the formation of their sexual cells and sometimes even in the formation of buds. The hybrids of uni-sexual crosses on the other hand are constant; so far as my experience goes, they do not segregate. If they are fertile they are, as a rule, as true from seed as their parents; but they may inherit the inconstancy of these (if, for instance, one of these belonged to an eversporting variety) and transmit it to their posterity.

A strong body of facts, which have been given elsewhere lead to the conclusion that crosses follow MENDEL'S laws if one of the two parents stands in the relation to the other of having arisen from it by retrogressive or degressive mutation. This means that the two parents of the cross possess exactly the same internal elementary characters; but that one or more of these occur in different conditions in the two parents; as for instance in the union of latent characters with active, of semi-latent with semi-active, and so forth. The opposite visible qualities, determined by the two different conditions of the same internal factor, constitute a so-called pair of characters. Therefore, racial forms which differ from one another by such qualities only, constitute a group for themselves, in the theory both of hybridization and of the origin of species.

Uni-sexual crosses are of an entirely different nature and lead to the direct production of constant hybrid races. They occur when at least one character is present in one parent, but is absent from the other. Hence MACFARLANE's name uni-sexual. In more complicated cases one form may have a larger excess of factors; or again, each of the parents may possess factors which are absent from the other. The excess will be unilateral, if one of the forms has arisen directly from the other, but bi-lateral if both have arisen in diverging directions from the same ancestor. In these cases vicarious pairs of characters do not exist in the two forms crossed, although externally they may seem to do so.

Therefore uni-sexual crosses correspond to progressive mutations; the Mendelian law, however, to the retrogressive and degressive forms of differentiation. Conversely we may infer that characters which, when crossed, behave in a Mendelian way, are in a latent or semi-latent condition, and that in the uni-sexual crosses an internal factor is entirely absent on one side. I willingly admit that the main purpose of this discussion is to make my principles clear, and to show how the two great branches of the theory of mutation may, in spite of the vast difference in their points of departure, ultimately lead to the same theoretical conception of the nature of elementary characters. This conception may in the present state of our knowledge be most conveniently formulated as follows:

Forms which have arisen by retrogressive and degressive mutation follow MENDEL's law, when crossed with their ancestors; whilst forms which have arisen by progressive mutation behave uni-sexually.

§ 3. THE THEORETICAL DISTINCTION BETWEEN SPECIES AND VARIETIES.

The idea of a fundamental difference between ancient and recent characters runs like a scarlet thread through the whole history of systematic biology. The nature and limits of this assumed difference have often furnished problems which the greatest investigators in this field have attempted to solve; and the answer has been a different one according to the information available at the time when it was attempted. From the transmutationists up to NÄGELI's well-known distinction between organic and adaptive characters there has been a long series of attempts to deal with these questions.

In ancient times the matter was easily settled by invoking supernatural causes. The higher systematic characters were assumed to have arisen by creation; the later ones by natural means; but in practice even this view led to confusion, because some authors regarded the genera, others the collective species, and yet others the constant elementary forms as the units which had been created.

Our discussions have led us also along several different lines, to the conviction that, as a matter of fact, there is a fundamental antithesis between ancient and recent characters, which finds its expression both in the phenomena of specific differentiation, and in those of artificial hybridization; for on the one hand forms can arise from others without the production of new factors, simply by the transference of factors already present into another condition, as from latent to active; and on the other hand, by the genuine appearance of new elementary characters. Progress in organization is due to the latter process, whilst the former is to a large extent the cause of the diversity of organic life.

If we compare this experimental result with the above theoretical considerations, we may assume that the distinction between older and younger characters consists in the difference between the formation of a new factor and the transposition of factors already present. On the basis of the doctrine of creation the origin of new units must be explained as being due to a supernatural cause, but no one has as yet applied this theory to the change in position of factors already present. Moreover in the light of the idea of evolution also the antithesis mentioned has its real and full significance.

It would lead me too far to analyze here the conceptions of other authors on these points, but such an analysis has led me to the conviction that the difference between the formation and the transposition of factors corresponds closely to the difference which the best systematists consider to exist between species and varieties.[1] A form which owes its origin to the production of a new internal factor is to be regarded as a species; a form which owes its peculiarity merely to the change in condition of a factor already present is to be regarded as a

[1] See also Vol. I, p. 185, and Vol. II, pp. 71-72.

variety;[1] or, as we have already expressed it in the first part of this volume (pp. 64, 71, etc.), the origin of new characters leads to specific differentiation, whereas the true derivative varieties arise by so-called retrogressive and degressive mutations without the formation of new characters. In my opinion this is, at any rate, the simplest view of the matter.

This, however, is purely theoretical, for in practice our definition can, at present, only rarely be applied. Here however the principles of hybridization come to our aid; for, expressed in hybridological terminology, our generalization runs: Forms, all of the characters of which follow MENDEL's laws in crosses, are to be regarded as varieties of the same species. This form of our generalization obviously admits of an immediate application in every case where the material can be dealt with experimentally.

Obviously this generalization is at present too sweeping; nevertheless the best investigators[2] have regarded the study of hybridization as an empirical foundation on which this distinction may be based. Moreover the difficulties are not really so great as they seem to be at first sight; for as soon as the body of evidence will have attained a certain extent, definite laws will be detected which will fit the majority of cases by judging from analogy.

The species, however, which I am here distinguishing from varieties are the smaller or elementary species; the delimitation of the larger or collective species is, from the very nature of the case, a question not for the experi-

[1] It should not be forgotten that varieties have been called incipient species and that from seed they are just as constant as species.

[2] See NÄGELI, *loc. cit.*, p. 396; FOCKE, *loc. cit.*, pp. 488, 502; NAUDIN, *loc. cit.*, p. 164; ABBADO, *loc. cit.*, p. 9, etc.

mental but for the comparative biologist.[1] The elementary species are demonstrably the existing units of the system; whilst the larger species are only aggregations of these. They will therefore be discussed in dealing with the question of the practical differences between species and varieties.

But, before I proceed to this, reference must be made to the more complicated but more common case in which two closely related forms differ from one another, partly by progressive and partly by retrogressive or degressive characters. To judge by the former they should be regarded as elementary species, by the latter however, as derivative varieties; and as they are hardly allowed to be in our system both at the same time, we must make a decision one way or the other

With a view to clearing up these difficulties let me deal with a particular instance, and select *Lychnis vespertina* and *L. diurna,* which are regarded by several systematists as belonging to one species, *Lychnis dioica*. If we regard these two forms as having been derived from a common original ancestor, and consider their individual characters, the difference in the color of the flowers stands out as the most striking distinguishing feature. The flowers of the original species must obviously have been red, and those of *L. vespertina* must have become white in the same way as those of other white-flowered varieties of red species. This view is supported by the fact that the colors of the flowers in these two species behave in exactly the same way in crosses as they do in many varieties, inasmuch as they conform to MENDEL's laws. Other differences between the two campions are

---

[1] I do not propose to enter here into the question of the desirability of a ternary nomenclature (see p. 65); it is entirely a question of convention.

the breadth of the leaves and the length of the flower stalks. But these characters do not segregate in the offspring of the hybrids. They are presumably to be regarded as results of progressive specific differentiation.[1] *L. vespertina* is, perhaps, a white-flowered variety of a red-flowered species which has disappeared. At any rate I do not think we shall go far wrong if we conclude that *L. vespertina* and *L. diurna* differ from one another partly by typical specific, and partly by varietal characters.

GÄRTNER has repeatedly expressed exactly the same view and has illustrated it by the same instance.[2] He says that any doubt as to the specific difference between closely related species, as for instance between *Lychnis diurna* and *L. vespertina,* can be most easily removed by crossing; for if such species give exactly similar hybrids with some other, i. e., with a third species, the difference between the two is of a varietal nature only. But if this does not occur we have proof that the essential nature of the species crossed, although they appear closely related with regard to their external features, is specifically distinct. For instance the two species of *Lychnis* just mentioned give wholly different hybrids with *Cucubalus viscosus*. On the other hand GÄRTNER lays stress on the fact that these species behave as varieties in regard to the color of their flowers when they are mutually crossed. Moreover *Lychnis vespertina* behaves as a variety with regard to the bending over of the teeth of the capsule, that is to say as a retrogressive variety of a species with the character of *L. diurna*.

[1] For a historical and critical treatment of the point, see a paper by R. ALLEN ROLFE, *Hybridization Viewed from the Standpoint of Systematic Botany,* Journ. Roy. Hort. Soc., April 1900, p. 197.

[2] GÄRTNER, *loc. cit.,* pp. 581-582.

We will now discuss the principle illustrated by this instance from a more general point of view. In the literature of the subject we frequently find the opinion that forms which are mutually fertile and produce a normal harvest of seed, giving fertile hybrids, are to be regarded as varieties of one and the same species. Forms, however, the union of which is followed by a diminution in fertility and the hybrids of which are less fertile than the species crossed, are regarded by the majority of the investigators as specifically distinct. These generalizations have served as criteria of relationship from the time of KÖLREUTER and GÄRTNER up to the present; and DARWIN himself relies on them in considerations of this kind.[1] Based as they were on extensive experience and on a profound systematic knowledge, they constitute principles which bid fair to become universally recognized. For these reasons they deserve to be placed in the foreground as a convenient point of departure for our discussion; and our object will be not to find out their weak points or to replace them by others, but rather to give them the more definite forms required by our present knowledge of hybridization.

Therefore we will start from the oft-cited proposition that varieties are only small species.[2] This means that the difference between species and varieties is not of a fundamental nature but rather of a gradual or even a conventional kind. Moreover we will start from the conception enunciated in the first part of this volume, according to which the forms which compose the collective species are mainly of two kinds (p. 64): *Homonom-*

---

[1] See FOCKE, *Die Pflanzenmischlinge*, pp. 436, 446-502, etc.
[2] See Vol. I, p. 171 and above p. 580.

ous forms or elementary species and *derived forms or true varieties.*

As we concluded in § 7, Part I, p. 65, the origin of elementary species is due to the formation of new elementary characters, i. e., to their actual numerical increase. True varieties differ from the species to which they belong by the latency of certain characters, which may either be active as in the type of the species, or latent as in the variety, or which may occur in a latent or semi-latent condition in the former, and become active or semi-active in the production of the variety. In other words, we may say that elementary species arise by progressive mutations, but derivative varieties by retrogressive and degressive ones (p. 71).

If we now compare the principles derived from the study of hybrids with these conclusions, we see that the two main types of hybrids are in essential agreement with these two systematic groups. Mendelian hybrids correspond to retrogressive and degressive specific differentiation, and consequently to true varieties; uni-sexual hybridizations correspond to progressive specific differentiation and consequently to elementary species.

There can hardly be any misconception as to the signification of this important conclusion. But it only indicates the principle and not its application to particular cases; and, as a matter of fact, it is only another form of the generalization enunciated above relating to the fertility of crosses and hybrids. For Mendelian crosses have as a rule the same fertility as the pure parent forms, and fertility does not diminish in the subsequent hybrid generations. In uni-sexual crosses, however, fertility diminishes and it does so in proportion as the relation between the two forms crossed becomes more remote.

The truth of this conclusion has become more obvious owing to the attention which has been paid to the subject since the re-discovery of MENDEL's work. The parallelism between the two groups of hybridization and the two types of systematic subdivisions has been most exhaustively dealt with by TSCHERMAK, who attempted to base upon it a principle of distinction between the specific hybrids and varietal mongrels.[1]

We will, then, regard the principle in its new form as demonstrated, and examine the question why the criteria which it supplies are not sufficient for universal application. In doing so I shall, for various reasons, leave the mutation crosses, which TSCHERMAK also regards as specific hybrids, out of consideration; and shall denote the Mendelian hybrids as bi-sexual in accordance with MACFARLANE's terminology, employing also the term uni-sexual in the sense in which it is used by that author. Expressed very briefly, therefore, *bi-sexual crosses produce varietal hybrids, uni-sexual ones specific hybrids.*

But some limitation is necessary; and herein lies the difficulty of the question, which is felt by every one who endeavors to apply the conclusions drawn from the study of hybrids to taxinomic problems. This limitation is, that the criterion really applies only to monohybrids; for di-polyhybrids, however, only in so far as they can be compared with these.

We have given the name of monohybrids to those mongrels whose parents differ from one another in a single elementary character only. Obviously they occur both in uni-sexual and in bi-sexual crosses. But, of

---

[1] E. TSCHERMAK, in the third appendix to his edition of MENDEL's *Versuche über Pflanzenhybriden*, p. 58.

course, any given monohybridic cross can only belong to one of these two groups. If it is bi-sexual and behaves in a Mendelian fashion we may immediately infer that the two parents are to be considered as varieties.[1] If it is uni-sexual they are elementary species, of which the one must have been derived from the other.

The di-polyhybrids are mongrels whose parents differ from one another in respect of two or more elementary characters. Two cases must be distinguished. Let us confine ourselves first to the dihybrids. In some cases the two points of difference may belong to the same category and therefore follow the same laws in the crosses and their products. If each of them, considered by itself, would lead to the conclusion that the parents were related as varieties, the same conclusion will obviously hold good for the combination. So, for instance, is *Papaver somniferum polycephalum Danebrog* to be regarded as a variety; so also, if it is allowed to judge by analogy, *Calliopsis tinctoria pumila purpurea* (Vol. I, p. 197). For the same reason the compound colors, which may be split into their components by means of crossing and can be reconstructed out of these, fall within the category of varietal characters.

Similarly forms, of which one has arisen from the other by two successive mutations in the progressive direction, and whose crosses. therefore, conform to the laws of uni-sexual unions, would most certainly have to be regarded as elementary species.

---

[2] And this independently of the nomenclature chosen. For instance according to the principles enunciated above, *Chelidonium laciniatum Mill.* will have to be regarded as a variety of *C. majus*, even when this more convenient name is retained. See Part I, p. 65. And from a practical point of view it would be very desirable to drop the attempt to correlate the nomenclature with the ever-changing systematic conceptions.

It may, however, also happen that in forms which differ from one another in two points only, one of these would give a bi-sexual union, whilst the other would lead to a uni-sexual cross. Such is the often cited instance of *Lychnis vespertina* × *diurna*. One of the characters would follow MENDEL'S law in crosses, whilst the other would tend to produce a constant intermediate form.[1] In this case, according to the former character, one parent would be related to the other as a derivate variety, but according to the latter as a homonomous elementary species.

Exactly the same, though to a greater extent, must be true of tri-polyhybrids. The points of difference between their parents can be all uni-sexual or all bi-sexual, or some of them uni-sexual and others bi-sexual. In the first case the parents are to be considered as elementary species, in the second as varieties; in the third, however, the principle affords no decision.

It is just this case which appears to be the commonest in nature. In experiments in hybridization, we must, if we wish at all to elucidate the laws to which the results conform, confine our attention to certain points of difference and leave all the rest out of consideration as of subordinate importance. MENDEL did this in his experiments with peas, and the same has to be done in crossing Maize, the races of which do not differ from one another exclusively in varietal characters; so also in *Lychnis vespertina* and *diurna* and in many other cases.

It would take us too long to continue this discussion further, and to accumulate examples; my meaning will, I believe, be sufficiently clear. It may be expressed in

---

[1] I call to mind the *Oenothera Pohliana* (*O. lata*×*brevistylis*), one of my crosses, the *lata*-character of which behaves as a mutational character in a cross; whilst the shortness of the stamen behaves as a Mendelian character.

its simplest terms, if we call those qualities which in crossing conform to MENDEL's law, varietal characters, and those which give uni-sexual unions, specific characters. This would conduce to a more general form of the thesis based on the experiments with *Lychnis* (p. 582), viz., that two related forms can differ from one another simultaneously in varietal and in specific characters.

Are they therefore to be regarded as varieties or as species? Here we arrive at the boundary which separates facts from conventional terminology. Here lies the point to which GOETHE refers in his well-known lines:

> "Dich im Unendlichen zu finden,
> Musst unterscheiden und dann verbinden."

The process of distinguishing is an objective one, that of combining subjective. The former is the immediate result of inquiry; in our case, of systematic studies on the one hand, and of experiments in hybridization on the other. The combination is partly a question of taste; it has to serve special aims; and above all it must facilitate general conceptions and mutual understanding.

It is not my task to go more deeply into the question of systematic subdivisions or to make any definite proposals.[1] My sole object is to place the actual facts in as clear a light as possible. This attempt, however, again leads to the conclusion that here also this insight can only be obtained on the basis of the theory of mutation. It is only by attempting to analyze the species into its component factors, the elementary characters, that we

---

[1] If two forms were found to differ from one another exclusively in varietal characters, but the number of these were very large, they would probably have to be separated as species. Here also the distinction between species and variety is an arbitrary one, and as a matter of fact, many of the larger groups and sometimes even whole families have among their distinguishing marks some which do not really differ from "varietal characters."

can hope to arrive at a conception of species which shall be both in accord with the facts, and justified by experiment.

Of course, I am well aware that the experiments which have been carried on up to the present are by no means sufficient and that much remains to be done. Numerous experiments in hybridization are necessary before they can serve as a foundation for systematic distinctions. But the leading principle in these researches must always be the attempt to determine the elementary characters.

§ 4. THE PRACTICAL CONCEPTION OF SPECIES.

Both collective and elementary species are called species; and this twofold significance of the word has thrust its roots so deeply into the history of descriptive biology, that it will probably never be wholly eradicated Linnaeus himself confused the two ideas; and whilst some readers derive from the study of his works the conviction that in his mind the collective forms were the true species,[1] others come to a different view of his attitude, and believe that in formulating the conception of species, he was considering the real units of the system.[2]

The fundamental conception from which almost all investigators start, is that species are the only real entities.[3] As to what these entities are, opinions differ. "On ne peut pas douter," says DE CANDOLLE, "que le groupe appelé espèce par l'illustre Suédois ne fut, dans

[1] See Vol. I, p. 20.
[2] S. BELLI, Observations critiques sur la réalité des espèces en nature au point de vue de la systématique des végétaux, 1901.
[3] C NÄGELI, Entstehung und Begriff der naturhistorischen Art, 1865, p. 31.

sa manière de voir, une association de formes voisines."[1] On the other hand JORDAN, as is well known, based his conception of the smaller or elementary species as the real species on the same fundamental proposition. The wish to see in the species something real, has always played a prominent part; but the reality as it appeared to the descriptive biologist has been very different from that in the mind of the experimental investigator.

It would certainly be desirable to agree to call only one of the two groups species; it is only the question which. The older view and the popular idea limit this term to the larger groups, and give the name of subspecies to the smaller ones.[2] But the term subspecies, as it is now in use does not signify a unit, but a group of units which is also compound and merely differs from the species itself in being smaller (Part I, p. 60). The modern tendency is to regard the smaller types as species, and wherever the criterion is of an experimental kind, like that employed by JORDAN, this view will predominate. Its importance to descriptive biology has recently been demonstrated in a most clear and convincing manner by BELLI;[3] and there seems every prospect of its being recognized by the best systematists.

It has been proposed to denote collective species by a special name, and the word "stirp" has been suggested. This term has been applied in this sense by several systematists,[4] and BELLI has adduced a long series of histor-

[1] ALPH. DE CANDOLLE. *Archiv. des sc. de la bibl. universelle*, Geneva, Febr. 1878, Vol. LXI, p. 4.

[2] In what follows, I shall leave varieties, in the sense in which this term has been used in the foregoing sections, out of account.

[3] S. BELLI, *loc. cit.*

[4] For instance, H. LÉVEILLÉ. *Monographie du genre Oenothera*, 1902, I, pp. 72, 106, etc.

ical and critical arguments for this practice. The word stirp would perhaps correspond most closely to the German word *Sippe*,[1] although this word has been employed by various authors in a different sense.[2] At any rate it is most desirable to make some distinction of this kind, and BELLI's suggestion might well form the foundation of such. But questions of nomenclature have little interest in this discussion and I prefer to leave the task to others.

We may now proceed to the practical delimitation of the conception of species.

Descriptive biology wants a definition, independent of the results of hybridization; in its commonest form it is based on the absence or presence of transitional forms, as was explained in the first part of this volume. Groups of individuals which are connected together by transitions are considered to belong to the same species. The limits between species correspond to gaps in the series.[3] Without some such convention the description of species occurring in collected material would be impossible; and this method has been employed by the best systematists since the time of DE CANDOLLE. It is only when direct experiments can be carried out that the problem can be dealt with in a different way.

Two difficulties, however, present themselves, which I shall now briefly deal with. In the first place, exactly

[1] *Die Mutationen und die Mutationsperioden bei der Entstehung der Arten.* Leipsic, VEIT & Co., 1901, p. 14.

[2] C. CORRENS, *Scheinbare Ausnahmen von der Mendel'schen Spaltungsregel für Bastarde,* Ber. d. d. bot. Ges., 1902, Vol. XX, Part 3 p. 170. See also the same author in Ber. d. d. bot. Ges., 1901, Vol. XIX, p. 77, note 1, and his *Monographie der Maisbastarde,* p. 1, and VON WETTSTEIN, *Grundzüge der geographisch-morphologischen Methode der Pflanzensystematik,* 1898, p. 3.

[3] See, for instance, *Année biologique,* IV, 1898, p. 470; V, 1899, p. 377 and elsewhere. Also BORRADAILLE, *On Crustacians,* 1901, p. 193; GRISEBACH, *Die Vegetation der Erde nach ihrer klimatischen Anordnung,* 1872, p. 9, and so forth.

the best varieties are, as a rule, not united by transitional forms with the parent species; in the second place, transgressive variability tends to obscure boundaries where they really exist. These limits are often overlooked in the descriptive method, and the search for them can only be carried out on experimental and statistical lines. With good right DE CANDOLLE speaks in such cases of provisioual species.[1]

§ 5. THE PARALLEL BETWEEN SYSTEMATIC AND SEXUAL RELATIONSHIP.

Since the doctrine of descent now commands general recognition, it is desirable that the systematic divisions should be an expression of the various degrees of relationship. Even before the appearance of DARWIN's works it was recognized that the task of systematic biology as a descriptive and classificatory science was different from the mere question of actual relationship. To investigate this and, where possible, to bring the divisions of the natural system into harmony with it, these were the ends which the pioneers in the study of hybridization had continually in view.

The result did not, however, correspond to this expectation. We have not, as yet, succeeded in bringing into harmony the study of hybridization with that of systematic biology. NÄGELI expressed this incompatibility most clearly by introducing his conception of sexual affinity. The degree of this affinity between two types was judged first by the degree of their fertility when crossed with one another, and, then, by that of the fertility of the hybrids thus produced.

[1] ALPH. DE CANDOLLE, *La Phytographie*, pp. 98, 167.

The simplest form to which it has been proposed to reduce the parallel between systematic and sexual relationship, is the following: (1) Plants which produce offspring when crossed with one another, always belong to the same genus; (2) Plants whose fertility is not diminished in crossing belong to the same systematic species (or collective species). Both generalizations are in great favor and are defended by prominent investigators. They have, however, a weak side, viz.. that they cannot be reversed.

Let us first examine the former proposition. It denies the existence of hybrids between distinct genera, or so-called generic hybrids. It originated amongst those philosophers who regarded the genera as having been created, the species, however, as having arisen from them by natural means. We have already dealt with the historical significance of these transmutationists in the first volume (p. 17). To them the view, stated above, is also due, that not only do species arise within the genera by a normal process of evolution, but that new forms may arise from these species by crossings. W. HERBERT is the most famous representative of this view,[1] which was later defended by GODRON. The latter investigator describes all genera, the species of which are fertile with those of related genera as artificial, and has collected a mass of evidence in support of this view.[2]

No fundamental objection can be brought against this view, and its adoption would lead in relatively few cases

[1] W. HERBERT, *Amaryllidaceae, With a Treatise Upon Cross-bred Vegetables,* London 1837, pp. 337 et seq. See also GÄRTNER, *loc. cit..* p. 152, and NÄGELI, *Sitzungsber. d. k. bayr. Akad. d. Wiss.,* Dec. 15. 1865, p. 400.

[2] A. GODRON, *De l'espèce et des races dans les êtres organisés,* 1859. Vol. I, pp. 225-236, and *Mém. Acad. Stanislas à Nancy,* 1862. pp. 206-298.

to an enlargement of genera. On the other hand it would put a check on the splitting up of genera which has been so much in vogue of recent years, and not less to the elevation of subgenera to the rank of genera. In practice, however, the application of HERBERT's suggestion has proved impossible. At that time there were only a few generic hybrids, but their number has undergone a considerable increase; partly no doubt through the arbitrary splitting up of genera mentioned above, but partly also through the rapid accumulation of experimental data. *Berberis* and *Mahonia* (e. g. *B. Neuberti*) could well be united into one genus. The suggested union of rye and wheat, a hybrid between which has been raised by RIMPAU,[1] into a single genus *Frumentum,* is not likely to win much favor; and the fact that BURBANK, in California, has raised a hybrid between *Nicotiana* and *Petunia* which he calls *Nicotunia,*[2] will hardly be regarded as a sufficient ground for a systematic union of these two genera. There are now about 150 bi-generic hybrids amongst the Orchids, especially between the genera *Laelia, Cattleya, Epidendrum* and *Sophronitis,* as also between *Zygopetalum, Colax* and *Batemannia.*[3]

The practical difficulties which stand in the way of HERBERT's proposal are, on the one hand, the fact that the limits to possible hybrid combinations are by no means definite, and on the other hand, the objection, which has so often been raised, that crosses are exceptional phenomena and that it would therefore be impossible to

[1] W. RIMPAU, *Kreuzungsprodukte landwirthschaftlicher Culturpflanzen.* Landwirthsch. Jahrb., 1891, p. 20, and Pl. VI, Fig. 58.

[2] LUTHER BURBANK, *New Creations in Fruits and Flowers* (Burbank's Experiment Grounds, Santa Rosa, California), 1893, with a figure of the *Nicotunia.*

[3] C. C. HURST, *Journ. Roy. Hort. Soc.,* Vol. XXIV, pp. 102, 125.

apply the principle in those numerous cases in which they fail. With regard to the former point, it should be noted that there are numerous natural hybrids which cannot as yet be made artificially, as for instance *Ribes Gordonianum*; in other words that many possible hybridizations do not succeed within the narrow limits of an experiment. The impossibility of a successful cross can therefore hardly ever be proved experimentally. With regard to the second point it suffices to cite the fact that in the great majority of genera no specific hybrids exist at all; and that therefore here the delimitation of genera according to this principle, would fail entirely.

We come now to the species. KÖLREUTER expressed the view that crosses within the limits of these are fertile and give fertile offspring; whereas crosses between species would either show a lessened fertility or at least would produce infertile hybrids. GÄRTNER and most of the more recent investigators have subscribed to this view, except that they regard diminished fertility, rather than the actual absence of it, as the index of the boundaries of the species.[1]

But even to these generalizations the exceptions are so numerous that unanimity in their application has not yet been reached. The parallel between sexual affinity and systematic relationship holds good in general, but fails only too often in particular cases.[2] NAUDIN regarded these deviations from the rule as exceptions,[3] and ABBADO and several other investigators have claimed the determining cause of these exceptions in individual cases to be the task of hybridological researches.[4]

[1] GÄRTNER, *loc. cit.*, pp. 163-164, 578-579, etc.
[2] See MURBECK, *Botaniska Notiser*, 1901, p. 214.
[3] CH. NAUDIN, *L'Hybridité dans les végétaux*, 1869, p. 145.
[4] ABBADO, *L'ibridismo nei vegetali*, 1898, p. 48.

These causes may fall into two entirely different categories; on the one hand, they may consist in insufficient systematic knowledge, on the other, in the insufficient number of experimental crosses. With regard to the former point it should be remembered that, although the systematist frequently takes latent characters into consideration, it is obviously by no means always possible to decide on systematic grounds whether a character which we do not see is really absent or exists only in a latent condition. Nevertheless, latency is often regarded as a retrogressive metamorphosis and therefore as the mark of a variety; whereas absence is considered as a phylogenetically older step and therefore as a specific character (see above p. 71).

From this discussion we see that we may cross a plant in which any given character is active either with one in which the internal factor for this character is absent, or with a species or variety in which it is present but in a latent or inactive state. Externally there is no difference between two such crosses, but fundamentally they are exactly opposite, and therefore it is to be expected that their results will differ. The cross, *active × absent* is a uni-sexual union and will presumably lead to a halving of the external characters of the parents in the hybrid, whereas the cross *active × latent* is a bi-sexual one, and follows MENDEL's laws, at least in ordinary cases. Many paradoxes, which at present seem to negative the parallel between systematic and sexual affinity may perhaps be explained by more exact investigation on these lines. FOCKE gives the following cases as instances:[1] "*Silene vulgaris*, and *S. maritima, Capsella rubella* and *C. bursa pastoris, Phaseolus vulgaris* and *Ph. multiflorus,* or the

[1] FOCKE, *loc. cit.,* p. 448.

species of *Diplacus* (*Mimulus*) do not seem to be morphologically more remote from one another than *Tropacolum majus* and *Tr. minus, Nicotiana latissima* and *N. Marylandica, N. rustica* and *N. Texana* or *Pisum sativum* and *P. arvense*. Nevertheless the results of crossing in the former cases exhibit all the characters of hybrids, but in the latter those of mongrels." FOCKE summarizes his conclusions on this subject in the thesis that "systematically probable" crosses often miscarry, whilst improbable ones sometimes succeed (*loc. cit.*, p. 457).

Even regarded from this point of view the difference between Mendelian and uni-sexual crosses offers itself as a criterion for distinguishing between species and varieties.

But sexual affinity does not always give reliable indications. In the first place GÄRTNER frequently found that fertility, as measured by the number of seeds ripening in a capsule, is different in reciprocal crosses. This shows that it is not merely determined by the degree of relationship, but obviously by some other causes besides. The best known of these is the length of the style; and the recent investigations of BURCK on the concentration and stimulating properties of the fluid secreted by the stigma have thrown much valuable light on this subject.[1] In extreme cases one of the crosses succeeds well, whilst the other does not at all, as for instance *Mirabilis Jalapa* × *longiflora, Geum urbanum* × *rivale, Sophronitis* × *Cattleya*, and so forth. In the second place some crosses do not succeed in spite of a very close apparent relationship, as for instance between *Anagallis arvensis* and *coerulea* (GÄRTNER).

[1] W. BURCK, *Over de beweging der stempels by Mimulus en Torenia*. Sitzungsber. d. Kon. Akad. d. Wet., Amsterdam, 1901, and in previous articles.

I do not propose to elaborate this theme further; it has often been dealt with and especially in great detail by FOCKE, who has presented it in a masterly way in his textbook on plant hybrids. The main conclusion, however, is that the majority of authors agree that systematic and sexual affinity, if properly understood, are essentially parallel; indeed, that they are really no more than two manifestations of one and the same thing, but that we have not yet succeeded in explaining the apparent exceptions to this parallel.

For our purpose, however, the important question is, whether the diagnoses of species and varieties will gradually come to be based on elementary characters as units, and whether sexual relationship will come to be judged by the number of differentiating units. GÄRTNER has already pointed out that those genera in which the largest numbers of hybrids have been produced are exactly those in which the number of very closely related species is the greatest (*loc. cit.*, p. 168). NÄGELI has elaborated this idea and SACHS has followed him in his *Lehrbuch der Botanik*. ABBADO, HURST, GILLOT and many others have also subscribed to this view.

The opinion expressed by these writers on the parallel between systematic and sexual affinity, may be summarized in the following thesis, viz., that the fertility of crosses and of the hybrids resulting from them, diminishes, on the average, as the number of points of difference (that is to say, that of the elementary characters, which constitute the differences) increases. But many more experiments are necessary before this suggestion can be examined critically or be regarded as resting on a sure foundation of experimental facts.

II. THE RANGE OF VALIDITY OF THE DOCTRINE OF MUTATION.

§ 6. THE SIGNIFICANCE OF THE AVAILABLE EVIDENCE.

Unlike the prevailing form of the theory of selection, the doctrine of mutation lays stress on sudden or discontinuous changes, and regards only these as active in the formation of species. The Darwinian form of the theory of selection regards both these and fluctuating variations as operative in the origin of new forms, whilst WALLACE favors the other extreme, according to which all formation of species goes by a slow and gradual process of change.

The two schools of thought naturally adopt different attitudes towards the doctrine of mutation. It is at once rejected by WALLACE's adherents, whilst those who incline to DARWIN's own form of the theory are less unreservedly inimical; many of them have even greeted it with open arms.

Inasmuch as these two lines of thought have been clearly manifested in the critiques which have been published of the first volume of this work, I propose to discuss them briefly here, in order to point out the fundamental questions which are involved in this controversy.

The extreme opponents of my theory maintain that there are no mutations at all: *Natura non facit saltus*, they say. What I have described as discontinuous changes, are, in their opinion, merely the extreme deviations brought about by ordinary variability; for the further these are from the mean, the rarer they are, and the greater are the intervals by which they are separated. The number of petals in *Ranunculus bulbosus semiplenus* oscillates around 9 or 10, frequently reaching 14, very seldom 20-23. Only in one case did I observe a larger number, which happened to be 31 (see p. 252). The gap between 23 and 31 is, however, not a discontinuous variation. It is perfectly normal, and quite a common occurrence in this part of QUETELET'S curves. In a general way, gaps of this kind in the curves of variation may be thus explained, and according to my opponents the so-called springs and jumps have to be explained in the same manner. They are assumed to be no more than the extreme variants of series which when investigated further, would prove to be continuous ones.

This view is chiefly maintained against my opinion by morpholgists[1] and statisticians.[2] It is, as KORSCHINSKY has lately shown, directly contradicted by horticultural experience;[3] and the absence of transitions and the stability of my new species of *Oenothera* prove that, in this case at any rate, true mutations do occur. The greatest obstacle in the way of agreement on this point, however, lies in the phenomena of transgressive variability, which to the morphological observer so often give

[1] Among my numerous critics I mention here CH. SCHRÖDER, *Die Variabilität der Adalia bipunctata L.*, Allgem. Zeitschrift für Entomologie, Vols. VI-VII, 1901-1902. The view taken by SCHRÖDER has since been proved to be erroneous by the experiments of A. G. MAYER on the colors of butterflies. See *Effects of Natural Selection and Race-Tendency Upon the Color-Patterns of Lepidoptera*, Museum Brooklyn Inst. of Arts and Sc., 1902, Vol. I, No. 2, p. 31.

[2] See the journal *Biometrika* and especially the articles in it by WELDON.

[3] S. KORSCHINSKY, *Mém. Acad. Imp. Petersbourg*, 1899, IX.

the appearance of transitional forms and only reveal their true nature when tested by breeding experiments.

Several of the critics who have expressed themselves more or less favorably on my theory, have pointed out that the greatest danger for it lies in this very point. In a very clear and concise summary of the doctrine of mutation MAC DOUGAL has expressed himself as follows: "The greatest misunderstanding which may likely arise in the consideration of these results will be that founded on the error of confusing fluctuating variability and mutability."[1]

The distinction between species-forming and fluctuating variability was first derived by DARWIN from his theory of pangenesis, and this may perhaps explain the antipathy which so many investigators bear towards it.[2]

The great majority of writers assume that fluctuating as well as discontinuous variability play a part in the formation of species.[3] This view of DARWIN, which under WALLACE'S influence gradually shifted into the background, has in latter years come again prominently to the front; and the various investigators concede here a less or there a greater share to discontinuous variations or mutations, according to their preconceptions and their experience in investigation. This long series of shades of opinion would seem to indicate that we are not concerned here with an independent principle, but with a gradual change of opinion from the prevailing theory

[1] D. T. MACDOUGAL, *The Origin of Species by Mutation*. Torreya, 1902, Vol. II, p. 99.

[2] See *Intracellular Pangenesis*, e. g., p. 214 (English ed.), and Ber. d. d. Ges., 1900, XVIII, p. 83.

[3] VON WETTSTEIN has published a useful summary of his views in the form of a lecture delivered to the Scientific and Medical Association at Karlsbad, and entitled *Der Neo-Lamarckismus und seine Beziehungen zum Darwinismus*, 1903.

to some other one; and especially amongst American investigators the tendency has been in recent years to proceed as far as possible in this direction.

If we look for a fixed point among these oscillating opinions we may well choose the view repeatedly expressed by DARWIN himself, that it is possible to imagine that characters may originate by a slow process, but may disappear all of a sudden.[1] In combining this with the distinction made in the first part of this volume between progressive, retrogressive and degressive formation of species, the proposition would run: Progressive formation of species may occur slowly and gradually, whilst retrogressive and degressive specific differentiation is due to mutations. Progressive differentiation consists in the formation of a new character which was not previously present; whilst retrogressive and degressive differentiation consists in the transference of internal factors, already present, from one condition to another. In the former case the active unit becomes latent; in the latter the latent becomes active, or the semi-latent semi-active; but the material vehicles of these characters remain fundamentally the same throughout; nothing new arises in the idioplasma.[2]

In horticulture, as we have seen, mutations are largely of the retrogressive or degressive kind. Discontinuous formation of species on the progressive line is much rarer. Nevertheless I believe that my researches with *Oenothera* have contributed instances which may demonstrate the occurrence of progressive mutations in this species at any rate. Obviously there is a great need of

[1] With reference to this point see the valuable critique by L. PLATE, *Ueber Bedeutung und Tragweite des Darwin'schen Selectionsprincips,* 1900, p. 37 and elsewhere.

[2] See below, §§ 9-11.

further investigations on this point, and these should not merely be concerned with new phenomena, but with the testing of results already obtained; for many instances of discontinuous origin stand in need of more convincing proof, and in other cases the progressive nature of a process which is interpreted as a mutation is often subject to doubt. In such investigations attention should be paid to the question whether the hypothetical premutations may perhaps be prepared gradually, whilst the new character which has been so developed in secret, might unfold suddenly. But it will take many years to decide these points.

Starting from general arguments KÖLLIKER[1] was the first to insist on the importance of mutations against DARWIN, indicating the process, which was then a purely hypothetical one, by the name heterogenic development. Others have expressed themselves favorably with regard to this view; especially K. E. VON BAER and BRONN, and also HAACKE, G. PFEFFER, DELAGE, CUNNINGHAM, WOLFF, DREYER, DRIESCH, EMERY[2] and many others. This doctrine has of recent years found its strongest champion in BATESON, whose views I have already dealt with above. Those authors too, who have made monographic studies of special genera and species have welcomed it; for instance WITTROCK, in his study of *Viola*, inclined to the view that species have originated discontinuously. Further, this doctrine is defended on purely speculative grounds by many prominent biologists, among whom I need only mention VON HARTMANN and also HAMANN and KERSTEN.[3] On the zoological side HU-

---

[1] KÖLLIKER, *Abhandl. Senckenb. Gesellsch.*, 1864. pp. 223-229.

[2] EMERY, *Biolog. Centralblatt*, 1893, No. 13, p. 723.

[3] See the careful and critical exposition in H. KERSTEN'S *Die*

BRECHT and MORGAN have expressed themselves in favor of the new view,[1] the latter on the ground of his studies in the regeneration of injured organs.[2]

Amongst practical agriculturists my views have been well received by EM. VON PROSKOWETZ and HJALMAR NILSSON. The former has conducted a long series of investigations on the transformation of the wild *Beta patula* into the sugar beet; and it was found that the changes do not by any means take place by imperceptible transitions but suddenly.[3] Each new character is brought to light at one stroke; it is not the product of selection but of internal processes, the nature of which we do not as yet know. Even such secondary characters as color, change in the same way. HJALMAR NILSSON, the director of the practical and experimental station for the improvement of seed at Svalöf in Sweden, has for many years been collecting a valuable mass of evidence, which promises to throw great light on the doctrine of mutation, but has not as yet, been grouped for that purpose. Judging from the oral and other communications which I have received from time to time from this investigator, I may state here that his results are in perfect harmony with the doctrine of mutation.[4]

idealistische Richtung in der modernen Entwickelungslehre, Zeitschr. f. Naturw., 1901, Vol. 73, p. 321.

[1] A. A. W. HUBRECHT, *De evolutie in nieuwe banen*, Utrecht, 1902.

[2] TH. HUNT MORGAN, *Darwinism in the Light of Modern Criticism*, Harpers Monthly Magazine, Feb. 1903, and in many other papers.

[3] EM. V. PROSKOWETZ JR., *Culturversuche mit Beta*, 1892-1901 in Oesterr.-Ungar. Zeitschrift f. Zuckerindustrie und Landwirthschaft des Centralvereins f. Rübenzuckerindustrie in d. Oest.-Ung. Monarchie, 1892-1902. The experiments of a given year will be found in the number for the following year, and in the number for 1892 will be found an account of the earlier experiments and the literature of the subject. For the mutations which occurred see especially the number for 1902.

[4] See the various numbers of the journal issued by the experi-

To the important observations already mentioned, made by HEINRICHER on *Iris pallida abavia*, by SOLMS-LAUBACH on *Capsella*, by WITTROCK on *Viola*, by BAILEY and WHITE on *Tomatoes*, and by many others, I have here to add the following. NOLL has described the sudden origin of a regular tendril in *Tropaeolum*, and draws conclusions from this in favor of the possibility of discontinuous changes and their significance for the theory of descent.[1] TRACY has observed the sudden origin of a dwarf variety of *Phaseolus lunatus*,[2] MACFARLANE has investigated the variability in the genus *Prunus*,[3] CARUEL has collected a number of cases in which direct transitions can be demonstrated and calls them "Euthymorphoses."[4] CARLSON has investigated the mutations of the forms of *Succisa* occurring in Sweden,[5] and LAURENT expresses himself in the same way with regard to several species of fruit trees.[6] Dr. J. W. HARSHBERGER sent me material of *Hibiscus moscheutos* and *Euphorbia ipecacuanha* from Pennsylvania, the extraordinary abundance of forms in which seems to indicate the occurrence of a period of mutation in these species; and Mr. L. COCKAYNE has given me information on some transformations of *Sarothamnus scoparius* and *Lupinus arboreus* observed by him in New Zealand. It may further be mentioned that

mental station at Svalöf, *Sveriges Utsädeförenings Tidskrift*, and particularly the *Arsberättelse under ar 1901*, in Vol. XII, 1902, No. 1, page 3.

[1] F. NOLL, *Das Auftreten einer typischen Ranke an einer sonst rankenlosen Pflanzenart*, Sitzungsber. d. Niederrhein. Ges. f. Naturk., Bonn, Jan. 14, 1895.

[2] W. W. TRACY, *American Naturalist*, 1895, XXIX, p. 485.

[3] J. M. MACFARLANE, *Publications of the University of Pennsylvania*, 1901, p. 216.

[4] T. CARUEL, *Bull. Soc. Bot. Ital.*, Florence, 1896, p. 84.

[5] *Bot. Not.*, 1901, p. 224.

[6] E. LAURENT, *De l'expérimentation en horticulture*, 1902, p. 12.

BORRADAILLE, working with decapods, came to the conclusion that great difficulty stands in the way of the explanation of specific differentiation by means of ordinary natural selection.[1] Mr. R. LAUTERBORN was so kind as to draw my attention to the appearance of the yellow *Atropa Belladonna lutea,* and to the evidence given on this variety by Dr. SCHÜZ.[2] Dr. RAATZ was also kind enough to send me some seeds of a most remarkable shiny brown variety of the sugar-beet which suddenly arose in the cultures of Klein-Wanzleben, and from which I obtained fine instances of the new character in my garden.

I have only given a selection from the long list at my disposal. My object was simply to show that the doctrine of mutation already finds adherents everywhere and is supported by a broad foundation of facts. This justifies the expectation that the difficulties which still stand in its way will ultimately be overcome.

§ 7. THE EXPLANATION OF ADAPTATIONS.

Ever since the belief in the common origin of organisms has been recognized as a basis for investigation and speculation, one aspect of the problem has aroused the special interest both of the author and of the student, viz., the explanation of adaptations. As a matter of fact this problem lies outside the scope of the present book, which is only concerned with the empirical foundations of the theory of descent. Nevertheless it seems to me that it

[1] L. A. BORRADAILLE, *Marine Crustaceans,* The Fauna and Geography of the Maldive and Laccadive Archipelagoes, Vol. I, Pt. 2, page 197.

[2] *Amtl. Bericht über die 33. Versamml. d. Naturf. und Aerzte,* Bonn, Sept., 1854 (Bonn, 1859), p. 139. A single specimen of the plant had been found a few years previously in the Black Forest.

The Explanation of Adaptations. 607

may be of interest to show that the prevailing view, according to which WALLACE's form of the theory of selection is the only one which will account for adaptations, is erroneous.

The view that all the characters of organisms vary in every desired direction, and that the slightest deviation may be subjected to the struggle for existence, and can be accentuated to, and finally fixed at, the necessary degree of development, is certainly an extremely convenient one. I willingly admit that almost anything can be squared with this theory in a very plausible way, and that explanations of this kind are very attractive to the student; but this is not science. The contradictions in such a system must be satisfactorily explained before it is accepted; and if we attempt to do this, we soon come to the conclusion that the hypothesis itself is not in harmony with the available evidence.

The limits of the applicability of the theory of selection, as applied to this question, are known to everybody; and without doubt they are extremely wide. How much the theory of mutation has to offer in this respect we do not know, because no attempt to estimate this has as yet been made, but everything points to the conclusion that this theory will explain adaptations just as completely, or rather just as incompletely, as the present view. It will, however, always have the special feature of emphasizing the hypothetical parts of the argument, rather than of dismissing them into the background.

At the present time the theory of selection has still the larger number of adherents; but amongst the younger investigators a train of thought is developing which, as we have seen above, ascribes a greater importance to discontinuous changes. For them fluctuating variability consists merely in an oscillation around a given point of equilibrium; whilst the formation of new species necessitates the attainment of a new equilibrium. Especially in America has this view made great progress, as has been described by CONN in an admirable exposition in his new work on evolution.[1]

Of the numerous writings which we owe to WALLACE'S school, I shall only mention here that on *Natural Selection and Tropical Nature,* of this author (1895), and PLATE's critical exposition of WALLACE's theory.[2] Although PLATE, at the conclusion of his clear and perceptive critique of the questions which pertain to this point, declares in favor of the theory of selection, his work, of all those of which I have made use, gives me the impression of the greatest objectivity; and I am convinced that the gulf between his views and those I hold will surely be bridged over some day. Therefore I refer the reader to his book for a closer study of these questions, and shall confine myself here to a few points which stand in the closest relation to those discussed above.

1. *The significance of fluctuating variability is very limited*, whereas the explanation of adaptations demands almost unlimited variability. In earlier days when the law of QUETELET was only known to apply to anthropology, almost all the changes of plants and animals were considered to be the consequences of ordinary variability, but now this is shown to be governed by laws which largely curtail its importance. In the first part of the first volume this theme has been discussed in detail, and I may here simply refer the reader to those chapters.

[1] H. W. CONN, *The Method of Evolution*, New York, 1900, p. 132.
[2] L. PLATE, *Ueber Bedeutung and Tragweite des Darwin'schen Selectionsprincips*, Leipsic, 1900. A very complete bibliographical list will be found on pp. 145-153 of this book.

A strong argument for my view was put forward by ROSA and CATTANEO.[1] According to these authors the extinction of large groups of species proves that the variability resident in them was powerless to adapt them to the changing conditions of life; and from this conclusion they infer that the ordinary variability, as it is *always* manifested, is not sufficient for this purpose. Obviously some other process is necessary.

II. *Fluctuating variability is linear*; it oscillates only in a *plus* and a *minus* direction, whilst adaptations demand a variability which will produce variations in all directions.[2] On this point also I have expressed my opinion in the first volume (p. 118). It constitutes, in my opinion, one of the strongest objections to the prevailing view; and it also shows more clearly than anything else how far DARWIN's adherents have departed from the views actually expressed by him. To DARWIN's mind the essential point was that the struggle for existence should have to select from material supplied by an indeterminate variability. Natural selection is a sieve. It creates nothing, as is so often assumed; it only sifts. It retains only what variability puts into the sieve. Whence the material comes that is put into it, should be kept separate from the theory of its selection. How the struggle for existence sifts is one question; how that which is sifted arose is another. In both respects, DARWIN's original view is still the best, but the point at issue has been often obscured by later writers. The meshes of the sieve are not such as to separate only the very best; on the contrary natural selection only throws out some part of the individuals, and amongst them the worst. i. e.,

---

[1] See below in § 12.
[2] GUSTAV WOLFF, *Der gegenwärtige Stand des Darwinismus*. 1896.

those the least adapted to the immediate external conditions. Selection is the elimination of the inferior, whilst the choice of the superior individuals should be called "election," and this leads to the stock-races (*elite* races) as in the selection of beets and cereals (Vol. I, pp. 99-118); or, as an admirable critic, A. KUYPER, says, "selection aims at the maintenance of species; election is the choice of persons."[1] The doctrine of the direct influence of the environment on organisms, as entertained by LAMARCK, is that against which DARWIN directed his hypothesis of indeterminate variability as being more in harmony with the demands of pure science. This old doctrine is repeatedly met with in modern times,[2] and this shows at least, in my opinion, that the prevailing form of the theory of selection does not find favor in those quarters.[3]

Thus the sieve of natural selection perpetually eliminates numerous individuals of inferior value; but how the differences between the individuals arise is another question. Linear variability provides differences only in two directions, by means of which selection can either increase or diminish, strengthen or weaken the various characters. It cannot effect more, unless material of another kind is provided by variation. The hypothesis of mutation meets this demand; for it necessarily assumes a variability in almost all directions, as I have shown

[1] A. KUYPER, *Evolutie*, Amsterdam 1899, p. 11.

[2] G. HENSLOW, *Does Natural Selection Play any Part in the Origin of Species*, Nat. Sc., XI, 1897, p. 166. WARMING, *Lehrbuch d. Oekologie*, p. 382. VON WETTSTEIN, *Ber. d. d. bot. Ges.*, 1900, Vol. XVIII, Generalversammlungsber. p. 184. STRASBURGER, *Ceratophyllum submersum*, Jahrb. f. wiss. Bot., Vol. XXXVII, p. 518, where a list of references to papers, dealing with the direct effect of the environment, will be found.

[3] R. V. WETTSTEIN, *Handbuch der systematischen Botanik*, 1901, p. 32.

in the first part of the first volume (p. 198), and our consideration of those species which are rich in subordinate forms, as well as the results obtained with *Oenothera Lamarckiana* have justified this claim. Thus we see that the current form of the theory of selection cannot supply the kind of variability which the theory demands, whilst the doctrine of mutation can supply it, as we know from actual observation.

III. *The first insignificant beginnings of new characters do not come under the operation of natural selection since they are of no significance in the struggle for existence.* This is the best known objection against the prevailing form of the theory of selection. It has been elaborated by many authors and admirably expressed by CONN in his work cited above, so that we need not deal with it further here. It ultimately leads every thoughtful investigator to the view that every organ must have its origin in a discontinuous variation.[1] The doctrine of mutation alone can overcome these difficulties although we must not forget that the objection is directed only against the present form of the theory of selection and not against DARWIN's own conception of it; for if the sieve of selection does no more than eliminate those of less fitness, and if its function is merely to increase the mean of those that remain, even the very slightest average progress must have a result, as DARWIN so frequently insisted.

In the doctrine of mutation, however, these slow transitions and these slight advantages have no place. Species-forming variability simply omits these, both in experiment and in horticultural experience, so that they constitute no obstacle to the theory.[2]

[1] CONN, *loc. cit.*, p. 134.
[2] It is in the explanation of instinct that the current form of the

IV. *The theory of selection explains the existence of useful characters, but does not explain that of useless or actually harmful ones.*[1] Whereas the doctrine of mutation assumes that specific differentiation does not take place in any definite direction, that mutations are produced independently of their adaptive value, and that they may survive, provided that they do not prejudice the existence or annul the fertility of the individual, the theory of selection cannot account for the origin of sterile forms of which, nevertheless, there are many. I have already mentioned instances of these in the first part of this volume, and will here only add a reference to a most remarkable sterile form of oats, recently described and figured by NOLL;[2] to the dates[3] and grapes[4] without pips, and to the highly branched and absolutely sterile variety of our gardens, called *Muscari comosum plumosum*, which belongs to a species characterized by a tuft of sterile flowers at the tip of the normal spike (Fig. 136).

Instead of giving a further discussion I shall content myself with referring to the curious case of *Mimulus* and *Torenia*, which BURCK has described, and which this

---

theory of selection is most evidently insufficient. See WASMANN, *Biol. Centralblatt*, Vol. XXI, Nos. 22, 23; also EMERY, *Gedanken zur Descendenz- und Vererbungstheorie*, Biol. Centralblatt, 1893, Vol. XIII, Nos. 13 and 14, p. 397; further W. WAGNER, *L'industrie des Araneina*, Mém. Acad. Imp. St. Pétersbourg, VIIth Ser., Vol. XLII, No. 11, 1894; and N. CHOLODKOVSKY, *Die Coniferenläuse*, Hor. Soc. Ent. Ross., XXXI, p. 43.

[1] For a list of these I refer the reader to DEMOOR, MASSART and VAN DER VELDE, *L'évolution régressive*, Paris, 1897, especially pp. 286-289.

[2] F. NOLL, *Sitzungsber. d. Niederrhein. Ges. f. Naturk.*, Bonn, March 4, 1901.

[3] CH. RIVIÈRE, *Société nat. d'acclimatation*, Paris, La Nature, 1901, No. 1477, p. 247. The tree in question grows near Hamma in Algeria.

[4] H. MÜLLER-THURGAU has given an exhaustive list of wholly or partially sterile varieties of grapes in *Experiment Station Record*, XI, p. 16, 1902.

The Explanation of Adaptations. 613

investigator claims to be of itself sufficient to disprove the doctrine of selection.[1] These plants contain four stamens in each flower, two large normal ones, and two

Fig. 136. *Muscari comosum plumosum.*

small abnormal ones. The latter contain fertile pollen, but never dehisce. The pollen is therefore absolutely

[1] W. Burck, *Kon. Akad. v. Wet.*, Amsterdam, 1901; Album d Natuur, 1902. See also the earlier writings of this author on facts which cannot be explained by the theory of selection.

useless, although if applied to the pistil, it is capable of producing an abundance of good seed. It is evident that this condition cannot have arisen either by gradual modifications or under the influence of the ordinary selection of useful characters.

All the difficulties which we have mentioned as besetting the current view, disappear if we substitute mutability for fluctuating variability, as the source of the origin of species; and there is no doubt that adaptations can be explained by mutability just as satisfactorily as by fluctuating variability.

§ 8. VEGETATIVE MUTATIONS.

In order to conduct an experimental investigation into the manner in which mutations arise, it is necessary to know when they occur. According to the prevailing opinion this moment is assumed to be that of fertilization. GALLESIO clearly expressed this view at the beginning of the last century, and advocated it against the belief in the direct influence of the environment which was common amongst breeders at that time.[1] Some association of species-forming variability with fertilization is generally regarded as being clearly demonstrated both in the animal and vegetable kingdoms, and especially for annual plants and those which are ordinarily multiplied by seed.

On the other hand we have the bud variations, or vegetative mutations as they should rather be called. They were well known to the older scientists, but DARWIN was the first to insist on their importance by collecting all

[1] G. GALLESIO, *Traité du Citrus; Teoria della riproduzione vegetabile,* Pisa, 1816; DE CANDOLLE, *Physiologie végétale,* II, p. 720.

the evidence bearing on them. Most of the later writers have agreed with him. DELAGE opposes the view that fertilization is the sole cause of variability, partly on the ground of the existence of bud-variations, partly because nothing new is produced in fertilization, in which nothing more occurs than a recombination of heritable characters already present.[1] SAVASTANO gives many reasons, derived chiefly from the study of woody plants, in support of the view that varieties usually arise from seeds and more rarely from buds.[2] BAILEY, on the other hand, lays greater stress on bud-variations. According to him bi-sexual reproduction is not a condition of variability, since many new varieties have arisen by vegetative means, such as several sorts of pine-apples, bananas, strawberries, apples, weeping willows, etc.[3]

KASSOWITZ goes furthest in this direction when he says:[4] "Even if there had never been any sexual reproduction, our earth would be peopled by beings differing widely from one another in their characters and in their functions; and there is no ground for the assumption that the differences between the most widely separated forms would have been any less without this (i. e., sexual reproduction) than it actually is."

The conclusion from this all too brief historical survey is that the importance of vegetative mutations is gradually obtaining wider recognition, whilst the attempt to associate species-forming variability with fertilization is coming to be regarded with less and less favor.

[1] Y. DELAGE, *L'hérédité*, 1895, p. 283.

[2] L. SAVASTANO, *La Varietà in arboricultura*. Annali d. R Scuola Sup. d'Agricoltura in Portici, 1899, I. 2, p. 63 and elsewhere.

[3] L. H. BAILEY, *The Plant Individual in the Light of Evolution*, Science, 1897.

[4] MAX. KASSOWITZ, *Allgemeine Biologie*, II, 1899, p. 247.

Before we proceed to a consideration of the facts let us first examine the question itself more closely. We may ask in which periods in the life of a plant the conditions for the appearance of mutation may be different. The life of a plant may be divided into the vegetative and the reproductive period, and there are therefore four possibilities to be taken into consideration. First we have the two periods themselves, then the two transitions from the one to the other, viz., the origin of the germ cells (the moment of the so-called numerical reduction of the chromosomes of the nucleus) and the fertilization, which latter is the beginning of the vegetative life. Confining ourselves first to general considerations, the processes involved in the origin of the sexual cells are obviously of a much more complicated nature than those involved in fertilization. On the other hand the sexual cells are usually regarded as more susceptible than the vegetative organs. For these reasons therefore, we might be inclined to assume for the moment of mutation not that of fertilization but some previous point of time. How much earlier, then becomes a further question.

In bi-sexual or Mendelian crosses segregation occurs during the maturation of the egg- and pollen-cells, and it is complete when these elements are being formed. For the egg- and pollen-cells of the monohybrids are no more of a hybrid nature, but entirely assume the one or the other of the two parental types. The same thing must hapen in mutation, for all the evidence seems to indicate that the egg- and pollen-cells have already mutated before they unite in fertilization.

Of course, as a rule the mutation will not find external expression until the germ develops, and thus the new character is only displayed after fertilization; but that the process should be its result, by no means follows from this. The moment of appearance evidently tells us nothing about the preparation which may have preceded it. This may have occurred during the sexual life, or may even have extended back into the vegetative stage.

The phenomena of sectorial variation, which are best known amongst striped flowers and variegated leaves (see page 114), but which have also been observed elsewhere, and especially in the sectorial segregation of hybrids support the latter view. I refer to the instance described above (page 276), of a variegated bud-variation in an oak. A variegated twig occurred on a bush whose leaves were otherwise quite green. But the point of insertion of the twig occurred on a variegated longitudinal stripe on the branch which produced it. The change therefore had not taken place in the actual origin of the bud, but long before. The term bud-variation is therefore, in such cases, not strictly applicable.

We can apply this instance to the appearance of mutations in general and say that the moment of the actual appearance of the character is preceded by a shorter or a longer period in which the change, although complete, was still in a latent condition. If, for instance, we are dealing with a transformation in flowers, a sectorial and a bud-mutation could occur without becoming externally visible. In the first part of this volume (page 123) I drew attention to the stamens with red stripes in striped flowers and dealt with the question whether the pollen grains themselves might not differ with regard to this mark, some of them possessing this character and others not. Obviously this question may be applied just as well to those characters which can not be seen in the stamens.

618 *Validity of the Doctrine of Mutation.*

I shall not pursue this train of thought further; but this much seems to be clear, that germinal variations may be the results of changes which have already taken place

Fig. 137. *Green Dahlia.* A branch whose terminal inflorescence b as well as the lateral one b' have become green, as in the remaining parts of the plant, whilst a branch has arisen at a from an axillary bud, bearing double red inflorescences of normal structure and without a trace of virescence. a', flowering and a'', a bud. See above, p. 92, Fig. 14 (1902). See also pp. 628-629.

in the pollen- and egg-cells; and that these changes themselves may have had their origin before the development of the sexual cells, and perhaps even before the origin of the flower itself. In other words:

Germinal variations may be regarded as a special case of vegetative mutations; and this possibility always remains open where the contrary cannot be proved.

Concluding these discussions I propose now to adduce a series of facts in which mutations have occurred vegetatively, that is to say, such facts as have hitherto been dealt with as bud-variations. It will be necessary to

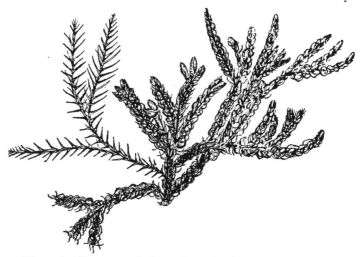

Fig. 138. *Cryptomeria japonica spiraliter falcata,* with an atavistic branch (see page 628).

consider three groups of phenomena separately: first, vegetative segregation in hybrids; secondly, vegetative atavism in eversporting varieties, especially as exhibited by striped flowers (Part I, Plate I), and thirdly, the true vegetative mutations which are usually of an atavistic nature (Figs. 137, 138), but sometimes may happen to be of a progressive kind.

*V*egetative segregations in hybrids are rare phenomena; but this may perhaps be due to the fact that in many instances of bud-variations the possibility of the plant in question being a hybrid has not been considered. A single case has fallen within my own experience; this is a hybrid between *Veronica longifolia* and its variety *V. l. alba.* The flowers of this hybrid are blue, and it may easily be kept in cultivation for many years by means of vegetative propagation. Ever since 1889, when I obtained the first specimen, I have had many thousands of stems in flower, amongst which I observed several cases of sectorial and of bud-variation, the last of them in the summer of 1902. The bud-variation sometimes occurred in the rhizome (1902); the whole shoot above the earth lacked the red pigment, both in its bark and its flowers, and so was easily recognizable before it bloomed. All the flowers were white, whilst those of the remaining shoots from the same rhizome were blue. Occasionally I found a raceme with white flowers arising as a lateral branch from a stem on which the rest of the flowers were blue (1894). The sectorial segregation is manifested in this hybrid in such a way that one side of the raceme had blue flowers, whilst those on the other were white.[1] The breadth of the longitudinal strip bearing white flowers is subject to variation; it may be either a half of the whole raceme (as in 1891), or a quarter (1898), or even less (1894, 1895). The seeds of flowers which have thus become white by vegetative mutation produce white-flowered offspring, so far as I am able to judge from some preliminary experiments.

As is well known, NAUDIN has crossed *Datura Stramonium* with *D. laevis,* and found amongst many hybrids with exclusively thorny fruits, three individuals which gave instances of vegetative segregation. These belonged

[1] *Ber. d. d. bot. Ges.,* 1900, XVIII, p. 86.

to the first generation and bore numerous fruits, the surface of which was partly thorny and partly smooth as in D. *laevis*. Sometimes one-half was smooth, but usually only a quarter or a smaller part. The valves without thorns had also reverted to the character of D. *laevis* in the fact that they were shorter than the thorned ones, and so did not fit into them properly. Fertilization had been left to insects, so that the results obtained by sowing the seeds would have been of no value. Such vegetative segregations appear, however, to be very rare in *Datura*. Other investigators do not seem to have obtained them; and I have myself frequently made extensive cultures of this hybrid in the hope of obtaining some, but as yet without success.

The records of instances of hybrid segregation are scattered through the literature of this subject, so that it may be worth while to give the following selection. SAGERET obtained a hybrid between *Brassica* and *Raphanus* with two types of pods.[1] DOUNET-ADANSON observed on an intermediate hybrid between *Abies Pinsapo* × *pectinata* a branch with the characters of *A. Pinsapo*.[2]

FOCKE mentions a case of the cross *Anagallis phoenicea* × *coerulea* which had red flowers but exhibited half a petal with the blue color of the latter species.[3] Vegetative segregations in the fruits of *Citrus* hybrids have frequently been described.[4] Flowers of different colors have been found occurring together on the same

---

[1] SAGERET, *Ann. Sc. Nat.*, 1826.

[2] *Bull. Soc. bot. Fr.*, 1899; ABBADO, *L'ibridismo, loc. cit.*, p. 26.

[3] W. O. FOCKE, *Nat. Ver. Bremen*, 1887, p. 422. See also GÄRTNER, *Die Bastarderzeugung*, p. 309, and FOCKE, *Pflanzenmischlinge*, p. 450.

[4] VERLOT, *La variabilité*, p. 14; KERNER, *Pflanzenleben*, II, pp. 559-560, etc.

stock in hybrids of *Helianthemum*.[1] Hybrid peas may be partly green and partly yellow[2] and similar segregations have been found by CORRENS and WEBBER on grains of hybrid maize.

But the best known instance of a vegetative segregation is furnished by *Cytisus Adami* (*C. Laburnum* × *purpureus*), which any one can observe in his garden or in parks. The hybrid is absolutely sterile, and all the specimens of it are grown from cuttings from one single individual. Whether the extraordinary tendency to segregate is peculiar to this hybrid as such, or to this particular individual, we cannot know.[3] The fact is, that this hybrid is intermediate between its two parents and besides this produces from time to time buds, of which some become strong branches of *C. Laburnum* with its large leaves and long racemes (Fig. 139 L), whilst from others the delicate, slender and bushy branches of *C. purpureus* arise, bearing their fruits and flowers either singly or in small groups (Fig. 140 P).

The current view is that *C. Adami* is a graft hybrid.[4] There is, however, neither an historical nor a physiological justification for this view.[5] The original raiser of the plant, the Parisian gardener ADAM, seems himself to have thought that this form which he put on the

[1] FOCKE, *Die Pflanzenmischlinge*, p. 473, where further instances will be found. See also BRAUN, *Verjüngung*, p. 336; DARWIN, *Animals and Plants Under Domestication*, etc. Similar facts are afforded by *Hieracium, Oxalis, Chamaedorea* etc.; see F. HILDEBRAND, *Jen. Zeitschrift*, Vol. XXIII, 1889, Plate XXV.

[2] F. WELDON, *Biometrika*, I, 2, 1902.

[3] See the remarks relating to *Datura* on the foregoing pages.

[4] *Intracellulare Pangenesis* (Ger. ed.), p. 206.

[5] I have given an historical account of *Cytisus Adami*, in Dutch, in the *Album der Natuur*, 1894, Part 7, under the title *Adam's Gouden Regen*. The original source of the historical data is found in *Annales de la Société horticole de Paris*, Vol. VII, 1830.

market only as a variety of *C. purpureus*, had arisen as a result of the grafting of *C. purpureus* on *C. Laburnum*; but his contemporaries did not agree with this, and CAMUZET maintains that he has seen the tree from which

Fig. 139. *Cytisus Adami*. A, A', A"; B, a branch of *C. Laburnum*, L, L', L", with numerous racemes bearing ripe pods.

ADAM had taken his buds for the purpose of making the grafts, and that this tree possessed all the characters of *Cytisus Adami*. The hypothesis that this form is a graft hybrid, originated with CASPARY (1865). No case of an undoubted graft hybrid has, however, as yet been produced experimentally and, consequently, the charac-

Fig. 140. *Cytisus Adami*. A, A', bearing at I a bunch of twigs of *C. purpureus*: P, H, and I.

ters which such hybrids would have if they existed are not known and can even hardly be guessed at. Therefore no conclusion concerning this question can be drawn from the characters of *C. Adami*.

The view that it is an ordinary hybrid which was afterwards grafted on *C. Laburnum* seems to be much more probable.[1]

In other data relating to graft hybrids this view has long been proved to be incorrect, as in LINDEMUTH'S ex-

Fig. 141. *Ulmus campestris variegata* with atavism by bud variation. A branch with larger green leaves has arisen at A.

periments with potatoes; or the grafts may have been made on hybrid stems, as LAURENT suspects to be the case in the *Néflier de Bronvaux*;[2] or again, a graft of

---

[1] The same view is adopted by E. LAURENT, *De l'expérimentation en horticulture*, Brussels, 1902, p. 16. The literature dealing with *Cytisus Adami* has been given so often that I need not go into further detail here.

[2] See the papers on this supposed hybrid between *Mespilus* and *Crataegus* in *Le Jardin* and *Journ. Roy. Hort. Soc.*, 1900. Vol. 24,

a hybrid may have been made on a normal plant, as WILLE has told me he suspects to be the case in the supposed graft hybrid consisting of a pear worked on a white thorn stock.[1]

The reader who is interested in the direct influence of the stock on the grafted bud is referred to the recent exhaustive studies by L. DANIEL.[2]

By pruning *Cytisus Adami,* BEYERINCK obtained very important results on the vegetative segregation of hybrids. He found that buds which, as a rule, are resting, but which can be made to develop by cutting off the higher branches, tend to produce the characters of *C. Laburnum* or of *C. purpureus,* so that we have it in our power to multiply the number of such segregations at will. More than one hundred instances were obtained by him on some few trees. Sectorial segregations of buds also occurred, sometimes transforming a longitudinal half of a shoot into *C. Laburnum,* whilst the other half remained *C. Adami.*[3] It is to be expected that the application of this principle to other cases will lead to the discovery of important facts.

Of the numerous instances of bud-variations described in the literature of this subject, many are, with-

---

p. 237. Also LAURENT, *loc. cit.*, p. 16. For a general review of graft hybrids see FRUWIRTH, *Züchtung landwirthschaftlicher Kulturpflanzen,* p. 72 ff.

[1] N. WILLE, *Mittheilungen d. biolog. Gesellschaft in Christiania,* Biol. Centralblatt, 1896, Vol. XVI, No. 3, p. 126. Perhaps this may be *Pyrus auricularis* (P. *communis* × *Sorbus Aria*) or a related hybrid. See DIPPEL, *Handbuch der Laubholzkunde,* III, p. 359.

[2] LUCIEN DANIEL, *La variation dans la greffe et l'hérédité des caractères acquis,* Ann. sc. nat. bot., 1899, VIIIth ser., Vol. VIII, pp. 1-226 and Plates I-X, and the subsequent publications of the same author.

[3] M. W. BEYERINCK, *Kon. Akad. v. Wetensch.,* Amsterdam, Nov. 1900.

out doubt, cases of such hybrid segregation, and therefore have no immediate bearing on the question of vegetative mutations. The same is true of the bud-variations of eversporting varieties (see Plate I, *Antirrhinum*) which

Fig. 142. *Rhus typhina.* A leaf of an otherwise green bush which was almost yellow from *a* to *b*; these leaflets have grown much smaller (on one side at *a*), *Doorn* (1886); collected by Mrs. WEBER.

have already been dealt with at sufficient length. Moreover the graded differences between the various branches of a plant belonging to an eversporting variety are not instances of mutations, and, as a rule, do not affect the hereditary properties of the seeds which they produce; as, for instance, in *Chelidonium majus flore pleno* (p. 336).

In the majority of cases a more detailed examination is urgently needed, before the true nature of even the commonest bud-variations can be properly understood. This is particularly the case in variegated plants, amongst which every one is familiar with the phenomenon, on shrubs and trees at any rate; but even here the process has not yet been exhaustively studied. On the one hand, some of these cases consist of bud-atavism, whole branches of a variegated variety reverting to the normal type of the species in their color as well as in their secondary characters (Fig. 141).[1] On the other hand, sometimes halves of leaves become green, or occasional branches with usually slightly, but sometimes finely, variegated leaves arise on green individuals (Figs. 142 and 143). On a large tree of *Morus nigra* in our garden the latter phenomenon occurs almost every year.

It is not until all these and similar cases have been excluded that bud-variations may be regarded as true cases of vegetative mutations. Even then we should require the proof that the deviating branches will reproduce their type from their seeds, after pure self-fertilization. In many cases, however, this is not possible because the bud-variations in question often bear no seeds, even when they occur regularly, as in *Cephalotaxus pedunculata fastigiata* (p. 109, Fig. 16) and in numerous other conifers, the bud-variations of which have been described by BEISSNER; as, for instance, in *Cryptomeria japonica spiraliter falcata* (Fig. 138, p. 619). Even from the green *Dahlia* I was, unfortunately, unable

---

[1] See above p. 111 and pp. 272-277.

to obtain any seed (Fig. 137, p. 618), because it flowered too late in the year. I have cultivated this plant, which was described and figured in the first section of this volume (p. 92), for many years by vegetative methods, because it is perfectly sterile. In the summer of 1902, however, it suddenly began to produce bud-varia-

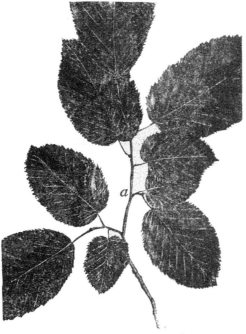

Fig. 143. *Carpinus Betulus.* At *a*, a partly variegated leaf on a tree which had otherwise only green leaves, Hilversum (1887).

tions, distributed in considerable numbers over the various main stems; they were apparently independent of one another, but seemed to arise in response to the same unknown external influences. The flowerheads of the atavistic branches were of the normal structure of the ordinary Dahlias, double, and with carmine red florets, at the tip of each of which was a white spot. From this we may conclude that this green *Dahlia* had arisen from the corresponding double-flowered variety.

I do not propose to adduce any further instances; those which I have given show that varietal characters may disappear in a vegetative way, the original characters becoming active again. For such a mutation neither the formation of sexual cells nor fertilization is necessary. Therefore the possibility that seed-variations may ultimately be derived from bud-variations cannot be denied; but many more experimental data will have to be collected before a final judgment based on a sufficient foundation of facts can be given. One of the best methods is that which I have already mentioned, as adopted by BEYERINCK, of artificially inducing bud-variations by pruning.

III. THE MATERIAL VEHICLES OF THE HEREDITARY CHARACTERS.

§ 9. DARWIN'S PANGENESIS.

The real meaning of the title of DARWIN'S book, *On the Origin of Species by Means of Natural Selection*, has often been largely misapprehended. In DARWIN'S mind the emphasis lay on the word "natural." He pleaded a natural origin of species against the accepted supernatural one. The whole object of his work was to show that the genetic relationships of animals and plants may be explained without invoking supernatural causes, and that their explanation by natural means is far more satisfactory and simple. For this conviction he succeeded in obtaining general credence, and thus laid a broad foundation for all future investigation in this sphere.

The modern tendency, however, is to shift the emphasis on the word selection and on the analogy between selection in nature and the artificial production of races in agriculture. In doing so it is often overlooked that half a century ago the science of the various forms of variability, and consequently of selection, was still in its infancy, and that we must not apply our present knowledge to the state of opinions which prevailed then. No distinction was drawn at that time, for instance, between mutability and variability; and it was DARWIN who first attempted in various cases to distinguish between these two types of variation.

Even now this contrast does not yet find so clear an expression in the available facts as to insure its immediate recognition. In DARWIN's time many more obstacles stood in its way, and it is probable that its real signification did not become manifest to him until after he attempted to deal with the phenomena of heredity in a theoretical way.[1]

As is well known, this attempt was made in his preliminary hypothesis of "pangenesis"; it is also known that he attempted to adapt his idea to other theories, prevalent at the time, by a series of subsidiary hypotheses which have now become superfluous; and that by doing so he did more harm than good to his theory. For, in combating these secondary hypotheses, most of his critics have overlooked the real value and significance of the main thesis.

In my book on *Intracellular Pangenesis* I have attempted to show how the importance of DARWIN's hypothesis can only be really appreciated if it is freed of these superfluous adjuncts.[2] In this essay I have also endeavored to prove that the germ of the theory reappears, in a more or less similar form, in the hypotheses of his successors; and that in these too, it is usually confused by useless or even erroneous suppositions. My object was to extract this essence and to bring it into as close relation with the available knowledge as was possible without the aid of too many auxiliary hypotheses.

It is not now my intention to give a review of the

[1] See *Different Kinds of Variability* in Darwin and Modern Science, pp. 66-74, 1909.

[2] *Intracellular Pangenesis,* translated into English by Prof. C. STUART GAGER, 1910.

enormous amount of literature which has since accumulated on this subject,[1] and with regard to the older theories, such as those of SPENCER, NÄGELI and HERTWIG, I need do no more than refer to my essay already quoted. My only task is to show that the evidence, brought forward in this book for the theory of mutation, affords a strong suport for the principle of pangenesis. All that is necessary to bring the results of observation into line with the doctrine of Pangenesis, is to substitute the idea of internal factors or material vehicles of hereditary characters for the empirical units of the visible qualities.[2] This view has been best worked out by JOHANNSEN in the section on the doctrine of pangenesis in his textbook of botany which has recently appeared; and this fact enables me to deal briefly with the topic.[3] • I propose to confine myself to a brief exposition of DARWIN's conception of pangenesis and to the modification of it which I suggested, without repeating all the observations on the subject which I have recorded in this book. I shall deal first with the essence of the hypothesis and then with the secondary hypotheses; and shall defer a discussion of the essence of the theory until later.

There are two essentially different views relating to the material vehicles of the hereditary characters of organisms. One view is that of SPENCER, according to

[1] Full lists of references are given in a large number of works of which the following are among the best: C. KELLER, *Vererbungslehre und Thierzucht*, 1895; H. MARLIÈRE, *Etudes sur l'hérédité*, 1895; E. B. WILSON, *The Cell in Development and Inheritance*, 1900; FRUWIRTH, *Die Züchtung der landwirthschaftlichen Culturpflanzen*, 1901, etc.

[2] *Ber. d. d. bot. Ges.*, 1900, XVIII, p. 83, and *Sur les unités des caractères spécifiques*, Revue générale de botanique, 1900, XII, p. 257.

[3] E. WARMING and W. JOHANNSEN, *Den almindelige Botanik*, 4th ed., 1901, pp. 675 ff. It is quite satisfactory to state here that several critics of the first volume of my book have anticipated this conception of the relation of the doctrine of mutation to pangenesis.

whom the character of every species constitutes an inseparable whole and every material vehicle, therefore, represents the sum of all its qualities. The adherents to this view are still in the majority. It is the very opposite of DARWIN's view, according to which the individual cells of the organism and the individual component elements within the cells each have their special representatives in the hereditary substance. Thus the material basis of inheritance is composed of as many different units as there are separate organs and types of cells.

NÄGELI has introduced the word idioplasma for this material basis; and for many reasons this term should be universally employed, especially since it can be used in speaking of the two opposite theories. To NÄGELI the idioplasm was a unit, but we may, as well, apply this term to the sum total of DARWIN's units.

The independent behavior of the individual hereditary characters both in the process of mutation and in hybridization, definitely proves, in my opinion, the correctness of DARWIN's assumption of separate material bases for each one of them; and the whole contrast between mutability and fluctuating variability can only be brought into harmony with the theory in the light of this principle.[1]

DARWIN's pangenesis may be summarized in the two following theses.[2]

The individual cells and organs of the whole organism are represented in every germ cell and every bud by

[1] The correctness of this view receives strong support from the fact that those of my critics who are partisans of WALLACE's form of the theory of selection, simply deny the distinction between mutability and fluctuating variability. See above, p. 599.

[2] DARWIN, *Animals and Plants under Domestication*, Vol. II, Chapter on Pangenesis; and my *Intracellular Pangenesis* (Engl. ed.), pp. 5 and 60.

definite material particles. These multiply by division and at cell division pass on from the mother cell to the daughter cells (doctrine of pangenesis).

Moreover all the cells of the body throw off these particles at various periods of their development. These reach the germ cells and hand over to them any characters of the organism which they may lack (hypothesis of transportation).

The multiplication of the material vehicles of heredity and their handing down, in the course of development, by the successive cellular and nuclear divisions, can be most clearly seen in those cases in which certain vicarious characters remain united during the greater part of the development, and do not become separated until the cell divisions have nearly come to an end. These units are then seen to be distributed after the manner of a mosaic. MACFARLANE was the first to draw attention to this significance of the phenomenon as seen in hybrids; and he has shown, especially in *Geum intermedium*, how the influences of both parents can remain combined in the individual cells, or can be recognized as separating out in these. *V*ariegated leaves often show these late separations very clearly,[1] often in large areas of the most widely different forms and shapes (Figs. 144 and 145), or in small groups (Fig. 146).[2] But our knowledge of a relation between these mosaic figures and development is not yet sufficiently complete to enable us to form a picture of these segregations of compound characters on the basis of the pedigrees[3] of the cells.[4]

[1] A. J. J. VAN DER VELDE, *Anatomie en physiologie der bonte bladen*, Handelingen, V. Vlaamsch Nat. en Gen. Congres, Bruges, Sept. 1901.
[2] MACFARLANE, *On the Minute Structure, passim*.
[3] See *Intracellular Pangenesis* (Engl. ed.), p. 88.
[4] Interesting particulars have been brought to light by the recent

636   *Vehicles of the Hereditary Characters.*

The second part of DARWIN's Pangenesis is met by insuperable difficulties, and has long since been regarded as superfluous. DARWIN himself admitted that it had only a very limited application to plants and corals, to

Fig. 144. *Ulmus campestris variegatus.* The variegated regions are developed in widely different extents on the different leaves.

Fig. 145. *Beta vulgaris saccharifera.* A chance variegated leaf. The yellow and green tissues are superimposed upon one another in such a way that the limits of the individual layers do not correspond.

investigations of BONNIER and FLOT. See G. BONNIER, *Formation des éléments du cilindre central*, Comptes rendus, Vol. CXXXI, p. 781 (Nov. 12, 1900), and *Sur la différenciation des tissus vasculaires, loc. cit.*, p. 1276 (Dec. 1900); and LÉON FLOT, *Sur l'origine commune des tissus dans la feuille et dans la lige, loc. cit.* (Dec. 1900).

worker bees and similar cases. If we let it drop, the first thesis stands out in a much purer light.

Fig. 146. *Quercus sessiliflora albovariegata.* A commercial variety. Green patches are scattered irregularly over a yellowish-white back-ground.

In this direction, GALTON and BROOKS have done great service to the theory of pangenesis. Both strongly support the main thesis, and attach little importance to the transportation hypothesis.

GALTON supposes that the various cells of the body are originally represented by special material vehicles. These vehicles combine to constitute the stirp which is practically the same as the idioplasma.[1] Many more vehicles must, however, be present in the germ than there are actual types of cells. The remaining latent particles play a still more important part in GALTON's theory than they do in DARWIN's, both in the explanation of ontogenetic development and of atavism. The germ substance is handed down unchanged in the cell divisions as well as in the multiplication of the individuals. It is only under certain circumstances that changes appear in it; an assumption, which is obviously necessary for the explanation of the transformation of species.

In an extraordinarily clearly written book on heredity W. K. BROOKS has also suggested a modification of the theory of pangenesis.[2] He does not reject the whole transportation hypothesis, but confines it to the transportation of a few particles in special cases, especially when the organism undergoes any change, whether from external or internal causes. A change in the environment of a cell induces it to throw off particles, and therefore to transmit to the offspring of the plant a tendency to deviate in the corresponding parts of the body in the same way (p. 83). The male germ cells are particularly adapted to the purpose of collecting these particles and transmitting them to the female germ cells. If a change has once become heritable through the medium of trans-

---

[1] FRANCIS GALTON, *A Theory of Heredity*, 1875.

[2] W. K. BROOKS. *The Law of Heredity, a Study of the Cause of Variation and the Origin of Living Organisms*, 1883. See especially pp. 80-100, 319, 327, etc.

ported particles, there is no further need for their participation, and they will simply be eliminated by natural selection.

BROOKS maintains that it is not necessary to assume special material vehicles for the hereditary characters of every leaf or of every flower or of every cell of the body. We must simply assume that all similar organs or cells are represented by the same unit. The otherwise highly complex conception of the structure of the idioplasm is thus considerably simplified.

The work of GALTON and BROOKS has contributed largely to freeing the theory of pangenesis of much useless ballast, and therefore to exhibiting its essence in a much purer light; but in regard to one important point they still adhere too closely to DARWIN's old conception of the theory. This point is the question whether the organs and the cells themselves are the units which we must think of as being represented in the idioplasm.

§ 10. INTRACELLULAR PANGENESIS.

In contrast to the opinion of the authorities cited in the previous section, I assume that the units are not the morphological elements, such as the parts of the body and tissues, nor the cells and their visible organs. On the contrary I assume them to be the internal elementary characters which determine the external features of the organism, and which must cooperate to build morphological elements (Fig. 147).

In his books, DARWIN is not always quite clear concerning what he means by a single heritable character. Often he speaks of his particles as representatives of cells; at other times other morphological elements or groups are regarded as units. Nevertheless, in some places, he has definitely expressed the view that every character which can vary independently of others must be represented by a discrete material vehicle.[1]

This principle has been largely supported by recent investigations. Morphological elements are coming to

Fig. 147. *Rubus fruticosus laciniatus.* The deep dissection of the margin of the leaf, which is the character of this variety, is seen to be expressed both in the foliage leaves and in the petals. Both phenomena are obviously the expressions of the same material vehicle of the internal character.

be recognized more and more as combinations of units, whilst independence in variation is regarded as the real sign of the existence of separate elementary units.

[1] DARWIN, *Variations*, 2d ed., 1875, II, p. 378. *Intracellular Pangenesis* (Engl. ed.), p. 19 ff.

But it is quite true that these elements are usually united into smaller and larger groups,[1] and that these then behave as units; in such cases the individual components of the group are either manifestly associated with one another or are influenced in the same way by external conditions of life, in the manner of their expression. The inflorescences of monœcious plants are typical instances of such groups of characters; other instances are furnished by stolons, and in general the development of organs may be dependent on causes which bring whole groups of characters into activity as such, keeping other groups in the latent condition.

With regard to the hypothesis of a transportation of the vehicles of the elementary characters, or the pangenes[2] as I call them, I agree with the view, expressed by GALTON and BROOKS, that this theory should be considerably limited, and I think that it would even be best to drop it altogether.

It is only within the cells that the assumption of such a transportation is necessary and only so far as regards the extrusion of the pangenes from the nuclei and their distribution through the protoplasm.[3]

According to my view the whole protoplasm consists of pangenes; and these alone constitute the living elements (*Intracellular Pangenesis* [Engl. ed.], p. 195). Following HERTWIG we assume that the hereditary char-

---

[1] *Intracellular Pangenesis* (Engl. ed.), p. 24 and elsewhere.

[2] In German there are almost as many names for these material vehicles as there are authors who have dealt with the subject; for instance, Mikroplaste, Archiplaste, Biomolecüle, Protobionten, Bioblaste, Elementarorganismen (ALTMANN), Plasome (WIESNER), Funktionsträger, Idioblaste (HERTWIG), Chonder (SCHNEIDER), etc.

[3] Where cells are mutually united by protoplasmic strings, the possibility of a transportation of pangenes from one cell to another seems still to be open.

acters are located in the nucleus, and that in cell-division they are transmitted from one cell-generation to the next. In the nuclei, however, most of the pangenes are inactive. To become active they must leave them or at least their framework and take up a position in the surrounding parts of the cell-body. A detailed consideration of the life phenomena of the cell has led me to the conclusion that it is indispensable to assume that the influence of the nucleus on the vital processes is a material one and that it will be found, on closer examination, that even the dynamic and enzymatic theories of this operation cannot make superfluous the hypothesis, that pangenes form the real substance of all protoplasm (*loc. cit.,* p. 202, 204).

The recent investigations by GERASSIMOW on cells of *Spirogyra* without nuclei or with two nuclei[1] strongly support this view, and from the zoo-physiological side DRIESCH and HANSEMANN have expressed themselves similarly.[2]

In the idioplasm of the nucleus the pangenes multiply by division. A part of those which have been formed remains in position and furnishes the vehicles of the hereditary characters for the next cell division. The other part, however, emerges from the nucleus and becomes active in the cytoplasm. Here they multiply so as to contribute considerably to the material out of which the several organs of the cell are built up, such as the chromatophores, the outer layer of the protoplasts, the walls of the vacuoles, etc. In this way they impose their

[1] J. J. GERASSIMOW, *Bull. Soc. Imp. Nat. Moscou,* 1901, Nos. 1 and 2; *Zeitschrift f. allg. Physiologie,* I, 3, 1902, p. 220; see also the literature cited there.

[2] H. DRIESCH, *Analytische Theorie der organischen Entwickelung,* 1895; D. HANSEMANN in *Virchow's Archiv,* Vol. CXIX, p. 315.

character on them, and this assumption furnishes an explanation of the fact that the functions of the cell-organs conform to the hereditary characters of the plants to which they belong. This migration from the nuclei is repeated at intervals,[1] and thus the body of the cell soon comes to consist almost solely of pangenes which have recently emerged from the nucleus.

§ 11. THE PANGENES AS BEARERS OF THE HEREDITARY CHARACTERS.

As an hypothesis, pangenesis serves a heuristic object; as a theory, it must serve as a basis from which a deeper insight into the nature of living substance may be obtained. I have not much to say here as to its heuristic value, since for myself pangenesis has always been the starting-point of my inquiries; at first only in a theoretical way, but afterwards also for the experimental investigations described in this book.[2]

Especially is it this hypothesis which has led me to search for mutations in the field,[3] because I hoped in this way to find facts which would throw a more immediate light on the bearers of hereditary characters, and thereby on the theory of heredity in general.

The doctrine of pangenesis only touches the kernel of the general theory of heredity and leaves the details to special theories; but experience has taught me that

---

[1] My belief that the transportation of the pangenes is largely brought about by the so-called streaming of the protoplasm and that this is a normal and general occurrence has not in the least been shaken by the arguments which have been urged against it.

[2] Se the bibliography at the beginning of this volume.

[3] I should like to insert here the following little coincidence. My *Intracellulare Pangenesis* was written during the summer holidays, spent near Hilversum in 1888, and the often described locality of *Oenothera Lamarckiana* was only about ten minutes walk away.

this kernel is a sufficient basis for experimental investigation, and that it is far more likely to lead to the discovery of new important facts than the elaborate tissue of hypotheses which have grown up around it. Moreover pangenesis is capable of much closer application than the opposite view that each of the units bears the whole of the specific characters. I confidently recommend DARWIN'S principle to any one in search of new lines of research in this field.

In the first place it has led to the proper distinction between the two main types of variability, viz., mutability and variability in the restricted sense. "Finally, we see," says DARWIN, "that on the hypothesis of pangenesis variability depends on at least two distinct groups of causes."[1] The first group embraces the failure, the over-production and the change in position of particles without their being themselves transformed in the process. These changes can explain a great deal of fluctuating variability. Into the other group fall the changes in the particles themselves producing new types which in multiplying will develop into new characters.

Into these categories fall three main types of variability, since the first group is obviously a twofold one, embracing in terms of my present view, on the one hand fluctuating variability, and on the other the regressive and degressive mutations. The former may be caused by changes in the number of the pangenes; the two latter, however, by the "transposition of gemmules and the redevelopment of those which have long been dormant." Besides these the origin of new forms of pangenes obviously corresponds to progressive mutability.[2]

[1] *Animals and Plants under Domestication*, II, 2d ed., 1875, p. 390.
[2] See also *Intracellular Pangenesis* (Engl. ed.), pp. 73 and 214.

Numerical changes of the pangenes are therefore the basis of fluctuating variability. Changes in the position of the pangene in the nucleus lead to retrogressive and degressive mutations, whilst to account for progressive mutation we must assume the formation of new types of pangenes.

The facts which have been described in this work, conform to this hypothesis so closely that they may be regarded as proofs of the truth of the principle. It seems desirable to deal further with this point without, however, elaborating subsidiary hypotheses.

MENDEL's discovery goes far to support the theory of independent bearers of hereditary characters. Their independence stands out more clearly in this case than in any other, with the exception of the process of mutation itself. In my first preliminary note on this subject I have pointed out the great importance of his laws in their bearing on the doctrine of pangenesis; and since that time CORRENS, BATESON, CUÉNOT and many other authors have more or less subscribed to this view. CUÉNOT, who, with BATESON, was the first to demonstrate the applicability of MENDEL's laws to the animal kingdom, calls the units which are concerned in these crosses *"particules représentatives."*[1] Whether these factors themselves are the pangenes of the nuclear threads, or whether these factors are composed of groups of similar units, is a highly important question which, however, can only be decided by means of future experiments.

For, as BATESON points out, it is still possible that the constant Mendelian hybrid races are not absolutely pure with regard to their individual characters; that is

[1] L. CUÉNOT, *La loi de Mendel et l'hérédité de la pigmentati*" *chez les souris*, Arch. zool. expérim. et générale, 1902. No 2

to say, in the formation of the germ-cells the dominant and the recessive characters may perhaps not separate fully, leaving, either always or only exceptionally, a trace of the dominant character in the germ-cell which has the recessive one, and *vice versa*. This trace may then be latent during the course of a number of generations, until at some later moment, and for some unknown reason, atavistic phenomena in such hybrid races awaken the memory of the original cross. Experience does not as yet support this view; it wants a much larger number of generations before a final verdict may be expressed. But it is obvious that an atavism of this kind, if it occurred, would suggest that the Mendelian units were of a compound nature.

These Mendelian factors maintain their independence during vegetative life and fertilization. According to previous conclusions, such crosses are always concerned with elementary characters which occur in a different condition in the one parent from that in which they occur in the other. There are mainly four distinct conditions: the active and the latent, the semi-active and the semi-latent. Their vehicles do not only separate in the formation of the sexual cells, but occasionally also in the vegetative life of the plant, as is demonstrated by the occurrence of so-called bud-variations in hybrids.[1] They are therefore in such cases only loosely associated and not blended together.

Fluctuating variability is due to variation in the number of equivalent pangenes; this explains why it is only linear (Vol. I, p. 118) and why it is manifested in two directions only. It goes in the *plus* direction by a multiplication and in the minus direction by a diminution of

[1] See pp. 619-620.

their number. Higher nutrition and favorable conditions of life effect an increase, whereas the opposite circumstances cause a decrease of this number. But the various kinds of pangenes are susceptible to these stimuli, on the one hand in a different degree, and on the other hand at different periods in the life of the plant, for some characters are highly variable, others not at all. In the first volume we have pointed out the existence of susceptible periods of variability. They teach us how it is possible that the different characters of organisms may react in different ways to the same external conditions. Correlative variability, in so far as it is not due to a coupling of pangenes by their association in groups, finds its sufficient explanation in this way.

The significance of normal fertilization appears in quite a new light when viewed from the standpoint of this conception. The conditions of life affect the several characters in a similar manner though in a different degree; but they cannot, so far as we can judge at present, combine in the same individual characters, which deviate in opposite directions. The only practical way in which this can be effected is by an exchange of elements, such as happens in fertilization and probably especially at the beginning of the formation of the sexual cells. In this way sexual reproduction can unite characters which vary in different degrees and directions, in every possible kind of combination; and it is left to natural selection to decide which of these combinations are the best in every individual case.

The theory of mutation assumes that the pangenes, or groups of similar pangenes in the idioplasm, may exist in various conditions and positions. The normal active condition is that in which they multiply at a definite period in the development of the organism and, in part, escape into the protoplasm, there to exercise their functions. Diametrically opposite to this is the latent condition, for in it this kind of multiplication is possible only to a very limited extent or not at all. In other positions two groups of dissimilar but homologous pangenes have a mutual effect upon one another which varies according as the one or the other obtains the mastery. This is seen in the case of the vicariating characters of the half races and eversporting varieties. Here the two elements are affected by external conditions in the same way but in vastly different degrees, the phylogenetically older one being scarcely at all susceptible, while the younger one is highly susceptible. If the latter retires into a latent state, as in the case of half races, the degree of their manifestation, that is of the migration of the material particles, from the nuclei into the protoplasm, is a limited one. If, however, they are in the semi-active condition, as in the case of the eversporting varieties, the result is the extraordinary variability which characterizes these races.

The nature of the difference between uni-sexual and Mendelian crosses is now obvious without further discussion. If each element finds its partner during the formation of the sexual cells of a hybrid, exchange takes place as in ordinary fertilization, and the Mendelian crosses become merely a special case of this. But if one or two or several elements do not find partners, the normal process will obviously be disturbed, since the two idioplasms do not fit one another exactly. And on the degree of this disturbance, that is to say on the number of differentiating elements, depend obviously in the first place the fertility of the cross, that is to say the capacity

of the hybrid to live, and in the second place the fertility
of the hybrid itself. If, however, they are fertile, the
unpaired characters probably simply divide in the primary
hybrids, at the moment of sexual reproduction, in a vegetative way, and this would explain the constancy of such
hybrid races.

Progressive mutations are due to the formation of
new pangenes. Dissimilar entities arise in the idioplasm
instead of only similar ones, and this is the process
which we have called pre-mutation. The pre-mutated
pangenes tend to be inactive at first, either because they
do not exist in sufficient numbers or for other reasons.
Obviously it is very probable that in different species
similar pangenes may lead to the origin of the same new
pangenes; and this might perhaps explain many phenomena of parallel progressive mutability.

Lastly we must suppose that the pangenes, or groups
of them in each of the conditions referred to, may be
more closely or more loosely associated with the remaining ones. If the association is a close one it will remain
the same through all generations, and the species or
variety is immutable with regard to the character in
question. If the equilibrium is an unstable one, the
character in question is mutable; and slight external influences may turn it into a stable condition and thus induce the visible mutations such as those of the *Oenotheras*.
Unfortunately, however, the nature of these influences
is still unknown. The stable condition, which in this way
arises out of the mutable one, can be either active or
latent.

It would be easy to extend this discussion further;
suffice it, however, to say that the relation of pangenesis to new discoveries is everywhere more or less ob-

vious, if closely examined. This demonstrates, in my opinion, the truth of the two doctrines of pangenesis and mutation; and opens an ever wider field for the investigation of hereditary phenomena.

IV. GEOLOGICAL PERIODS OF MUTATION.

§ 12. THE PERIODICITY OF PROGRESSIVE MUTATIONS.

The essence of the theory of mutation lies within the narrow limits of the Linnean collective species and agrees equally well with the theory of descent with modification and with the doctrine of creation. Its special province is the question how those smaller species originate which were supposed in pre-Darwinian times to have arisen by natural laws from the created types, i. e., from the collective species.

But the light shed by the new theory extends far beyond these narrow limits. Its full importance can better be estimated from a general point of view than by a reconsideration of the facts already given, and the final judgment will probably depend in a larger measure on its applicability to the broad questions of descent, than on the significance of the facts upon which it is based.

Therefore it seems desirable to show that the mutation theory is really in closer accord with present views regarding the phylogeny of plants and animals, in many and indeed in the most important points, than the prevalent form of the theory of selection.[1] In doing so I shall confine myself as much as possible to the opinions

[1] See my lecture delivered before the association of German naturalists and medical men at Hamburg in September 1900, *Die Mutationen und die Mutationsperioden bei der Entstehung der Arten* (Leipsic: Veit & Co., 1901).

of the best authorities; and shall not propose any new hypotheses, but merely point out the agreement between the doctrine of mutation and the theories which have been put forward by others. I shall be treading new ground and shall therefore be as brief as possible, referring the reader to the literature on the subject for information on special points without dealing with these in detail.

I will first discuss the conclusions which may be derived from a consideration of the mutation period in *Oenothera Lamarckiana,* and shall then attempt to show that these are in perfect harmony with geological and paleontological facts.

Starting from the fact that our *Oenothera* is at present in a condition of mutability, we naturally ask the question whether this condition has had a beginning or not. If it had, the plant must have had, at some previons time, immutable ancestors; if it had not, all its ancestors, back to the most simple organisms, were as mutable as it is now.

The former view agrees with that which was held about the middle of the previous century, before the spread of DARWIN's ideas. The general conception was, "que les espèces varieraient plus à certaines époques de leur existence qu'à d'autres."[1] This obviously leads, in our special instance, to the supposition of a period of mutation; and this is exactly the view expressed in the first volume of this book. It leads, further, to assume periodic mutations which have alternated with periods of immutability; for if all the various elementary characters whose accumulation has ultimately led to the origin of

[1] H. LECOQ. *Géographie botanique,* 1854. See also ALPH. DE CANDOLLE. *Géographie botanique raisonnée,* II, pp. 1100-1102.

The Periodicity of Progressive Mutations. 653

our species, have arisen suddenly, these changes must have been distributed more or less regularly over the whole line of ancestors of the *Oenothera*. How many steps are combined into a single period of mutation cannot be determined, and the question is obviously of secondary interest only. The available evidence seems to indicate that only one step in the same direction occurs at one time; but obviously this does not exclude the possibility of periods in which more numerous changes occur.

In order to apply the results obtained with our primroses to earlier hypothetical periods of mutation, I will repeat the empirical pedigree of the first volume (p. 224), but in a somewhat different form. I will indicate the lateral branches which arise from the main stem in successive years, that is to say the new species, in the form of radiating groups (Fig. 148). Each group denotes the mutations in a single generation. The main stem continues unchanged and successively produces the individual groups. Together, however, they obviously belong to one and the same period, inasmuch as each of them mainly consists of the same species and in approximately equal proportions.

In order to compare this period with previous ones the whole figure may be compressed to a single group. This has been done in the upper part of Fig. 149. The lateral branches do not arise here from a single point, and this is intended to indicate the fact that the figure embraces a series of generations in which the variations were repeated.

As stated above, we will now assume that the ancestors of our *Oenothera* have not always been mutable. Therefore our group must have a limit below, and must, so to speak, be borne by a stem without lateral branches.

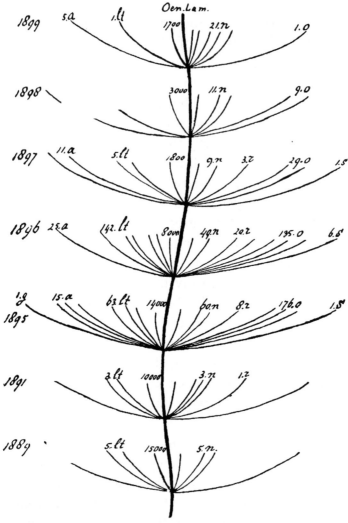

Fig. 148. Pedigree of *Oenothera Lamarckiana*, exhibiting the yearly origin of new species in my experimental garden in the years 1889-1899. *g, O. gigas; a, O. albida; lt, O. lata; n, O. nanella; r, O. rubrinervis; o, O. oblonga; s, O. scintillans*. The numbers preceding the letters are those in which the species in question arose. The numbers on the main stem show the extent of the yearly cultures.

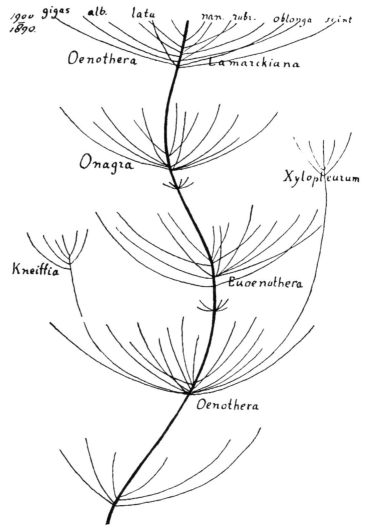

Fig. 149. Schematic pedigree of progressive formation of species; based on *Oenothera Lamarckiana*. The upper group is a reduced form of Fig. 148, and contains the same new species. *Onagra* is the sub-genus to which *Oenothera Lamarckiana* belongs. *Euoenothera*, *Kneiffia* and *Xylopleurum* are other sub-genera of *Oenothera*. The two small groups of lateral branches which have been intercalated are intended to represent the several intermediate periods of mutation. The figure can be continued downwards in a similar manner.

If now we follow this stem downwards, we must obviously sooner or later arrive at another mutation period, and of course one of which, although direct observation is no longer possible, so many products remain, that we may conclude with a high degree of probability its similarity with the period observed by me. I am referring to the differentiation of the sub-genus *Onagra,* and of its numerous species such as *O. biennis* L., *O. muricata* L., *O. cruciata* Nutt, etc. I have already dealt at length in the first volume (p. 440) with this hypothetical period, and therefore may now limit myself to representing this *Onagra*-period in Fig. 148 in the same way as the group above it, which relates to the variations now being produced by *Oenothera Lamarckiana.*

Obviously we may now continue our scheme downwards. We next reach the sub-genus *Euoenothera,* many of the species of which are very like those of *Onagra,* and have, indeed, sometimes been confused with them, as, e. g., *O. odorata* with *O. suaveolens.* From these we attain to the genus *Oenothera* itself, whilst other sub-genera form lateral branches, of which *Kneiffia* (Fig. 89, p. 458) and *Xylopleurum* have been selected as examples in Fig. 149.

I have so often made reference to the vestiges, left by other past, but relatively recent, periods of mutation[1] that I may now confine myself to mentioning the following: *Draba verna,* according to JORDAN and ROSEN (Vol. I, p. 173 and Fig. 3 on page 22); *Viola tricolor* (Vol. I, p. 23, Fig. 4) according to WITTROCK's researches, *Hieracium, Rubus, Rosa, Helianthemum* and many other genera with their numerous closely related species are

[1] It would appear, from WASMANN's beautiful investigations, that certain beetles (of the genus *Dinarda*) which live in association with ants are at present in a mutable state. *Biolog. Centralblatt,* XXI, Nos. 22 and 23, Dec. 1901.

instances of such groups. In such cases STANDFUSS, in conducting his well-known experimental investigations into the relations between closely allied species of insects, uses the expression, "changes like successive explosions."[1] Every genus rich in species gives him the impression of an explosion. It looks as if an original species burst into hundreds of forms, the smaller species, among which some survived and constitute the present species. The genus is obviously only this original or collective species.

Our Fig. 149 could be continued further downwards. From the elementary species we came to the collective species, and from these to the sub-genera and genera; in the same way to the more remote explosions would correspond the sub-families and families and the higher grades of the system. If the whole system were perfectly known to us, and if the pedigree had the form of an ordinary dichotomous table, each point of division would represent a period of mutation, from which, however, only the selected lateral branches, and not all those which had arisen, would be included in the picture.

So much then for the speculations to which an affirmative answer to the question proposed above (p. 652) would lead. In the following section we shall see how naturally these fit in with the results of paleontological investigation; but we must now discuss what are the results which would follow a negative answer to the same question.

Such a negative answer would imply the assumption that all the ancestors of our *Oenothera*, back to the first

[1] M. STANDFUSS. *Experimentelle zoologische Studien.* Neue Denkschriften d. allg. schweiz. Ges. f. d. ges. Naturw., 1898, p. 23. Further the articles of the same author in *The Entomologist*, May 1895, and in *Bull. Soc. entomologique de France*, 1901, No. 4. Also his *Handbuch der paläarktischen Grossschmetterlinge*, Zurich, 1896.

forms of life, have been mutable. Let us consider this view in relation to two important results of our investigation. In the first place it is obvious that *Oenothera* is not the only mutable plant. According to the researches of BAILEY and WHITE tomatoes are now almost certainly undergoing such a change, and cocoa-palms, since their introduction into the Indian Archipelago, must almost certainly have passed through such a period. Everywhere in the vegetable kingdom we come across vestiges of periods of mutation; and we should be led to the conclusion that the phylogeny of plants is represented by a richly-branched pedigree in which, downwards from the mutational groups now living, the lines must always be composed of mutable ancestors; for the assumption that mutability is an uninterrupted process is exactly the hypothesis from which we start.

But not all plants and animals are mutable at the present time; on the contrary, mutability is a very rare phenomenon. This circumstance can only be brought into harmony with the theory of the ever mutable main lines of the pedigree, by assuming that they have produced lateral branches in which the capacity for mutation has been lost. That such has often been the case we may confidently infer from the available evidence. According to the principles enunciated in the previous chapter, all that is necessary to bring this about, is that the representative elements be transferred from their unstable into a rigid condition.

The whole pedigree would then appear as a freely branched system of continuous mutable lines without gaps, and which are everywhere clothed, if I may so express it, by numerous immutable lateral branches. These would then stand in the same relation to the stem as the short foliage-bearing branches of our trees do to the long branches which form their crowns.

Every genus and every sub-genus would then contain at least one mutable species, from which the others have arisen, and this one might either still survive in their midst or have perished. In the former cases, presumably rare, these parent species would agree most closely with the supposed generic types of GÄRTNER which he regards as the central or original forms of the genus, on the ground of their behavior in crosses.[1]

It is easily seen that the contrast between the two pedigrees, to which an affirmative and a negative answer of our question respectively leads, is not of a very fundamental kind; and that the two can be reconciled if we regard the periodical mutability of the former and the large number of immutable branches of the latter as the two chief features.

Let us now consider the conclusions which the paleontologist DANIELE ROSA has drawn from his extremely important studies on the diminution in variability in connection with the appearance and extinction of species.[2] A detailed study of the phylogeny of extinct forms led him to conclude that the prospect of the survival of genera and families, and indeed of whole orders, was demonstrably correlated with their richness in forms. Cases like *Lingula* which have remained the same with very slight changes from Cambrian times up to the present

---

[1] GÄRTNER, *Bastarderzeugung im Pflanzenreich*, pp. 273-280.
[2] D. ROSA, *La riduzione progressiva della variabilità, e i su i rapporti coll' estinzione et coll' origine delle specie*, Turin, 1899. German translation by H. BOSSHARD, *Die progressive Reduktion der Variabilität und ihre Beziehungen zum Aussterben und zu der Entstehung der Arten*, Jena, 1903. G. CATTANEO, *I limiti della variabilità*, Rivista di Sc. Biolog., 1900, Vol. II, Nos. 1-2. See also E. D. COPE, "The Law of the Unspecialized," *Primary Factors of Organic Evolution*, Chicago, 1896.

day, are extremely rare.[1] On the contrary we usually see that the smaller groups sooner or later die out, whilst it is only in those cases in which variability, that is to say the production of new forms, has been most active that the groups continue for longer periods of time. Incapacity to vary dooms a group to death; only those who can most easily and quickly adapt themselves to changing conditions of life can survive. Species-forming variability is therefore not a universal capacity for variation, but only the result of quite special conditions which may often be absent from certain groups.

If we assume that the mutability in the main lines of the pedigree is an uninterrupted condition, and that this power, once lost, cannot be regained, it is clear that every branch of the pedigree, i. e., every larger or smaller group, is doomed to extinction as soon as the mutable species in it become extinct from some cause or another. On the other hand it is easy to see that the more numerous the mutable types are, the greater is the species-forming capacity of the whole group and, consequently, the greater its prospect of maintenance throughout geological ages.

Without giving a definite expression of opinion, it does not seem to me to be likely that mutability has continned throughout geological times without interruption. Therefore I think it more probable that there has been alternation between mutable and immutable periods. This latter view, moreover, is in agreement with the conclusions arrived at by ROSA.

[1] Further instances are given by HUXLEY, *Proceed. Roy. Inst.*, III, p. 151; and by POULTON, *Brit. Assoc.*, 1896, *Zool. Section*, Presidential Address. For the *Foraminifera* see CARPENTER, *Introduction to the Study of the Foraminifera*, 1862, p. xi, etc.

§ 13. ITERATIVE FORMATION OF SPECIES.

There is a great deal of evidence to show that species arise in groups, and that they originate discontinuously in the geological strata.[1] For various groups of animals and plants the exhaustive studies of KOKEN have shown that this mode of the origin of new forms in the geological strata is the usual one.[2] He calls this phenomenon the iterative formation of species. According to him a persistent species produces "varieties" which appear in swarms at certain periods; these periods are separated by more or less long phases of rest. He observed this first among the more ancient gastropods; but cases of the iterative formation of species have been described also amongst the Craniadae and Pectinidae.

It does not seem to me to be going too far to argue that the conclusions derived, in the foregoing section, from the actual observation of the process of mutation fit in with these results of paleontological investigation in a perfectly simple and satisfactory way, whilst the old theory of selection can only account for this periodicity by the help of special hypotheses. WHITE, who has thorougly investigated these phenomena from a paleontological point of view,[3] has recently pointed out the agreement of my views with his conclusions.[4] Our Fig. 149, (p. 655) could be used as a schematic representation of

[1] W. O. FOCKE, *Die Pflanzenmischlinge*, 1881, p. 509.

[2] E. KOKEN, *Paläontologie und Descendenzlehre*, Jena, 1902, and the literature cited here. See especially pp. 12-13. See also W. B. SCOTT, *On Variations and Mutations*, Am. Journ. Sc., Vol., XLVIII, p. 355.

[3] CHARLES A. WHITE, *The Relation of Biology to Geological Investigation*, Report of the U. S. Nat. Mus., 1892, p. 245.

[4] The same, *The Saltatory Origin of Species*, Bull. Torrey Bot. Club, Aug. 1902.

KOKEN's conception. In each of the several periods the new forms appear in a swarm, whilst the periods themselves are separated by phases of rest. According to the theory of selection the species themselves should be transformed into new ones; but according to the theory of mutation the original species does not disappear, whilst the extremes press forward. In the case of *Oenothera Lamarckiana* the main stem continually multiplies with undiminished vigor. Its derivate species have the greatest difficulty in maintaining themselves in competition with it in the natural state. Even so, as KOKEN points out, paleontology recognizes numerous cases in which the type species persists alongside those to which it has given rise, and it may even sometimes persist after these have disappeared.

The genetic association of the individual types can be demonstrated by experiment; but in paleontology conclusions relating to this point must obviously be based on considerations of a comparative nature. Apart from this, everything seems to be exactly the same. "The swarms of varieties and species succeed one another like the stories of a house. Similar forms recur by being produced at various times by the conservative guardians of the race, but not by one giving birth to the other."[1] Paleontology has the great advantage of directly demonstrating the stories which follow one another, as such. Comparative biology, on the other hand, has to infer them from classification, whilst experiment will probably always have to confine itself to a single story.

According to the precedent set by WAAGEN, the several forms of a group which follow one another in the course of time, and by means of which a type is gradually

[1] E. KOKEN, *Jahrb. d. k. k. geol. Reichsamts*, 1896, p. 40.

changed, are usually designated, amongst paleontologists, as mutations.[1] The name varieties is applied by them to those forms which live side by side at the same time and constitute the rays of a fan or the units of a swarm in our diagram in Fig. 149. In this sense my *Oenotheras* are related to one another as varieties, but to the parent form as mutations. In experimental science, however, these paleontological terms would prove very inconvenient; and the older meaning of the word "mutation," as it was used by botanists long before WAAGEN, is greatly to be preferred. What varieties are, will for a long time remain a topic of discussion.[2]

§ 14. THE BIOCHRONIC EQUATION.

The characters of organisms are not unlimited in number. However complex the structure of a higher plant or animal may seem, and however much the characters which compose them may give the impression of being unlimited, no one will deny that, when more closely examined, their organization will appear, although not simple, at least a great deal simpler than it seemed to be at first sight.

COPE states that for the 28,000 species of vertebrates there are only a few hundred organs on which their variation and diversity rests.[3] If we examine the dichotomous tables for the identification of species in the most various groups of animals and plants, we are astounded at the

---

[1] See H. E. ZIEGLER, *Ueber den derzeitigen Stand der Descendenzlehre in der Zoologie*, Jena, 1902; and the same in *Zool. Centralblatt*, 1902, Nos. 14-15.

[2] See above § 3, p. 578. SAGERET defined mutations as "*Variétés qui se forment sous nos yeux.*" *Ann. Sc. nat.*, 1826, p. 299

[3] E. D. COPE, *The Primary Factors of Organic Evolution*, 1896.

small number of characters necessary for such identification. If, in the case of single forms, we look at a diagnosis of species, genus, family or order, we find only a small series of characters referred to. If we attempt to describe a higher plant as completely as possible, it becomes difficult to prolong the list for over more than a few hundred characters; and even if we have regard to internal structure,[1] latent characters, and so forth, it is very difficult to attain to thousands of characters. The significance of this difficulty is best illustrated by the fact that such a description of a single form would cover over a hundred pages of print.

The structure of our eye is infinitely wonderful; the series of intermediate stages between it and a simple spot of pigment is immeasurably great; and a period of millions of years would be needed on the theory of selection for the attainment of the present high degree of organization from those first beginnings by means of ordinary variability.[2] But MURPHY, BROOKS and many others have pointed out that these considerations do not necessitate the conclusion that it must have happened in this way.[3] On the contrary, the extraordinarily long time which the theory demands, leads us to suspect that there is some weak point in the argument.

It is perhaps here that the theory of mutation, regarded from a general point of view, manifests its greatest advantages over the prevailing form of the theory of selection. In the first part of the first volume I have attempted to show that it agrees with results of experi-

[1] A. GRAVIS, *Rech. anat. sur les org. végét. de l'Urtica dioica*, Mém. sav. étr. Acad. Belge. Vol. XLVII, 1884; and the same author, *Rech. anat. et phys. sur le Tradescantia virginica*, ibid., 1898.

[2] DARWIN, *Origin of Species*, p. 143.

[3] W. K. BROOKS, *Heredity*, 1883, 2d. ed., p. 283.

ment far better than that theory. We now see that when applied to the great problems of life it is free from those insuperable difficulties, which so many investigators have found to stand in the way of the theory of selection.

The theory of selection demands almost unlimited time for the evolution of organisms; for the mutation theory, on the other hand, the time which the physical geologists grant to life, is amply sufficient. This view was first clearly expressed by BROOKS, in accordance with HUXLEY, when he showed that all the difficulties which beset the theory of selection and which, according to many investigators, needed 2,500,000,000 years for the whole process of evolution, would disappear if we assume relatively sudden and discontinuous changes to take place from time to time.[1]

The most distinguished investigators demand a period of about 24 million years, to cover the duration of life on the earth. If the ancestors of our *Ocnothera Lamarckiana* have produced, once in every 4000 years, a mutation which made them richer by a single character, our plant would now be composed of 6000 such characters, a number far higher than comparative and systematic science can by any means accumulate in its description.

This rough calculation shows at any rate that the demands made by the theory of mutation are not so exorbitant as those made by that of selection. Neither the number of mutation periods passed through, nor that of the characters acquired in them, is beyond our powers of comprehension. On the contrary the phenomena viewed from this standpoint are such that they indicate the possibility of a much closer investigation.

[1] W. K. BROOKS, *loc. cit.*, p. 286.

Let us now proceed to consider the relation between the degree of organization and the speed of the evolution from a more general point of view. For this purpose I shall deal with the several factors as briefly as possible, and propose to begin with biological time.

Many investigators have attempted to reach an approximate estimation of biological time, i. e., the duration of life on the earth. Proceeding on entirely different lines, the best of them have arrived at results which agree in a most remarkable way. From this fact we may infer that the calculations, which from their very nature must be more or less vague, probably represent a fairly close approximation to the truth.

I take the following,[1] partly from LORD KELVIN's famous researches, and partly from the clear exposition given by W. J. SOLLAS in his address as President of the geological section of the British Association in the meeting of 1900; and further from the recent investigations of DUBOIS.

LORD KELVIN based his first calculations on the increase in temperature in the successive depths of a mine.[2] This increase, however, has been shown by more recent investigations to vary considerably. The older determinations gave from 25 to 37, or sometimes as much as 50 meters for each degree Centigrade. In the neighborhood of the North American lakes, however, in a shaft of 1396 meters, an increase of 1° C. per 122 meters has

[1] An exhaustive review of the subject can be found in *Album der Natuur*, Sept. 1901; and the matter is also dealt with by H. CHARLTON BASTIAN, *Studies on H erogenesis*, London, 1901, pp. i-x. See also *Nature*, Sept. 1900 and *Revue scientifique*, April 1901.

[2] SIR WILLIAM THOMSON (afterwards LORD KELVIN), *The Secular Cooling of the Earth*, Transact. Roy. Soc. Edinburgh, 1862, Vol. XXIII.

been observed, and near Przibram in Bohemia, an increase of 1° C. for every 69 meters. Inasmuch as these two latter records have been made in districts which are further removed from local sources of special high temperature than the older mines, we must conclude that the earth has already cooled down much further than was previously thought, and that the period of 20 to 40 million years arrived at by LORD KELVIN, is by no means too high an estimate.

GEORGE DARWIN calculates that the moon separated off from the earth at least 56 million years ago, and GEIKIE put as the maximum for the existence of the earth's crust, a hundred million years. The general view is that the formation of the sea occurred fairly soon, geologically speaking, after the formation of the crust, and that no great period of time was necessary for a cooling of the water, sufficient to render life possible.

Further data for similar calculations are furnished by the action of rivers. These carry certain dissolved salts to the sea. From the mean proportion of sodium chloride which they contain and from the total volume of water which is poured into the sea every year by all of them, we can calculate how much the saline contents of the sea must increase from this cause. The total amount of dissolved salts in the ocean can also be calculated and we can then estimate the number of years necessary for the accumulation of this quantity. From these data JOLY calculated the age of rivers to be 90 million years; but it is highly probable that the continents were originally far richer in salt than now, and that the rivers have more or less exhausted them, that is that they carried more salt to the sea in former times than they do now. Applying this qualification, SOLLAS has reduced JOLY's result to fifty million years at the most.[1]

EUGÈNE DUBOIS has made use of the calcareous contents of rivers as the starting-point of his calculations.[2] He starts from the fact that carbonic acid is the source of plant food, and that the process of assimilation is the only one on this earth by which oxygen arises on a large scale. His arguments led him to conclude that the total amount of oxygen in the atmosphere has become free in this way. Now, carbonic acid is contributed to the atmosphere by the action of volcanoes. Once arrived here, it is partly decomposed by plants, and partly itself acts on rocks, and especially in combination with lime and magnesia forms salts which are washed out by the rain and carried by the rivers to the sea. Here, however, these salts are again laid down in coral banks, shells, and so forth, and in this way arise the enormous calcareous strata which constitute so large a portion of the hard crust of the earth. The volume of these layers can be approximately calculated, and the figure thus obtained, when divided by the annual contribution, gives some idea of the duration of the whole process. In basing his calculation on the chalk only, DUBOIS arrives at an estimate of 45 million years; but if magnesia is included as well, obviously a much smaller figure must be arrived at, viz., 36 million years.

I have still to mention briefly two further methods of arriving at this result. HELMHOLTZ found that the

---

[1] For a further discussion of these calculations see E. DUBOIS, *Kon. Akad. v. Wet. Amsterdam,* Jan. 1902, p. 503.

[2] E. DUBOIS, *ibid.,* p. 495, and also *loc. cit.,* June and August, 1900. The same author, *Over den Kringloop der stof op aarde,* Leyden, 1899; and *Over den ouderdom der aarde.* Kon. Ned. Aardryksk. Genootsch., 1900.

sun can have shone for only about 20 million years with approximately the same energy as that with which it shines now; and since this is the first condition of life on the earth, we must assume that its duration has been about the same as this period. The most authoritative estimate for the total thickness of the geological strata, and of the speed with which they have been laid down, is that of 80 kilometers, laid down at a rate of 30 centimeters per century, and this leads to an estimate of 26 million years for the whole period.

Therefore, about 20 to 40 million years is the period of the duration of life upon the earth; and LORD KELVIN, who a few years ago subjected the data, on which this estimate is based, to a critical reconsideration, came to the conclusion that the duration of life on the earth may provisionally be put at about 24 million years.[1]

We will therefore now adopt this figure as a basis for our further arguments.

The second question is this: How quickly have the individual periods of mutation followed on one another? We have very few data which enable us to arrive at any conclusion on this point. As is well known, the parts of plants which have been preserved in the sepulchers of the pyramids along with the mummies, and in other monuments of the same period, such as flowers, leaves, fruits, cereals, straw and weeds of the fields, prove the great antiquity of many species which are still existing. Numerous species are no doubt older than the pyramids, and have therefore remained unchanged for a period of at least 4000 years. The remains of lake dwellings,[2]

[1] See the review in the *Phil. Mag.*, Jan. 1899.
[2] Osw. HEER, *Die Pflanzen der Pfahlbauten*. Schweiz. Naturf. Gesellsch., 1866, No. LXVIII, with one plate; also C. SCHRÖTER and J. HEER, *Lebensbild von Oswald Heer*, Zurich, 1885.

the drawings on Roman coins, and many other facts of the same kind conduce to similar estimates.[1] On the other hand, the rarity of mutable plants in comparison with immutable ones, and also the small number of genera and other groups rich in species, as compared with the ordinary types of the European and American floras, lead by an entirely different chain of argument to conclusions which mainly support those reached above.

We may therefore assume as a provisional conclusion that a few thousand years elapse on the average between two successive periods of mutation. Of course, it is extremely probable that the speed of the process of evolution has not at all times been the same. On the one hand we must suppose that at first it was more rapid than it is at present.[2] On the other hand there must have been periods of greater mutability and periods of relative stagnation, possibly in the whole animal and vegetable kingdom, but certainly in special lines of descent owing to which some have reached a high degree of differentiation in the same period of time in which the progress in other lines has been relatively small. The Cambrian period divides biological time into two approximately equal parts, no fossil remains from pre-Cambrian times are known In Cambrian times members of all the more important groups of invertebrates suddenly appear, and among plants the Algae are richly represented. It almost seems that only those lines of descent which have made their evolution on the continents have begun in post-Cambrian times.

In a very attractive essay BROOKS has shown how

---

[1] Instances of the ages of certain plants are given by DE CANDOLLE, *Géographie botanique*, II, 1063-1068, 1086 etc.

[2] On this point see my lecture cited above, pp. 52-57.

this transformation from the non-fossiliferous to the fossiliferous period may be imagined to have taken place.[1] In the beginning life was chiefly confined to the upper levels of the sea; and extended to only those depths to which the rays of the sun penetrated, thus supplying the source of energy for the nutrition of the smaller Algae. These latter were almost the only source of nourishment for the animals which therefore had not left this region yet; consequently they were mostly small and of delicate structure, and without such parts as could become fossil. Afterwards it was the discovery, as BROOKS calls it, of the possibility of life on the gloomy bottom of the sea, on the dead remains of the swimming organisms sinking there, which extended the distribution of life and furnished a new and most variable abode for living beings. Thus was started the rapid and abundant evolution in the numerous directions which now constitute the main lines of organic descent.

Besides this period of rapid evolution, BROOKS, together with other writers, assumes that there have been other special periods of great variability; for instance at the time when land-animals and again when man originated (*loc. cit.*, p. 217). The distribution of fossils also indicates the existence of periods in which the formation of species has been especially rapid.[2]

The question arises: Were the individual mutations greater in such periods, or did they only follow more rapidly upon one another?[3] This question is one of comparative anatomy and of systematic science. Some

---

[1] W. K. BROOKS, *The Foundations of Zoology*, 1899, pp. 215-237.

[2] *Die Mutationen und Mutationsperioden*, p. 56; also W. K. BROOKS, *Foundations of Zoology*, p. 218; CHAS. A. WHITE, *The Relation of Biology*, p. 296, etc.

[3] E. KOKEN, *Paläontologie und Descendenzlehre*, 1901, p. 30.

investigators hold one of these views, others the other. If we assume that the individual mutations in such periods were changes of a greater amplitude, they might be designated by a special name, for instance by the one suggested by SCHNEIDER, the "descenses."[1] There is no fundamental difference between these and mutations, and the same changes may, according to SCHNEIDER, in some lines attain to the magnitude of descenses, whilst in others they may remain of merely subordinate importance.

At present, however, I am concerned merely with an approximate and average estimate, and the knowledge at our disposal suggests that an estimate of a few thousand years fairly closely represents the truth.

A third question relates to the number of elementary characters of which one of the higher animals or plants is composed. According to the theory of selection an almost unlimited number of complications would be possible. In my *Intracellular Pangenesis* I have shown that, quite on the contrary, the number in question cannot be so inordinately great; for we repeatedly see the same characters recurring in different organisms, many of them in systematic groups widely remote from one another, as for instance in the higher plants and the higher animals. I need only mention the close similarity between the chemical processes involved in digestion in the stomach and in the leaves of insectivorous plants. Tendrils and climbing plants, submerged or swimming water-plants, heterostylic and cleistogamous flowers, parasitism and saprophytism and numerous other instances could be adduced. Everywhere nature has built up the whole

---

[1] K. C. SCHNEIDER, *Lehrbuch der vergleichenden Anatomie*, Jena 1902, pp. 244, 248.

extraordinary richness of its forms, from a relatively small number of elementary units, for almost every individual character is found in numerous species, and it is to their different grouping and their combination with the rarer factors that the extraordinary diversity of living forms is due.[1]

As we can easily see, this view effects a considerable simplification of the problem. Many authors have expressed their agreement with it, and only last year SCHNEIDER stated his opinion clearly enough, when he said that the number of histological characters is by no means a very large one.[2]

The question is, however, how many elementary characters an angiosperm or a higher animal possesses, on the average. In the former case I have attempted to draw up lists of characters. Every such list consists of two parts. The first half embodies the characters which had been acquired up to the time when the systematic group, to which the plant belongs, originated, and this part is therefore the same for every species within the group. The second half contains the later characters, viz., those of the phylum and orders down to the species and varieties. In drawing up such lists, it is easy to reach the number of a few hundred characters; but then the task becomes more arduous, and finally insuperable difficulties are encountered. But it is evident that even a complete list would scarcely embrace more than a few thousand characters for any single plant.

Our conclusions may therefore be summarized in the following theses:

[1] *Intracellulare Pangenesis* (Ger. ed.), p. 7; English translation by Prof. C. STUART GAGER, 1910, p. 8.
[2] *Loc. cit.*, p. 248.

1. The number of elementary characters of a higher plant, that is to say, the number of mutations through which its ancestors have passed from the beginning, is probably not more than a few thousand.

2. The average intervals of time between two successive periods of mutation are similarly to be estimated at a few thousand years.

3. From this we may conclude that a period of some millions of years is sufficient for the whole development of the animal and vegetable kingdom; or, in other words,

4. The doctrine of mutation does not demand a longer period for the duration of life than that which has been given by LORD KELVIN, viz., 24 million years.

These theses may be most simply summarized in the statement that the product of the number of elementary characters of an organism, and of the mean interval of time between two successive mutations of its ancestors, is equal to the extent of biological time.[1] If we call the former magnitude M (the number of mutations), the length of the intervals L and biological time BT, we have the expression

$$M \times L = BT.$$

I have called this the biochronic equation.[2] It will, I hope, help to demonstrate the importance of the doctrine of the elementary units of organisms, and thus bring this doctrine prominently before the eyes of the general reader, as well as of the trained investigator. This is my main object in enunciating it.

[1] The mutations referred to in these paragraphs are of course progressive mutations.

[2] *Die Mutationen und die Mutationsperioden,* p. 63.

INDEX.

ABBADO, 595.
ABBINK-SPAINK, 572.
Abies excelsa, fasciated, 492.
Abutilon, 30.
Acacia cornigera, 311; *diversifolia*, 23; *verticillata*, 310.
Acer Pseudo-Platanus, tetracotyls, 360.
Achillea Millefolium, 35; *rosea*, 321.
Adaptations, explanation of, 606.
Aesculus Hippocastanum, 201, 225, 370.
Affinity, sexual, 592.
Agave vivipara, 23.
Agrostemma coronaria bicolor, 62, 85; *Githago*, 15, 501; *G. nicaeensis*, 84; *G. pallida*, 65.
Aloë verrucosa, 70.
Alpine plants, 59.
Amarantus speciosus, 361; tricotyls, 398; tri-radiate, 497.
Amphi-syncotyly, 457.
Anagallis arvensis coerulea, 84; *phoenicea coerulea*, 621.
Analysis of organisms, 567.
Anemone coronaria plena, 13.
Anthemis nobilis, 93.
Anthyllis Vulneraria, 59.
Antirrhinum majus, 289, 345, 460; peloric, 224; *striatum*, 120; terminal leaves of, 376; tricotyls, 432.

Aquilegia, 30; *chrysantha*, 83.
Arabis alpina, 274; variegated, 287.
Arnica montana, 272.
Arum maculatum immaculatum, 62.
Arundo donax, 268.
Artemisia Absynthium, fasciated, 507.
ASCHERSON, 95.
Asperula azurea setosa, tricotyls, 358.
Aspidistra elatior, 268.
Aster Tripolium, 35, 498; annual, 299; fasciated, 512.
Atavism, 44, 71, 104; by bud-variation, 110, 625; mutational, 109; phylogenetic, 107; physiological, 107; vegetative, 619.
Atavists, 514; offspring of, 561; significance of, 554.
Atropa Belladonna lutea, 63, 606.
Aurea forms, 282.

BAILEY, 102, 615.
Ballota nigra, 284.
Bananas, red, 112.
Barbarea vulgaris variegata, 283.
BATESON, 8, 645.
Beet, annual, 291; sugar, 333.
Begonia Sedeni, 322; *semperflorens*, 35, 321.
BEISSNER, 106.

BELLI, 589.
Beta patula, 604; *vulgaris*, 60; *vulgaris saccharifera*, 636.
Betula alba, 112.
BEYERINK, 626.
Biastrepsis, 535.
Bidens grandiflora, 35, 195; *tripartita*, 78.
Biennial plants, 291.
Biochronic equation, 663.
Biology, systematic, 567.
Biscutella laevigata glabra, 62.
BLARINGHEM, 110.
Bluebottle. See *Centaurea Cyanus*.
Boehmeria biloba, 69.
BONNIER, 59; and FLOT, 636.
BORRADAILLE, 591, 606.
Boskoop, 91.
Brassica Napus oleifera, 298.
BRAUN, 112, 326, 531.
BRIEN, on beets, 294.
BROOKS, 670.
BRUYNING, 335.
Buds on leaves, 70.
Bud-variation, 111, 122, 131, 138, 154, 617; on variegated plants, 273.
BURBANK, 594.
BURCK, 612, 597.
BURKILL, 330.

Cactus Dahlia, 16.
Calceolaria, 225; peloric, 323.
Caltha palustris, 22, 28.
Camellia japonica, 22.
Campanula Persicifolia alba, 83; *pyramidalis alba*, 83; *rotundifolia*, 33, 310.
Cannabis sativa, 359; tricotyls, 426.
Capsella Bursa Pastoris apetala, 97; *Heegeri*, 96, 331.
Carlina acaulis, 59.

CARLSON, 605.
Carnation, wheat-ear, 92.
Carpinus Betulus, 629.
CARRIÈRE, 11, 57, 111.
CARUEL, 605.
CASPARY, 70, 506.
Castanea vesca, 572; *variegata*, 274.
Casuarina quadrivalvis, twisted, 539.
Catacorolla, 15.
Catananche coerulea alba, 84.
CATTANEO, 609.
Caulescens, 59.
CELAKOWSKY, 28.
Celosia cristata, 33, 337, 489, 497, 517; *variegata*, 116.
Centaurea Cyanus, 117, 330.
Centranthus macrosiphon, 460, 458.
Cephalotaxus pedunculata, 109.
Cereals, 335.
Characters, antagonistic, 7; beginnings of, 611; hereditary, 567; latent, 18, 148; semilatent, 19; specific, 588; useless, 588; varietal, 588; vicariating, 648.
Chelidonium majus, double, 324; *laciniatum*, 86; *latipetalum*, 86.
Chenopodium album, tricotyls, 386.
Chlorotic branches, 281.
Chromosomes, 616.
Chrysanthemum coronarium, 82, 85, 195; *C.album*, 84; *inodorum*, 161, 284; *inodorum plenissimum*, 184; ostrich-feather, 16; *Parthenium*, 270; *segetum*, 161, 289.
Citrus hybrids, 621.
Clarkia pulchella, 144, 289; *carnea*, 85; tricotyls, 429.

Clover, crimson, 229.
Cochlearia Armoracea variegata, 278.
COCKAYNE, 605.
Cockscomb, 33, 517.
Coffea arabica, terminal leaf of, 377.
Combs, 494.
CONN, 608.
Convolvulus tricolor, 120.
COPE, 663.
Coreopsis tinctoria, 15, 82.
Coriandrum sativum with pitcher, 465.
Corn marigold, 161.
Correlation of anomalies, 234.
CORRENS, 591.
Corylus Avellana, 112.
Cotton, 30.
Crepis biennis, fasciated, 509.
Crosses, fertility of, 648; reciprocal, 597; types of, 576; unisexual, 577.
Crossing, 16.
Cryptomeria japonica monstrosa, 495; japonica spiraliter falcata, 619.
Cucumis sativus, 70.
CUÉNOT, 645.
Curve, dimorphic, 165; many-peaked, 523.
Cyclamen persicum, 22.
Cynips Kollari, 269.
Cynoglossum officinale bicolor, 62.
Cypripedium caudatum, 220.
Cytisus Adami, 622; Laburnum, 225.

Dahlia, 116; cactus, 16; green, 91; striata nana, 82; variabilis, 458; variabilis fistulosa, 100; variabilis viridiflora, 90.
DANIEL, 626.

Daphne Mezereum album, 63.
DARWIN, GEORGE, 667.
Datura Stramonium × D. laevis, 620.
Daucus Carota, annual, 299.
DAVENPORT, 8.
DE CANDOLLE, ALPHONSE, 32, 61, 589.
DE CANDOLLE, CASIMIR, 69.
Degressive, 71; mutations, 575.
DELAGE, 615.
Delphinium, 113.
DELPINO, 66, 558.
Dianthus barbatus, 285; barbatus torsus, 550; Caryophyllus spicatus, 92.
Digitalis lutea, 497; parviflora, 83; purpurea monstrosa, 222.
Dihybrids, 586.
DINGLER, 495.
Dipsacus, fullonum, 562; laciniatus, 541; sylvestris annual, 295; sylvestris torsus, 529.
Discoidea, 79.
Double flowerheads, 194.
DOUNET-ADANSON, 621.
Dracocephalum moldavicum, 369, 386; speciosum, 368.
DRIESCH, 642.
DUBOIS, 668.
Duration of life, 666.

Election, 610.
Elementary species, 65.
Eléments de l'espèce, 61.
Elite races, 610.
Empetrum nigrum, 63.
ENSINK, 550.
Epilobium hirsutum, 328.
Equation, biochronic, 603.
Equisetum Telmateja, twisted, 538.
Ericaceae, 95.

Erigeron bellidiflorus, 498.
ERNST, 94.
Erodium cicutarium album, 84.
Eucalyptus Globulus, 311.
Euphorbia exigua, 60; *ipecacuanha*, 605.
Euthymorphose, 605.
Eversporting Variety, 21.
Evonymus japonica, 504.
Explosions, 657.

Fagus sylvatica, 370; syncotyls, 462.
Fasciated growing point, 493.
Fasciation, 463; in eversporting varieties, 508; in half races, 502; plane of, 518; radiate, 497; ring, 496.
Ferns, crested, 337.
Fertilization, 616.
Ficus religiosa, 18.
Flax, 85.
Flecked leaves, 267.
FLOT, 636.
Flowers, double, 195.
Fluctuating Variability, 645.
FOCKE, 596, 621.
Forma alpestris, 59; *genuina*, 60.
Formation of species, 661; degressive, 71.
Forms, derived, 64; homonomous, 64.
Frumentum, 594.

Galeobdolon luteum, 226.
Galium Aparine, 537; twisted, 531; *verum*, 537.
GALLESIO, 614.
Galls, variegated, 269.
GARJEANNE, 31.
GÄRTNER, 582.
GEIKIE, 667.
Genista Germanica, 62.

Gentiana punctata concolor, 62.
Geological periods, 651.
Geranium molle fasciatum, 522; *pratense*, 119.
GERASSIMOW, 642.
Gesnera Geroltiana, 377.
Geum intermedium, 635; *urbanum*, 280; variegated, 287.
Gleditschia sinensis inermis, 101.
Gloxinia superba, 15, 22.
Godetia amoena, 84.
GOEBEL, 26, 71, 92, 106.
GOETHE, 588.
Gold-green Variety, 270.
Gooseberry, 63.
GRAVIS, 664.
GROOMBRIDGE, 116.
Gymnosperms, 495.
Gypsophila paniculata, 553.

Half race, 21; twisted, 541.
HANSEMANN, 642.
HARSHBERGER, 605.
Hedera Helix variegata, 284.
HEEGER, 96.
HEINRICHER, 28, 107.
HEINSIUS, 166, 285.
Helianthemum, 622.
Helianthus annuus syncotyleus, 466; *tuberosus*, 505; Variegated, 287.
Helichrysum bracteatum, 115; tricotyls, 430.
HELMHOLTZ, 668.
Helwingia rusciflora, 69.
Hemi-syncotyly, 457.
Hemi-syncotylous race, 476.
Hemi-tetracotyl, 346.
Hemi-tricotyl, 346, 359.
HERBERT, 593.
Hereditary coefficient, 545.
Hesperis matronalis, 136.
Heterogenic development, 603.

HEUZÉ, 298.
Hibiscus moscheutos, 605.
Hieracium umbellatum, 498.
HILDEBRAND, 83.
Hilversum, 643.
HOFFMANN, 76.
HOFMEISTER, 14, 83, 201.
HOLMBOE, 81.
Hornmeal, 453.
Hortensia, 112.
HUBRECHT, 604.
HUXLEY, 665.
Hyacinthus orientalis, 70.
Hybridization, 72; in twisting, 562.
Hybrids, generic, 593.
Hyoscyamus pallidus, 85.
Hypericum perforatum, 33.
Hypocotylous buds, 70.
Hyssopus officinalis albus, 84.

Idioplasma, 634.
Impatiens balsami, 115.
Inermis, 62.
Instinct, 611.
Intermediate races, 7, 25.
Iris pallida abavia, 107; *Pseudacorus*, 82; *xiphioides*, 115.

JOHANNSEN, 633.
JORDAN, 590.

KASSOWITZ, 615.
KELLER, 633.
KELVIN, 666.
Kerria japonica, 112; *variegata*, 274.
KICKX on pitchers, 323.
KLEBAHN, 531.
KOKEN, 661.
KÖLLIKER, 603.
KORSCHINSKY, 600.
KRASAN, 58.

KRELAGE, 198.
KUYPER, 610.

LAGERHEIM, 498.
Lamium album maculatum, 226; variegated, 287.
Larkspurs, 113.
Latency, 66.
Latent characters, 19.
LAURENT, 605.
LAUTERBORN, 606.
Lavandula Spica, 60.
Leaves, peltate, 69; split, 556; terminal, 377; variegated, 265.
LECOQ, 652.
LEMOINE, 16.
LE MONNIER, 562.
LENECEK, 32.
Leonurus Cardiaca, 225.
Lilium candidum plenum, 89.
Lime-tree, 32.
Limosella aquatica, 59.
Linaria vulgaris, 31, 289; *peloria*, 201.
LINDEMUTH, 625.
LINDLEY, 61.
Linear variability, 609.
LINNAEUS, 60.
Linum usitatissimum album, 84.
Lobelia syphilitica, 84.
Lolium perenne ramosum, 32.
LUDWIG, 8, 167.
Lunaria biennis, variegated, 284.
Lupinus arboreus, 605; *luteus* 225; twisted, 322.
Lychnis chalcedonica alba, 84. *dioica*, 581; *vespertina*, 97.
LYNCH, 82.
Lysimachia vulgaris, 312.

MAC DOUGAL, 601.
MAC FARLANE, 577, 605, 635.
Madder, 493.

Madia elegans, 85.
Magnolia obovata, pitchers, 27, 323.
MARLIÈRE, 633.
Material vehicles, 631.
Matricaria Chamomilla discoidea, 81; flore plenissimo, 187.
MAYER, 600.
Medicago lupulina, 232.
Melampyrum pratense, 225; tetracotyls, 375.
Melilotus coerulea monophylla, 87.
MENDEL, 576, 645.
Mentha aquatica, 220.
Mercurialis annua, 460, 572; spiral torsion, 464; syncotyl, 463; tricotyls, 428.
Middle race, 21.
Mimulus, 612.
MOLL, 541.
Monocotyledons, 66.
Monohybrids, 585.
MORGAN, 604.
MORREN, 320.
MÜLLER, FRITZ, 30.
MÜLLER, H., 82.
MUNTING, 187.
MURR, 81.
Muscari comosum plumosum, 612.
Mutation, degressive, 569; period, 652; progressive, 569; retrogressive, 569; vegetative, 619.
Myosotis alpestris, 86; *alpestris compacta*, 271; *azorica Victoria*, 330.
Myosurus minimus, 330.

NÄGELI, 578.
NAUDIN, 620.
Néflier de Bronvaux, 625.
Neo-Lamarckism, 601.
NESTLER, 495.

Nicotunia, 594.
NILSSON, 604.
Nitella syncarpa, 94.
Nitrates, manuring with, 453.
NOLL, 605, 612.
Nutrition, increased, 307.

Oats, sterile, 612.
Oenothera Berteriana, tricotyls, 415; *glauca*, tricotyls, 454, 458; *hirtella*, 346; *hirtella* tricotyls, 423; *Lamarckiana*, sectorial variegation, 285; *Lamarckiana*, variegated, 273, 285; *rubrinervis*, tricotyls, 384; *rubrinervis*, variegated, 285.
Oil plants, 298.
Onagra-period, 656.
Origin, polyphyletic, 233.
Orobanche Galii, 255.

Pangenesis of DARWIN, 631.
Pangenes, 641, 643; groups of, 647; new, 649; new types of, 645.
Papaver Argemone, 30; *commutatum*, 33, 243; *commutatum polycephalum*, 575; *nudicaule*, 115; *nudicaule aurantiacum*, 200; *Rhoeas*, 289; *Rhoeas*, tricotyls, 358, 435; *rupifragum*, yellow seedlings, 289; *somniferum Danebrog*, 85; *somniferum polycephalum*, 14.
PAUL, WILLIAM, 12.
PEARSON, 8.
Pedicularis palustris, 507.
Pedigree, 653.
Pelargonium, *zonale*, green, 92.
Peloria anectaria, 203.
Pelorias, heritable, 220.
Penstemon gentianoides, 359.
Pentacotyl, 346, 357.
PENZIG, 230.

Index. 681

Periodicity, 44, 323, 651.
Periods, susceptible, 647.
Petalomania, 92, 243.
Petalomanous types, 7.
Petunia, 338; double, 200.
PEYRITSCH, 225.
Phacelia tanacetifolia, tricotyls, 436.
Phaseolus coccineus, 293; *lunatus*, 605; *multiflorus*, 70, 293, 501.
Phlox Drummondi alba, 84.
Phosphates, manuring with, 453.
Phylogeny, 651.
Picris hieracioides, 458.
Pinus Abies aclada, 63, 93.
Pinus sylvestris, 327.
Pitchers, 18, 148, 464.
Plantago lanceolata ramosa, 148; *lanceolata*, split ears, 506; *major f. bracteata*, 32; *major rosea*, 33.
PLATE, 602, 608.
Poa alpina vivipara, 23.
Podocarpus Koraiana, 110.
Polemonium dissectum album, 84.
Polygonum Convolvulus, tricotyls, 389; tri-syncotyls, 461; *Fagopyrum*, 289.
Polyhybrids, 586.
Potato seeds, 309.
Potentilla anserina, 318; *arenaria*, 58.
PRAIN, 18.
PREHN, 83.
Premutation, 571, 649.
Primula sinensis, 22.
Progressive mutations, 576.
PROSKOWETZ, VON, 604.
Prunus, 605.
Pyrethrum Parthenium, aureum, 7; *roseum*, 199.

Quercus sessiliflora albo-variegata, 637.

RAATZ, 606.
Races, balanced, 8; half, 574; inconstant, 574; intermediate, 8, 574; non-isolable, 227; spirally twisted, 543; thoroughbred, 422.
Ranunculus aconitifolius, 59; *acris*, 93; *acris petalomana*, 7; *arvensis*, 62; *arvensis inermis*, 98; *auricomus*, 32; *bulbosus*, fasciated, 490; *bulbosus semiplenus*, 243.
Raphanus Raphanistrum, 15; cotyl-pitcher, 460.
Red berries, 63.
REINKE, 106.
Reseda odorata, 13.
Retinospora, 106.
Retrogressive, 576; mutations, 645.
Reversions, 111.
Rhus typhina, 627.
Ribes Gordonianum, 595.
RIMPAU, 94, 292.
Robinia Pseud-Acacia, 62.
ROSA, 609, 659.
Rubia tinctorum, 493; spiral, 540
Rubus fruticosus laciniatus, 640; variegated, 275.
Rye, perennial, 294; split ears of, 489; sterile, 94.

SAGERET, 621, 663.
Sagittaria sagittifolia, 310.
Salix aurita, 323.
SALTER, 12.
Salvia sylvestris alba, 84.
Sand bed culture, 305.
Sarothamnus scoparius, 605.
SAVASTANO, 615.

Saxifraga decipiens, 320; umbrosa, 70.
Scabiosa alba, 83; atropurpurea, 372, 553.
SCHÜZ, 606.
SCHRÖTER, 63.
Scirpus lacustris, 70.
Scrophularia nodosa, 289; tricotyls, 407.
Sectorial plants, 145; variation, 620; variegation, 276.
Sedum reflexum cristatum, 513.
Seedlings, yellow, 289.
Seeds, choice of, 332; of atavists, 516.
Segregation, vegetative, 619.
Selection, double, 545.
Semi-latent characters, 19.
Senecio Jacobaea, 81.
Sexual relationship, 592.
Sieve of natural selection, 610.
Silene Armeria alba, 84; *Armeria rosea*, 85; *conica*, tricotyls, 389; *conoidea*, tricotyls, 390; *inflata*, tricotyls, 437; *noctiflora*, 390; *noctiflora*, variegated, 287; *odontipetala*, tricotyls, 357.
Sinapis alba, 460.
Single variations, 6.
Sippe, 591.
SOLLAS, 666.
SOLMS-LAUBACH, 96.
Sonchus palustris, 506.
Sophora japonica pendula, 101.
Species and varieties, 578; collective, 61, 589; conception of, 567; elementary, 58; incipient, 9; origin of new, 71; provisional, 592; sub-progressive formation of, 67.
Specularia speculum, 327.
SPENCER, 633.
Spinacia oleracea, tricotyls, 391.

Spiraea sorbifolia, 505.
Spiral torsion, 368, 463, 527.
Spirogyra, 642.
STANDFUSS, 657.
Stellaria graminea aurea, 271; *Holostea apetala*, 96.
Stirp, 590, 638.
Strawberry, "Reus van Zuidwyk," 101.
Striped sorts, 267.
Sub-progressive formation of species, 67.
Subspecies, 60.
Succisa, 605.
Sugar-beet, brown, 606; fasciated, 507.
Svalöf, 605.
Sycios angulata, 458.
Syncotylous intermediate races, 466; race, isolation of, 469.
Syncotyly, 457.

TAMMES, 502.
Taraxacum officinale, 496.
Taxus baccata, 111; *fastigiata*, 98.
Teratology, 21.
Tetracotyl, 346.
Tetragonia expansa, 32, 84, 502, 514; forked, 374.
Tetrapoma, 331.
Teucrium Polium, 60.
Thoroughbred races, 423.
Thymus serpyllum, variegated, 271.
Time, biological, 666.
Tomatoes, 102.
Torenia, 612.
Torsions, heritable, 527; local, 558; rare spiral, 537.
TRACY, 605.
Tradescantia repens, 279.
Transgressive variability, 466, 600.

Transitional forms, 601.
Transportation hypothesis, 635.
Tricotylous intermediate races, 393.
Tricotylous races, isolation of, 422.
Tricotyls from bought seed, 380.
Tricotyly, partial variability of, 444.
Trifolium incarnatum, 289; *incarnatum quadrifolium*, 227; *pratense*, 289; *pratense*, pinnate leaf, 231; *pratense quinquefolium*, 36; *repens perumbellatum*, 317.
Tropaeolum, tendril, 605.
Tulips, striped, 115.
Twisted plants, 527.

Ulmus campestris variegata, 625, 636.
Uni-sexual, 577.
Units, number of, 598.
Uropedium Lindenii, 220.
Urtica urens, 553.

Vaccinium, 95.
Valeriana alba, 458; *officinalis*, 527.
Variability, fluctuating, 608; increase in, 9, 14.
Variations, germinal, 619; parallel, 67; sectorial, 122, 201; taxinomic, 69.
Variegated leaves, 265; plants, 272, 628.
Variegation, sectorial, 285.

Varietas aurea, 7, 270; *bicolor*, 62.
Varieties, derived, 60; eversporting, 8, 18; golden, 7; horticultural, 58; sterile, 88; systematic, 58.
Variety, 57, 64, 580, 584.
VAN DER VELDE, 635.
VERLOT, 57.
Veronica Buxbaumii, 330; *longifolia*, 494, 496, 620; *scutellata pubescens*, 63.
Vicia Faba, 501.
Vicia lutea hirta, 63.
VILMORIN, on striped flowers, 113.
Viola tricolor maxima, 337.
Violet, Dames', 136.
Viscaria oculata, twisted, 551.
VROLIK, 90.

WAAGEN, 662.
WALLACE, 599, 608.
WASMANN, 656.
WEEVERS, 537.
Weigelia amabilis, 28, 537.
WELDON, 600.
WETTSTEIN, VON, 56, 293, 591, 601.
WHITE, 661.
WILLE, 626.
WILLIAMSON, 13.
WILSON, 633.
WITTROCK, 82, 603.

Xanthium canadense, 336.

ZEINER LASSEN, 505.

Lightning Source UK Ltd.
Milton Keynes UK
UKHW050425211218
334320UK00022B/601/P